THE SCIENCE BOOK

WEIDENFELD & NICOLSON

Contents

First published in the UK in 2001 by Cassell & Co

This edition first published in the UK in 2003 by Weidenfeld & Nicolson

Design copyright © Cassell & Co 2001
Text copyright © Cassell & Co 2001, with the exception of: *Nature's Numbers* © 1995 Ian Stewart (The Calculus); *The Periodic Kingdom* © 1995 Peter Atkins (Mapping the Elements); *The Origin of Humankind* © 1995 Richard Leakey (The First Humans); *Just Six Numbers* © 2000 Martin Rees (From Newton to Einstein); *Words and Rules* © 1999 Steven Pinker (Words and Rules); *The Pattern on the Stone* © 1998 W. Daniel Hillis (Turing Machines); *River out of Eden* © 1995 Richard Dawkins (The Digital River); *Why is Sex Fun?* © 1997 Jared Diamond (A Unique Species); all published in the UK by Weidenfeld & Nicolson and in the US by BasicBooks.

A CIP catalogue record for this book is available from the British Library.

ISBN 13: 978-1-84188-254-3 Paper
ISBN 10: 1-84188-254-2

Editor: Peter Tallack
Project Editors: Peter Adams and Tim Whiting
Design Director: David Rowley
Designer: Clive Hayball
Picture Researchers: Lesley Grayson, Barbara Izdebska
Production Manager: Rosanna Dickinson
Additional Editors: Stephen Guise, Kathryn Duke, Matt Milton, Rebecca Lal, Sarah Bercusson, Annabel Blandford, Peter Frances, Sasha Heseltine
Indexer: Dr Christine Boylan

Printed in China.

Weidenfeld & Nicolson
Wellington House
125 Strand
London WC2R 0BB

A note to the reader: *Dates are often obscure, particularly early on when only an approximate decade is given for lives or events. These are based on the consensus of science historians. Where an entry covers a series of achievements, the choice of person(s), event or date to peg it on is often somewhat arbitrary. Organizations rather than individuals are occasionally credited for international or collaborative efforts. SI and metric units are used throughout; and a billion is taken as one thousand million (10^9). The publisher will add missing dates of scientists to any future edition if supplied with them.*

3

Preface
Susan Greenfield

Science is increasingly part of our everyday lives: more than in any previous era, science today is influencing what we feel, what we eat, how much leisure time we enjoy or suffer, and how we choose to communicate in an ever-accelerating information age.

Enormous progress is being made in both the physical and the biomedical sciences, and in the interaction between the two. If we are to harness these changes for the benefit of society, it is vital that we are all scientifically literate.

This book goes a long way to introducing the general reader to the astonishing wide world of science. Not only can we read each bite-sized entry independently to gain immediate insight into the individual discoveries that have shaped our world and lives, but we can also dig deeper into the book to explore the background science from which these discoveries have sprung. As a completely innovative step, the book adopts a marvellous organizational framework that documents in chronological order the great episodes in the development of all branches of science.

This historical perspective, as well as the fabulous and arresting artwork, should appeal not just to people seeking facts, but also to those wishing to appreciate these facts in their wider aesthetic and logical contexts. Indeed the book might also stand as a testament to the much-needed bridging of the stultifying gulf between the arts and the sciences: perusing the pages of this volume will be as enjoyable and illuminating as reading any work of history, literature or philosophy.

The Science Book does more than provide an easy way of accessing the facts of individual discoveries. It also presents the basic ideas behind the science, and, most importantly, allows us to see how personalities give rise to these ideas. Readers will gain, in a succinct and beautiful fashion, a true education in science that will equip them for the challenges that science will pose in the twenty-first century and beyond.

Foreword

Simon Singh

The Science Book consists of just 250 episodes from the history of science. At first glance, it would seem that a couple of hundred snapshots could not hope to capture any of the richness and subtlety of the scientific quest to understand the universe. But perhaps by extrapolating between entries we can begin to appreciate the scientific method, the route by which truth is sought, established and overturned. Take for example the tortuous and complex path that brought us to our current model of how planets orbit the Sun.

Although Greek astronomers had previously considered a heliocentric system, the work of the Alexandrian astronomer Ptolemy established an Earth-centred view of the universe that persisted for over 1,000 years. In 'The Mathematical Collection', re-titled 'The Greatest Composition' (*Almagest*) by Arabic translators, Ptolemy began with two axioms: that Earth is the mathematical centre of the universe and that the divine objects of the heavens move in uniform and circular motions. But observations showed that the planetary orbits could not all be described by simple circular motion, because planets would occasionally halt their usual progression from west to east, travel backwards from east to west, then resume their usual eastward orbit. Today we know that this occurs because Earth is not at the centre of the universe. Rather, it orbits the Sun, and the apparent backwards motion of other orbiting planets is the result of observing orbits from an object that is itself orbiting.

But Ptolemy could still resolve the observations with his axioms. The Earth was fixed, and the planets did indeed follow steady circular paths, except that the centre of each planetary circle (epicycle) followed its own circular path (the deferent). In fact this idea was based on the work of Aristotle and others, but Ptolemy gave it a firm mathematical foundation. It seemed the entire orbiting cosmos could now be described. Ptolemy's approach was right in one respect: a successful theory must fit with the observations. But a successful theory is not just descriptive. It should also explain the underlying reality of the phenomenon observed, reflecting the actual principles responsible. Ptolemy's model fails on this count.

Ptolemy's model survived for centuries largely because it concurs with our common-sense view of a static Earth. But as Albert Einstein said, 'Common sense is nothing more than a deposit of prejudices laid down by the mind before you reach eighteen'. Reluctant to

Foreword

abandon common sense, astronomers instead added more cycles and epicycles to the Earth-centred model in a desperate effort to make it fit the real world. Eventually, at the start of the sixteenth century, Nicolas Copernicus re-formulated a model based on a heliocentric system. His theory of the Solar System was simpler, more elegant and even more accurate than the Ptolemaic view. (Simplicity and elegance have always proved to be valuable in formulating scientific theories. Aesthetics is a truer guide than common sense.) A century later, Johannes Kepler, drawing on the detailed observations of Tycho Brahe, rejected Copernicus's circular orbits and replaced them with ellipses. The refined model led to an even greater accuracy between theory and observation, and the foundation was firmly set for Isaac Newton to formulate his law of gravity, which outlined the forces behind Kepler's elliptic orbits.

What emerges is an interplay between theory and observation. When there is a conflict, the choice is either to adapt the theory in some ad hoc way or to question its fundamental validity. In general, scientists prefer to tinker with their cherished theories rather than abandon them. But great scientists dare to question the orthodoxy and have the audacity to construct a new theory that better describes the observations and more closely reflects the underlying truth. As the sociologist Robert K. Merton put it, 'Most institutions demand unqualified faith; but the institution of science makes scepticism a virtue'.

In the most extraordinary episodes of science, a combination of scepticism, creativity and rational thought can spontaneously give rise to a new theory before a significant discrepancy between the previous theory and observation arises. In the twentieth century, Einstein's theory of general relativity overtook Newton's law of gravity. According to Einstein, Newton's laws were a rough approximation to his own deeper theory, but they had been sufficient to have agreed with all previous observations. Einstein's theory, however, made predictions that challenged astronomers to make new highly accurate measurements that would reveal the inadequacy of Newton's theory and the precision of his own. In 1919 Sir Arthur Eddington's eclipse measurements showed the extent to which the Sun could bend starlight, exactly as predicted by Einstein, but not by Newton.

In a way, *The Science Book* is a tribute to scepticism. The latter pages describe the currently accepted truths in various fields, but even though the theories are admired, they too should be questioned and tested. The danger is that today's theories can become so established that they attain the quality of intellectual common sense, tempting scientists to take on board prejudices that prevent them from seeking an even better understanding of the universe.

Introduction

Peter Tallack

This book traces the story of the achievements that have changed the course not only of science itself but also of whole areas of thought. It covers the traditional natural sciences (physics, chemistry, biology, astronomy, Earth sciences), as well as psychology, archaeology, paleaoanthropology, medicine and mathematics. Technology features only when it has led directly to scientific advances, as with the telescope, the microscope or the computer.

The question of what to include and what to omit vexed me for months. Although the book acknowledges the early philosophers whose investigations heralded the dawn of scientific thought, the selection tends to favour discoveries, theories or methods that solved a long-standing problem, opened up entire new areas of enquiry or changed our view of the world. Some were initially misunderstood but later accepted. And a handful have been chosen that had a profound impact on a particular area of science even though they eventually turned out to be wrong.

This provisional nature of scientific knowledge is especially clear in the modern selections, which have been the most difficult to make. Around nine-tenths of all scientists who have ever lived are alive and working now. While it is relatively easy to assess with hindsight the achievements of previous generations, it is often hard to compare the significance of recent contributions and to know whether they will stand the test of time – though they do convey a sense of the breadth and diversity of current science, and hint at discoveries yet to come.

Although science does not develop in a linear way, the chronological structure of the book provides a unique insight into the general trends and influences across the centuries, as well as the cross-fertilizaton of ideas among different branches of science. But science is not just a series of facts. The book also explores the fruitful mix of imagination, creativity, competition, muddle, intuition, ingenuity and mistakes that characterizes science as a very human activity.

Finally, in celebrating the fascination and wonder of science, I have been forced to omit many important scientists and scientific milestones. A selection such as this is inevitably subjective – if not 'foolhardy', to use the words of one eminent physicist I consulted. Even science can be a matter of taste.

Origins of counting

Swaziland *c.*35,000 BC

Like all other human activity, the origins of counting are shrouded in historical haze. Bones serving as tally sticks are the earliest archaeological evidence. The most ancient is a baboon's fibula inscribed with 29 notches, found in Swaziland, southern Africa, and dated to about 35,000 BC. It may represent the lunar cycle as it resembles 'calendar sticks' still in use in Namibia. A wolf bone with 55 notches was found in Moravia, Czech Republic (30,000 BC), and the shores of Lake Edwards in Zaire yielded the Ishango bone, a tool-like object with a quartz insert (9000 BC). The tally marks are often arranged in various groups, although what they represent remains a mystery.

A faint echo of the tally system continues in some modern numerals: our own numeral one, as well as the Roman and Chinese numerals for one, two and three, are merely tally marks, I, II and III. But counting in ones is limiting, and civilizations soon began to group together single digits to create more flexible number systems based on 5, 10, 12, 20 and 60. Most seem to have their origins in a variety of finger-counting methods.

The first evidence of writing in general and numbers in particular comes from Mesopotamia, the land between the Tigris and Euphrates rivers. The Sumerian base-60 system demanded special symbols to represent 1, 10 and 60 and was the first place-value system invented – the value of a digit depended on its position. It originated in about 2500–3000 BC, but survives today in our time calculations and angle measurements. Although written records were kept on clay tablets, calculations were often performed with different-shaped pebbles to denote various goods. By about 2000 BC this method had been replaced by an abacus. The first writing system based on an alphabet was developed by the Phoenicians at the beginning of the first millennium BC, and with it the representation of numbers by letters.

See also Astronomy before history pages 12–13, Zero pages 34–35, Algebra pages 38–39, Deciphering hieroglyphics pages 136–137

Right A Sumerian clay tablet recording the bill for the sale of a field and a house paid in silver. It is written in cuneiform script and dates from *c.*2550 BC.

Astronomy before history

Egyptians *c.*3000–1000 BC, Babylonians *c.*2000–1000 BC, Chinese *c.*2000–1050 BC

Science started with astronomy. Our ancestors looked at the sky and saw a combination of constancy and variability. The 4,000 or so easily visible stars congregated into memorable unchanging patterns. Seven wanderers moved in front of this constellatory background. The Moon and Sun were discs. Their paths lay in a zodiacal band.

Night by night the phases of the Moon changed. But every 29.5305882 days it returned to the same shape. The constancy of this interval defined the month and probably spurred the ancients to count. The Sun also obeyed fixed rules. Its rising and setting positions on the horizon changed systematically, as did its noon-time altitude. These changes coincided with changes in ambient temperature, length of daylight and animal and plant behaviour. Soon the year was defined and then divided into seasons. Midwinter, spring and midsummer became important festivals. Monuments such as Stonehenge in southwest England were built to point towards the rising Sun on the day of the summer solstice. The Egyptian pyramids were oriented with their sides along the north–south and east–west axes. They also contained meridional shafts that pinpointed the transit of stars such as Sirius and Thuban. That the numbers of days and months in the year were not whole numbers kept astronomers busy organizing their religious calendars.

Of the wandering planets, Mercury and Venus never strayed far from the Sun. By contrast, Mars, Jupiter and Saturn generally moved from west to east around the zodiac. But annually they appeared to stop, moved backwards for a time, and then resumed their normal motion. These 'retrograde' loops varied in size from planet to planet, as did the celestial speeds and brightness. Explaining this variability provided the scientific astronomers with a complex puzzle. Luckily, the puzzle was not so complex that they abandoned their endeavours, but it did take the works of Copernicus, Kepler and Newton to solve it.

See also Origins of counting pages 10–11, Music of the spheres pages 14–15, Celestial predictions pages 26–27, Earth-centred universe pages 30–31, Sun-centred universe pages 42–43, A new star pages 48–49, Laws of planetary motion pages 52–53, Heavens through a telescope pages 54–55, Newton's *Principia* pages 78–79.

Right Prehistoric observatory: Stonehenge embodied the precise astronomical knowledge of its Neolithic builders, who identified the heavens as a source of supernatural power.

Music of the spheres

Pythagoras *c.*580–500 BC

Pythagoras was born in Samos in Greece in 580 BC and then settled in Kroton in the south of Italy. While travelling in Egypt and Babylon he learnt a great deal, bringing this knowledge back to the West. He is credited with the suggestion that the Earth is round, although it is difficult to distinguish his discoveries from those of members of his philosophical school, a kind of secret sect that flourished for 200 years. As evidence he cited the way the Pole Star appears higher in the sky as one moves progressively north; a ship's stern disappears before its sails as it goes over the horizon; and Earth's shadow, during a lunar eclipse, is always circular.

Earth was supposedly surrounded by a series of crystalline transparent spheres (the 'perfect' shape), each rotating uniformly (the 'perfect' motion) and carrying a celestial body. Later the Pythagorean school introduced a spinning Earth to account for the daily motion of the stars, Sun, Moon and planets. As justification they pleaded simplicity: rotating a small Earth was easier than rotating a huge sky.

Pythagoras regarded numbers as 'the very essence of things'. As a scientist and a musician, he was fascinated by the relationship between the pitch of a note and the length of its harp string, and he extended these harmonic proportions to astronomy. The planets were not randomly placed: their distances and speeds were thought to fit simple number proportions. Hippolytos remarked that Pythagoras was the first to reduce the motions of the seven heavenly bodies to rhythm and song.

Although the name Pythagoras is most closely associated with the Pythagorean theorem

Aristotle's legacy

Aristotle 384–322 BC

Aristotle grew up and spent most of his life close to the seashores of Greece. There, among the warm rock-pools of the intertidal zone, he could see a great concentration of animal types: starfish, sea anemones, crabs, worms and fish. His writings show that he observed these animals closely, and the questions he asked about them have dominated biology ever since.

Aristotle aimed to distinguish the different kinds of animals, and his writings mention 560 species in all. He thus began the research programme to catalogue Earth's biodiversity – a programme that proceeds apace today. Although Aristotle did not classify plants, a pupil of his did. Botany began when one of Aristotle's pupils, Theophrastus, learned about the plants that soldiers had encountered during the expeditions of another of Aristotle's pupils, Alexander the Great.

Aristotle founded anatomy, embryology and physiology. He described the internal anatomy of animals such as crabs, lobsters, fish and squid – his detailed description of the internal mouthparts of the sea urchin was almost perfect, and the organ is still called 'Aristotle's lantern'; he was particularly interested in blood, and described the heart (though mistaking its function) and blood vessels; and he studied embryos, such as the chick. Aristotle also wrote influentially about mathematics, physics and cosmology – his laws of motion were widely accepted until Galileo and Newton – and was the inventor of formal logic.

Aristotle had moved (on the advice of the Delphic Oracle) in his youth to Athens to become a student of Plato. Plato and Aristotle dominate subsequent intellectual history, but it is Aristotle who is the founder of science. Plato's thinking was religious, mystical and poetic. Aristotle was never less than equivocal about supernatural explanations, and did not use them in his physics and biology. Scientists since his time have added the experimental method, but they have continued with his observational method and his use of deductive reasoning.

See also Birth of botany pages 18–19, Human anatomy pages 44–45, Natural magnetism pages 50–51, Circulation of the blood pages 58–59, Falling objects pages 60–61, Boyle's *Sceptical Chymist* pages 70–71, Naming life pages 88–89, Eggs and embryos pages 140–141, Crop diversity pages 292–293, Genetics of animal design pages 460–461, Diversity of life pages 468–469

Right Leading the way: Aristotle was one of the truly great scientists even though his rational thinking often led him to the wrong conclusions.

Birth of botany

Theophrastus *c.*372–287 BC

Theophrastus was a pupil of Aristotle, with whom he founded the Peripatetic school of philosophy ('the Lyceum') in Athens. After becoming head of the school in 322 BC he produced two botanical works, *Historia Plantarum* ('Natural History of Plants') and *De Causis Plantarum* ('Reasons for Vegetable Growth'), which remained the most authoritative treatises on the subject for more than 1,500 years. They acquired a wider readership when Theodore of Gaza (after whom the plant genus *Gazania* is named) translated them into Latin in 1483, and dedicated them to Pope Nicholas V.

Both works contain wide-ranging observations on plants, including aspects of morphology, anatomy, pathology, seed germination, grafting and propagation, crop cultivation and medicinal use. They also include the first description of the pollination of date palms and discuss the sexuality of plants – an aspect of plant reproduction that was not proven experimentally until the work of Rudolph Jakob Camerarius in 1694. Theophrastus's sources of botanical information seem to have come from the garden at the school – possibly the first botanic garden in history – and for less parochial botanical matters he might have gleaned information from soldiers returning from campaigns with Alexander the Great.

The books take the form of long descriptions of parts and processes, to which Theophrastus attempted to apply names based on words from everyday Greek, and in them developed an early example of the technical jargon that remains characteristic of scientific discourse. Derivations of *anthos* (his word for a flower) and *pericarpion* (the pericarp is the tissue forming the wall of a fruit) are instantly familiar to botanists today. In his books he classified almost 500 plants, establishing some taxonomic groupings that still persist and gaining the admiration of the great eighteenth-century taxonomist Carolus Linnaeus, who described him as the 'father of botany'.

See also Aristotle's legacy pages 16–17, Medicinal plants pages 28–29, Plant sex pages 82–83, Naming life pages 88–89, Nitrogen fixation pages 208–209, Crop diversity pages 292–293, Photosynthesis pages 344–345, Green revolution pages 428–429

Right Theophrastus's excellent descriptions of plants, their germination and their parts influenced botany for centuries. Pictured here are six medicinal herbs, from a 13th-century Arabic manuscript.

جنطيان

دهن بلسان

زاج مشوي

فلفل اسود

سادج هندي

فو

Euclid's *Elements*

Euclid *c.*325–265 BC

The most influential work in the history of mathematics is undoubtedly the *Elements* by Euclid, a mathematician who worked in Alexandria during the reign of Ptolemy I. Little is known of Euclid's life, but his book became so popular that it supplanted all other similar works for more than 2,000 years. It is not a wholly original work, but rather a textbook that distills all previous knowledge of elementary mathematics – the basic building blocks of geometry and arithmetic. But Euclid's skills as a teacher shine through in its concise, logical and coherent structure. He became the standard-bearer of Greek mathematics as various translations, commentaries and editions of the *Elements* were dispersed throughout the Mediterranean, Asia and Europe.

The *Elements* consists of 13 books of gradually increasing complexity. It begins with some basic definitions of, for example, a point, a line, a surface and a circle. Euclid then sets out several fundamental concepts ('axioms' and 'postulates') that cannot be proved but which can be accepted as true by intuition, such as that non-parallel lines meet at a point. Reasoning from these starting points, he then constructs by deductive logic a series of theorems or 'propositions'. Further theorems are then built up on the foundations of earlier ones. Although most of the treatise deals with geometry, it also looks at ratio and proportion and the theory of numbers. Among the highlights are Euclid's proof that for a right-angled triangle the area of a semicircle along the hypotenuse is equal to the sum of the areas of the semicircles on the other two sides (a generalization of Pythagoras's theorem); that the number of primes is infinite; and that the square root of two is irrational. In the final three books, Euclid proves that there can only be five regular, or Platonic, solids: the tetrahedron, the cube, the octahedron, the dodecahedron and the icosahedron.

Not until the nineteenth century was it realized that Euclid's axioms were not absolute truths. This led to the development of a new kind of geometry, which came to form the basis of relativity and quantum mechanics.

Above A 13th-century French representation of geometry, showing a mathematician using a pair of compasses.

See also Music of the spheres pages 14–15, Non-Euclidean geometry pages 144–145, General relativity pages 278–279, Limits of mathematics pages 308–309, Fractals pages 446–447

Right The first printed edition of Euclid's *Elements* issued in Venice in 1482. It has since been through more than 1,000 editions, making it the most successful textbook of all time.

Unctus est cuius ps nõ est. ℂ Linea est
lõgitudo sine latitudine cui⁹ quidé ex/
tremitates sr duo pũcta, ℂ Linea recta
é ab vno pũcto ad aliũ breuissima exté/
sio i extremitates suas vtrũq; eo℞ reci
piens. ℂ Supficies é q̃ lõgitudiné 7 lati
tudiné trĩ h3:cui⁹termini quidé sũt linee.
ℂ Supficies plana é ab vna linea ad a/
liã extésio i extremitates suas recipiés
ℂ Angulus planus é duarũ linearũ al/
ternus practus:quã expãsio é sup sup/
ficié applicatioq; nõ directa. ℂ Quãdo aut angulum ptinét due
linee recte rectiline⁹ angulus noiaf. ℂ Qñ recta linea sup rectã
steterit duoq; anguli vtrobiq; fuerit eq̃les:eo℞ vterq; rect⁹erit
ℂ Lineaq; linee supstãs ei cui supstat ppendicularis vocaf.ℂ An
gulus vo qui recto maior é obtusus dicif.ℂ Angul⁹ vo minor re
cto acut⁹appellaf.ℂ Termin⁹é qõ vniuscuiusq; finis é.ℂ Figura
é q̃ tmĩo vl'termis ptinet.ℂ Circul⁹é figura plana vna qdem li/
nea ptéta:q̃ circũferentia noiaf:ĩ cui⁹medio pũct⁹é : a quo⁹oés
linee recte ad circũferétiã excuntes sibiinicé sũt equales. Et hic
quidé pũct⁹cétrũ circuli dr.ℂ Diameter circuli é linea recta que
sup ei⁹centrũ trãsiens extremitatesq; suas circũferétie applicans
circulũ i duo media diuidit.ℂ Semicirculus é figura plana dia/
metro circuli 7 medietate circũferentie ptenta.ℂ Portio circu/
li é figura plana recta linea 7 parte circũferétie ptéta: semicircu/
lo quidé aut maior aut minor.ℂ Rectilinee figure sũt q̃ rectis li/
neis cõtinent quarũ quedã trilatere q̃ trib⁹rectis lineis: quedã
quadrilatere q̃ q̃tuor rectis lineis. q̃dã mltilatere que pluribus
q; quatuor rectis lineis continent. ℂ Figurarũ trilaterarũ:alia
est triangulus hñs tria latera equalia. Alia triangulus duo hñs
eq̃ha latera. Alia triangulus triũ inequalium laterũ. Max iterũ
alia est orthogoniũ:vniũ.i.rectum angulum habens. Alia é am
bligonium aliquem obtusum angulum habens. Alia est origoni
um:ĩ qua tres anguli sunt acuti. ℂ Figurarũ auté quadrilatera℞
Alia est q̃dratum quod est equilaterũ atq; rectangulũ. Alia est
tetragon⁹long⁹:q̃ est figura rectangula : sed equilatera non est.
Alia est helmuaym: que est equilatera : sed rectangula non est.

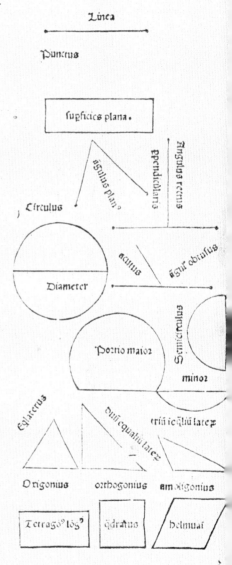

De principijs p se notis: 7 pmo de diffini-
tionibus earandem.

Linea

Punctus

supficies plana.

ppendicularis

Angulus rectus

ãgulus plan⁹

Circulus

acutus

ãgul⁹ obtusus

Diameter

Semicirculus

Portio maior

minor

Eqlaterus

duõ equalia latez

triũ ieq̃liũ latez

Origonius

orthogonius

am bligonius

Tetrago⁹ lõg⁹

q̃dratus

helmuaj

Moving the world

Archimedes *c.*290–212 BC

The Greek mathematician Archimedes was born in Syracuse, Sicily, and studied in Alexandria, before returning to live in his native town. In his lifetime he became famous as much for his practical discoveries as his mathematical theories. He invented what is still known as 'Archimedes' screw', which is used as a pump to raise water, as well as a system of compound levers in pulley form, which he demonstrated (so Plutarch claims) by pulling single-handedly a fully laden ship on to shore. And at the request of King Hieron II, he is supposed to have built military machines to defend Syracuse against the Romans, such as mechanical cranes that could pick up enemy ships with an iron 'beak', swing them around in the air and then dash them against the cliffs. These devices relied on the principles of the pulley and the lever as worked out by Archimedes. In fact the lever was such a powerful idea that Pappus, a fourth-century Alexandrian, wrote: 'Give me a place to stand and I will move the Earth.'

Such public displays of Archimedes' ingenuity were in contrast to his belief in the purity of a life devoted to impartial study. The archetype of the absent-minded professor stems largely from Archimedes, and his running naked through the streets of Syracuse shouting *Eureka!* ('I have found it!') is merely one of many legends his behaviour gave rise to. What sparked this particular exhibition was the law of hydrostatics: when a body is immersed in a fluid, its loss of weight equals the weight of the fluid displaced.

Aware that his fame had spawned plagiarists, Archimedes baited them by distributing false theorems. And there was much to be copied, including formulae for the areas and volumes of geometrical figures and ground-breaking work in calculus. His death at the hands of an over-zealous Roman soldier was greeted with anguish by the whole of the Mediterranean world.

See also Falling objects pages 60–61, Newton's *Principia* pages 78–79

Right Calculated death: the killing of Archimedes by a Roman soldier, as depicted on a mosaic.

Circumference of the Earth

Eratosthenes *c.*276–194 BC

Because the distance from the Earth to the Sun is about 12,000 times the diameter of our planet, one can assume that all sunbeams are parallel, even at locations far removed from each other. This enabled Eratosthenes to calculate the diameter of Earth.

Eratosthenes of Cyrene (now Shahat, Libya), a polymath, geographer and director of the great library at Alexandria, read in a papyrus book that at Egypt's southern frontier outpost of Syene (modern Aswan), near the first cataract of the Nile, vertical sticks cast no shadow at all at noon on the summer solstice, 21 June. At this precise time the Sun could be seen reflected in the water at the bottom of wells. He worried why, at exactly the same moment, a vertical stick at Alexandria, some 800 kilometres due north, did cast a shadow. The most sensible conclusion was that Earth was spherical, not flat. By measuring the length of the shadow, he calculated that the sunbeams at Alexandria made an angle of one fiftieth of a circle (just over seven degrees) to the vertical at the specific time. So the distance between Alexandria and Syene was one fiftieth the circumference of a great circle around the Earth. As this circle has a circumference of $2\pi R$, he could easily calculate a value for R, the radius of Earth.

Eratosthenes hired a man to pace out the distance between Alexandria and Syene. The value obtained was 5,000 stadia, a unit not known precisely today (J. L. E. Dreyer, the Irish historian of astronomy, equated 1 stadium to 157.7 metres). But it is clear that the value obtained for the radius of Earth is within a few per cent of the correct value. Eratosthenes was the first person to measure accurately the size of a planet, and his measurement was widely believed by his contemporaries.

See also Music of the spheres pages 14–15, Planetary distances pages 74–75, π pages 92–93, 'Weighing' the Earth pages 112–113

Right On the horizon: proof of the curvature of the Earth, 1764. Pythagoras was the first man known to us who taught that the Earth was spherical.

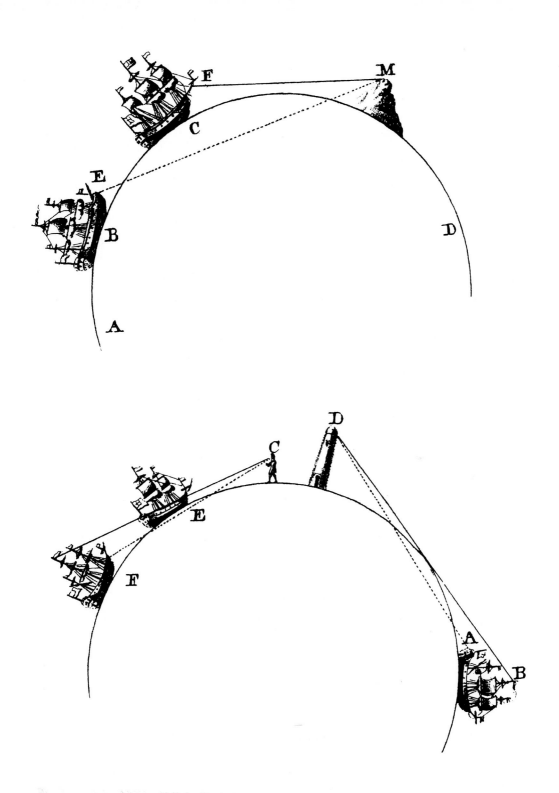

Celestial predictions

Hipparchus *c.*190–120 BC

Hipparchus was the greatest astronomical observer of antiquity, responsible for converting Greek mathematical astronomy from a descriptive into a predictive science. He measured the length of the year to be 365 days 5 hours and 55 minutes, but realized that the seasons had different lengths – spring, summer, autumn and winter lasting 94.5, 92.5, 88.125 and 90.125 days respectively. Using geometry, he then calculated (developing trigonometry on the way) that the distance from the Earth to the Sun was at its shortest around 4 January and longest on 4 July. Typical of his accuracy is the value he obtained for the length of the month. His 29 days 12 hours 44 minutes and 2.5 seconds was short by only 1 second.

Hipparchus was also responsible for combining Babylonian and Greek astronomical traditions, and adopting the 360-degree circle into Western mathematics. By translating Babylonian works, he drew up a list of lunar eclipses in the previous 800 years. Noting that the star Spica had changed coordinates over a period of 150 years, he went on to discover the 'precession of the equinoxes'. The First Point of Aries (the Sun's position on the first day of spring) was found to move by 1 degree per century (its actual rate is 1 degree per 71.7 years). This slow westward shift of stellar coordinates is caused by the Earth's spin axis wobbling around like a decaying top in a cycle that takes 25,800 years.

Hipparchus thought that the positions of the stars might be changing. Inspired by seeing a new star in the constellation of Scorpio in 134 BC, he produced the first detailed stellar catalogue. The positions of 850 stars visible to the naked eye were carefully noted. Innovatively, he divided these into varying classes of importance, the brightest being of first importance and the faintest sixth. This 'magnitude' system is still used today.

See also Astronomy before history pages 12–13, Music of the spheres pages 14–15, Euclid's *Elements* pages 20–21, Earth-centred universe pages 30–31, Sun-centred universe pages 42–43, A new star pages 48–49, Heavens through a telescope pages 54–55, Planetary distances pages 74–75, π pages 92–93, Distance to a star pages 150–151, Climate cycles pages 276–277

Right Star trails: as the Earth rotates, the heavens seem to revolve around the poles during the night. With a long-exposure photograph it is possible to catch this apparent motion.

Medicinal plants

Pedanius Dioscorides *c*.20–90

Until the rise of the pharmaceutical industry in the late nineteenth century, most medical remedies came from plants. Medical students learned botany as a routine part of their studies, and popular recipe books looked to the botanic kingdom as the main source of help in time of plant disease. The roots of this approach arose in antiquity, the classical source for which was *De Materia Medica*, by Pedanius Dioscorides, a surgeon with the Roman army who, like Galen, wrote in Greek.

Dioscorides's work both consolidated and extended the ancient knowledge of medical treatment. Although few biographical details survive, he was said to have lived 'a soldier's life', and certainly travelled widely in the Mediterranean region, accumulating information about local plants and knowledge on how to gather, store, prepare and use them in the best possible way. He recorded detailed information about his plants, including habitats, physical characteristics and which parts of them were best exploited in treatment. Because so many ancient texts have been lost, it is impossible to say how many 'new' remedies Dioscorides introduced, but his monograph described a multitude of medicinal plants, among them cinnamon, belladonna, juniper, lavender, almond oil, ginger and wormwood. Fewer than a quarter of his plants had been mentioned in the Hippocratic writings.

Dioscorides's *De Materia Medica* was not confined exclusively to plant remedies: a minority of its preparations were of animal or mineral origin. By covering the whole range of available medicinal products using a simple and empirically informed approach, he established a standard that lasted for over 1,600 years. His work was copied and extended throughout the Middle Ages and the Renaissance, with the consequence that he remained a 'living' author. Through the *Herbal* of the Englishman Nicholas Culpeper, and others, his influence still persists, especially in the belief that plant remedies are more 'natural' and therefore safer.

See also Birth of botany pages 18–19, The body in question pages 32–33, Mauve dye pages 172–173, Regulating the body pages 188–189, Aspirin pages 226–227, Crop diversity pages 292–293

Right An opium poppy, from an illustrated edition of *De Materia Medica* prepared by a Byzantine artist in Constantinople 400 years after Dioscorides compiled the work.

Earth-centred universe

Claudius Ptolemy *c*.90–168

Claudius Ptolemaeus (Ptolemy), the Alexandrian astronomer, geographer and mathematician, was dogmatic, and his *Almagest* (which extends to 500 pages in modern translation) was taken seriously for nearly 1,400 years. He didn't merely suggest that the heavens appeared spherical; he insisted bluntly that they were exactly spherical. Because the night sky described a perfect hemisphere, the Earth must be at the centre of the universe – its natural place according to Aristotle's cosmology. Further, Ptolemy's Earth wasn't spinning; birds and clouds would be left behind if it were.

The *Almagest* revisited much of the work of Hipparchus 300 years previously, including measurements of the angle between the plane of the Earth's equator and the plane of its orbit, estimates of the distances of the Earth to the Moon and the Sun, and a catalogue of stars. It also listed 44 constellations to which Ptolemy gave the names we still use today (such as Orion and Leo).

Ptolemy's most influential contribution was his mathematical theory of planetary motion. Luckily, the planets all have nearly circular orbits around the Sun, so his Earth-centred system could be used to predict their positions with reasonable accuracy. In a nutshell, Ptolemy's planets moved steadily around a perfect circle (the epicycle), and the centre of that circle moved steadily around another perfect circle (the deferent), which had the Earth at its centre. But Ptolemy introduced awkward complications such as the 'equant point' to account for the varying planetary speeds and motions and their actual elliptical orbits. Still, Ptolemy's *Almagest* was a mathematical *tour de force*, with all European astronomers up to the time of Tycho Brahe believing that the spheres carrying the heavenly bodies were physically real.

Ptolemy also wrote extensively on astrology and geography. In fact, his geographical treatise was as prominent in classical geography as his astronomy treatise was in classical astronomy.

See also Astronomy before history pages 12–13, Music of the spheres pages 14–15, Celestial predictions pages 26–27, Sun-centred universe pages 42–43, Laws of planetary motion pages 52–53

Right As astronomer and geographer, Ptolemy held supreme sway over the minds of scientific men down to the 17th century.

The body in question

Galen *c.*130–201

The Greek physician Galen compiled a system of medical knowledge that dominated the Western world until the Renaissance. Although his doctrines were revised in the seventeenth and eighteenth centuries, his notions of 'phlegmatic', 'melancholic' and 'choleric' personalities remain with us today – quite an achievement for a former doctor at a Roman gladiatorial school.

Brought up in Pergamon, near the Aegean coast of modern Turkey, Galen was influenced by the medical thinking of Hippocrates, the anatomical and physiological teachings of Herophilus and Erasistratus, and the philosophy of Plato and Aristotle. He believed that a good doctor must embrace not only the natural sciences but also the principles of logic underlying rigorous medical investigations. His prolific writings amount to a comprehensive and rational system of medical philosophy.

In Galenic medicine, the body was thought to be permeated by four vital fluids ('humours') – yellow bile, blood, phlegm and black bile. These were modified by four elements (fire, air, water, earth), four qualities (warmth, moisture, coldness, dryness), the four stages of life (childhood, youth, adulthood, old age) and the four seasons. In general, disease was ascribed to disturbances in the balance of the humours by, for example, diet and climate – hence the therapeutic practice of blood-letting. Also pervading the body were three kinds of vapour or spirit, formed in the three principal organs – heart, liver and brain.

Galen's greatness lay in his observations, not his theories. He was an active experimentalist who held public anatomy lectures and demonstrations. But prohibition of human dissection meant he was forced to use animals for research. Not everything he saw applied to human anatomy, so he occasionally got things wrong. But he made important contributions to our understanding of digestion, nervous impulses, the spinal cord, blood formation, respiration, the heartbeat and the arterial pulse – which he was the first to use as a diagnostic aid.

See also Aristotle's legacy pages 16–17, Human anatomy pages 44–45, Circulation of the blood pages 58–59, Boyle's *Sceptical Chymist* pages 70–71, Regulating the body pages 188–189, Nervous system pages 210–211, Conditioned reflexes pages 242–243, Nerve impulses pages 366–367

Right A page from Roger of Salerno's *Book of Surgery*, showing patients queuing up to receive treatment. By the 12th century, the Italian city of Salerno had established Europe's first medical school.

Zero

Brahmagupta *c.*598–665

Zero as a concept is fundamental to mathematics, and zero as a symbol seems to us now as familiar as the other nine numerical symbols, yet it took many centuries to make zero work in a mathematically precise way. The Babylonians, Egyptians, Greeks and Romans largely worked without a recognized zero symbol. In about AD 130 Ptolemy added the letter 'omicron' as a zero to the Sumerian number system based on 60, but this fell into disuse.

The starting point for the zero in use today was in India. In the seventh century Indian mathematicians often used a word to denote the absence of a number in their place-value decimal system (so avoiding confusion of, say, 305 with 35 or 350). This was represented as a dot, which developed into a recognizable zero symbol. Although the first indubitable use of this symbol is in an inscription on an Indian stone tablet dating from AD 876, earlier writings attest to the struggles to make zero a part of the number system. The Hindu astronomer and mathematician Brahmagupta tried to define the arithmetic operations involving zero 200 years earlier. Addition and subtraction were not a problem and any number multiplied by zero obviously gave zero. But division proved to be trickier. He stated incorrectly that zero divided by zero is equal to zero, and left fractions such as $^0\!/_2$ and $^3\!/_0$ as they were without finding an answer, although he did add that the first of these might be taken as zero.

About 200 years later the Jain mathematician Mahavira wrongly claimed that a number remains unchanged when divided by zero, although he rightly said that the square root of zero is zero. Even in the twelfth century, the leading Indian mathematician Bhaskara stated that dividing by zero gives a quantity, 'as infinite as the god Vishnu'. Despite these hiccups the Hindu decimal system spread westwards to Persia, the Arabian empire and Europe, as well as eastwards to China. A modern resolution of the pivotal role of zero in the mathematics of the infinitesimally small and the infinitely large awaited the work of the German mathematician Georg Cantor in the late nineteenth century.

See also Origins of counting pages 10–11, Algebra pages 38–39

Right Indian mathematics and astronomy developed hand in hand. Here astronomers observe stars using a theodolite while consulting Sanskrit texts on astronomy and trigonometry.

Unweaving the rainbow

Abu-'Ali Al-Hasan ibn Al-Haytham *c.*965–1039

The history of optics is intimately linked to one of our most loved natural phenomena – the rainbow. According to Aristotle a rainbow was formed by clouds acting as a huge lens and reflecting sunlight. Ptolemy later experimented with refraction – the bending of light as it travels from one medium to another (such as air, water or glass) – but ignored it altogether when considering the rainbow.

Around 1025 the Arab physicist al-Haytham, better known in the West as Alhazen, wrote the hugely influential *Treasury of Optics*. He explained how lenses worked, described the construction of the eye, made parabolic mirrors (now used in telescopes) and gave experimental values for the refraction of light, stating correctly that on entering a denser medium, light bends towards the perpendicular because it slows down. He also discussed the rainbow, but on this he unfortunately followed Aristotle's theories.

The crowning achievement in medieval optics was *De Iride* ('On the Rainbow') by the German professor of theology Theodoric of Freiburg. In 1304 he used a spherical flask filled with water to simulate water droplets. Combining experimentation with geometry, he discovered that a rainbow is formed from a ray of light being refracted as it passes from air to water, reflected internally within the droplet and finally refracted again as it re-emerges into the air. He also gave the correct angle of about 42 degrees between the centre of the rainbow and its halo. (As often happens in science, thousands of miles away two Arab scientists came up with the same conclusions at about the same time.)

What Theodoric could not work out was how a secondary rainbow was formed or why its colours were reversed. Three hundred years later the French philosopher René Descartes showed that the secondary rainbow is due to two internal reflections within the droplets.

See also Perspective pages 40–41, Newton's *Principia* pages 78–79, Spectral lines pages 130–131, Greenhouse effect pages 184–185

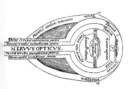

Above Anatomy of the eye, from a 16th-century edition of Alhazen's *Treasury of Optics*.

Right A 17th-century explanation of the role of the eye as part of the seeing process. Here the lens, the Sun and the eye all play crucial roles in the mystery of optics.

Artis et Naturæ, Tubi et Oculi, in speciebus solarib9 præsentandis consensus. N.° 3

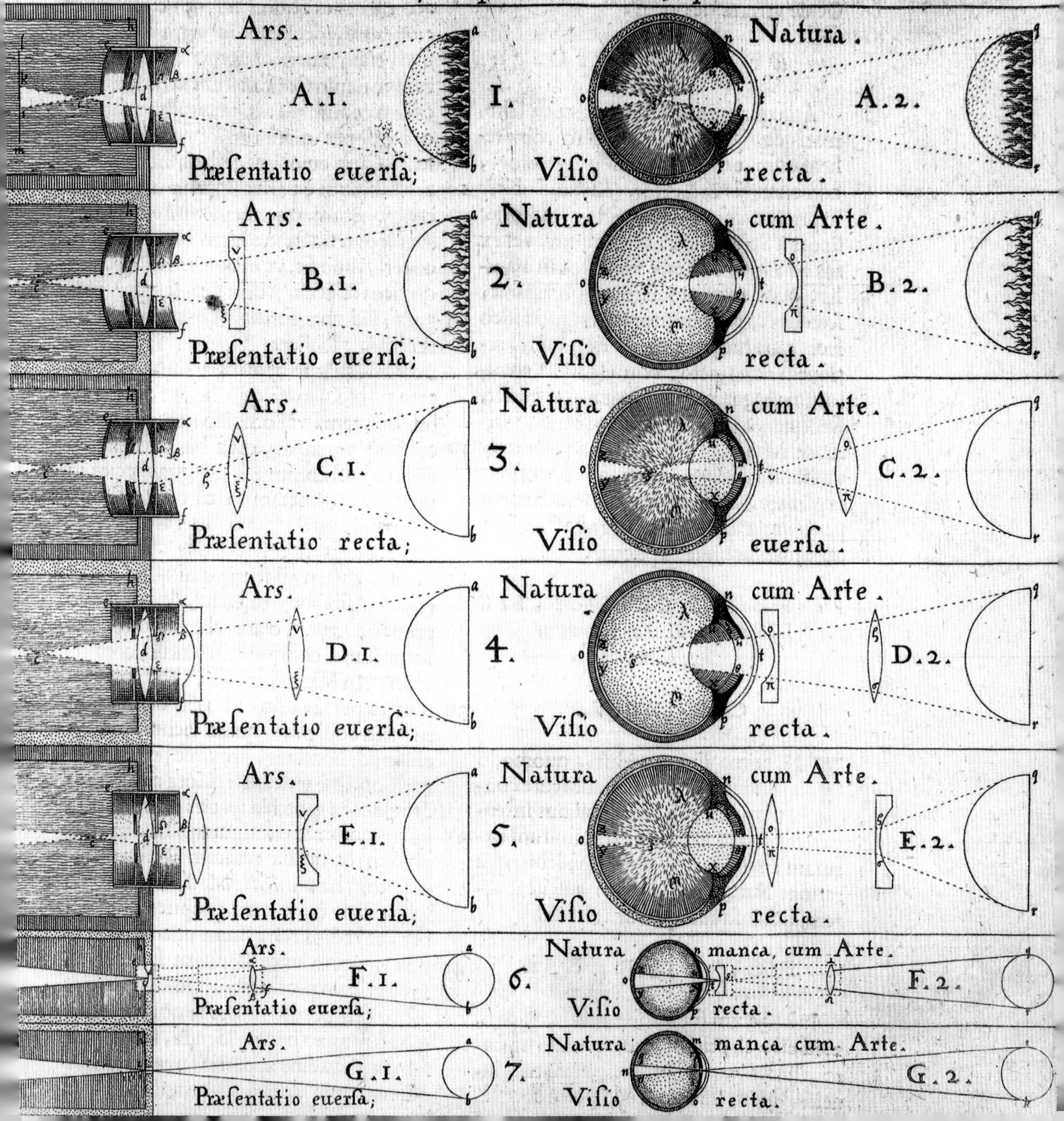
Ars.
A.I.
I.
Natura.
A.2.
Præsentatio euersa;
Visio
recta.
Ars.
B.I.
2.
Natura
cum Arte.
B.2.
Præsentatio euersa;
Visio
recta.
Ars.
C.I.
3.
Natura
cum Arte.
C.2.
Præsentatio recta;
Visio
euersa.
Ars.
D.I.
4.
Natura
cum Arte.
D.2.
Præsentatio euersa;
Visio
recta.
Ars.
E.I.
5.
Natura
cum Arte.
E.2.
Præsentatio euersa;
Visio
recta.
Ars.
F.I.
6.
Natura
manca, cum Arte.
F.2.
Præsentatio euersa;
Visio
recta.
Ars.
G.I.
7.
Natura
manca cum Arte.
G.2.
Præsentatio euersa;
Visio
recta.

Algebra

Fibonacci *c.*1170–1240

Our familiar ten numerals, from 0 to 9, have their origins in India, were adopted by the Muslim world and later transmitted to Europe during two centuries of feverish translation in Moorish Spain and Sicily. Twelfth-century Toledo was a particularly fertile place for scholars, with the world's classics of mathematics, science and literature translated between Arabic, Greek, Hebrew, Latin and Catalan. Many mathematical words come from this period. For example, 'algorithm' comes from the name of the ninth-century Arab mathematician al-Khwarizmi, and 'algebra' from the Arabic *al-jabr*.

In 1202 the Italian mathematician Leonardo of Pisa (nicknamed 'Fibonacci') published his *Liber Abbaci*, which popularized the new arithmetic of the 'nine Indian numerals' and the 'zephirum', or zero. Fibonacci devoted considerable space to mercantile mathematics, and an industry for professional 'calculators' developed to handle financial calculations. The first printed book on accountancy was Luca Pacioli's *Summa Arithmetica* (1494), and books on financial and navigational calculations became very popular in ports throughout Europe.

Hand in hand with the new numbers came the emergence of algebra. Most of mathematics up until this period was written in words, with problems and their solutions given rhetorically, numbers being largely for purposes of calculations. The Arabs extended Greek work on solving equations and initiated the move to a 'syncopated' style, with formulae given partially symbolically and partially in words. (Pacioli's own book was a mixture of the rhetorical and the algebraic.) Progress was slow and every book adopted its own algebraic symbolism. We find the '=' sign first used in England, and the '+' and '–' signs coming from Germany. By the time of Descartes, algebra seems very similar to our own, despite his incongruous use of '∞' for '='. And certainly by Newton the use both of the new numerals and of algebra was truly well-established.

See also Origins of counting pages 10–11, Zero pages 34–35

Right Representation arithmetic, from Gregor Reisch's *Margarita Philosophica*, 1508. The transition from the Roman abacus to the use of Hindu-Arabic numerals was surprisingly slow, with centuries of rivalry between the two systems.

Perspective

Leon Battista Alberti 1404–72, Piero della Francesca c.1412–92

The Renaissance revival of classical architecture and proportions was inspired by *On Architecture* by Vitruvius, a Roman architect who wrote in the first century BC. But the Renaissance was also the great age of realism in art. Italian painters were interested in giving their two-dimensional surfaces the illusion of three-dimensional reality. To achieve this they had to understand the laws of perspective.

The first Renaissance writer on perspective was the artist Leon Battista Alberti. His *La Pittura* appeared in Latin in 1435, while a later version, in Tuscan vernacular, was dedicated to the architect Filippo Brunelleschi, who used perspective in his work. The standard problem was the construction in perspective of the chequered floor (*pavimento*) seen in many paintings of the time. Alberti's explanation in *La Pittura* lacked the detail required for any artist to put it into practice, and was probably aimed at impressing his wealthy patrons. The first work to show a thorough understanding of the mathematical rules for achieving perspective was by Piero della Francesca. Completed around 1488, *De Prospectiva Pingendi* ('On Perspective for Painting') was never published but existed in manuscript form, with sections finding their way into later printed works such as Luca Pacioli's *De Divina Proportione* (1509), illustrated by his friend Leonardo da Vinci. From the standard *pavimento* problem, Piero proceeded to deal with more complicated shapes such as the human figure.

But constructions for such realistic figures were beyond most people and an industry developed for the manufacture of various mechanical aids such as pinhole cameras and reference grids to guide artists. Many examples appear in the work of the German artist Albrecht Dürer. He completed his own treatise on proportion in 1532 but thought the mathematics too hair-raising for his readers and published a simplified version two years later – the first-ever mathematics book printed in German.

See also Euclid's *Elements* pages 20–21, Unweaving the rainbow pages 36–37, Non-Euclidean geometry pages 144–145, Fractals pages 446–447, Quasicrystals pages 482–483

Right 'The Flagellation of Christ' by Piero della Francesca, which shows many features from his treatise on perspective, including the chequered *pavimento* and architectural elements.

Sun-centred universe

Nicolas Copernicus 1473–1543

Nicolas Copernicus was born in 1473 in Torun, Poland. In 1496 he went to study law and medicine in Italy. There he became interested in astronomy. Late-fifteenth-century Earth-centred astronomy was in trouble. For a start, the calendar was out of step. What's more, Ptolemy's 'equant point' was thought to be an 'unnatural' complication; his lunar orbit had the Moon's apparent size changing greatly throughout the month, which it clearly did not; and his elaborate overall approach to planetary orbits was incoherent. That each planet's orbital period was related to the Sun's year also posed a problem.

In 1503 Copernicus returned to Poland to serve as canon, under his uncle, at the cathedral at Frombork (now Frauenberg, in Germany). His untaxing ecclesiastical duties allowed him to concentrate on astronomy. He reformed the subject at a stroke by removing the troublesome equants and relocating the Sun. The Sun ceased to be a member of the seven moving astronomical bodies, and took a position at the centre of the system. The Earth was demoted, becoming the third planet out from the Sun, and the Moon ignored the Sun and stayed in orbit around the Earth.

Copernicus divided the planets into two sensible groups: those inside the Earth's orbit and those outside. He fixed the ordering of the planets, whereas Ptolemy's ordering had been arbitrary. The distances of the planets from the Sun and the orbital periods of the planets could be calculated, and were found to be harmoniously linked. The movement of the Earth easily and simply explained the backward ('retrograde') motions of Mars, Jupiter and Saturn.

Rheticus (Georg Joachim von Lauchen) published a summary of Copernicus's work in 1539 and supervised the publication of the complete *De Revolutionibus* in 1543, just as Copernicus died. This was immediately adopted as a manual for those interested in working out planetary positions.

See also Astronomy before history pages 12–13, Music of the spheres pages 14–15, Celestial predictions pages 26–27, Earth-centred universe pages 30–31, A new star pages 48–49, Laws of planetary motion pages 52–53, Heavens through a telescope pages 54–55, Planetary distances pages 74–75

Right The Copernican system. This 1660 engraving dramatizes the centrality of the Sun, and cleverly portrays the apparent cycle of the zodiac through the daytime sky and the night sky.

Human anatomy

Andreas Vesalius 1514–64

The rebirth of investigative science during the Renaissance had a profound impact on anatomy and physiology. Once again people started to question nature rather than relying on knowledge based on ancient authority or superstition. In the hands of a young Belgian physician, Andreas Vesalius, the increasingly respectable practice of human dissection yielded new knowledge of the body, which he transmitted to the Latin literate in his towering masterpiece *De Humani Corporis Fabrica* ('On the Structure of the Human Body'), published in Basle, Switzerland, in 1543.

Born in Brussels, Vesalius studied medicine at Louvain (in what is now Belgium) and in Paris. He obtained his medical degree in Padua in Italy at the age of 23, and stayed on at the university as a professor of anatomy. In typically unorthodox style, he performed his own anatomical demonstrations rather than delegating the dissections to a menial. In 1539, his fame growing, he was given the bodies of executed criminals for his research as well as cadavers for teaching purposes. Four years later he presented his findings in *Fabrica*, seven superb books of text and illustrations.

The value of *Fabrica* lies not just in the correction of Galen's anatomical errors, such as the belief that a bile duct opened into the stomach as well as the duodenum or that the human mandible was composed of two bones. Vesalius also set himself the task of making anatomy the foundation of all medicine. In his work, natural science was divorced from philosophy, perhaps for the first time. Anatomical knowledge gained from direct contact with the human body was shown to be essential for the progress of medicine. He challenged the popular disdain of doctors to performing their own surgical work and inspired students to learn anatomy by doing their own dissections. His accurate illustrations of muscle-and-nerve men posing naturally among ancient ruins are icons of the birth of a new medicine.

See also The body in question pages 32–33, Circulation of the blood pages 58–59, Nervous system pages 210–211

Right Vesalius relied on scrupulous dissections and the latest in artistic techniques and printing for the more than 200 woodcuts in his book.

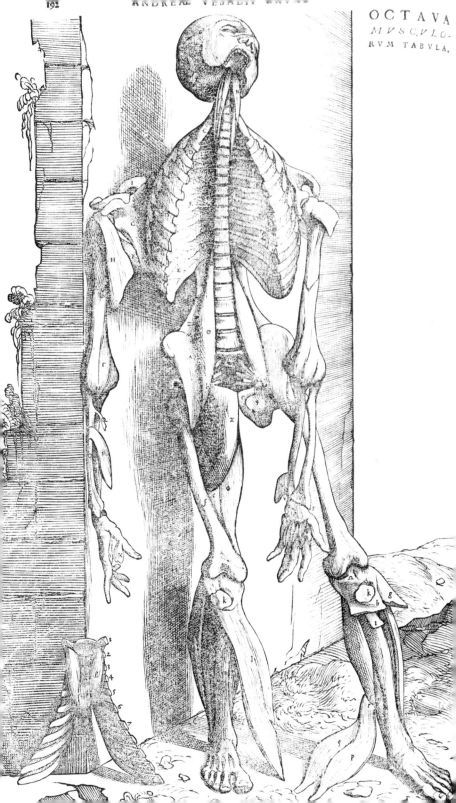

Fossil objects

Conrad Gesner 1516–65

The sixteenth-century Swiss naturalist and physician Conrad Gesner has been acclaimed as the greatest naturalist of his time. In total he published 72 works, and left 18 in progress. In 1565, the year of his death in Zurich from the plague, he completed his innovative *De Rerum Fossilium* ('On things dug up from the Earth'), which marked the start of the emerging science of palaeontology. Also notable are his *Historia Animalium*, in which he attempted to describe all the animals then known, and his *Bibliotheca Universalis*, which contained the titles of all the books then known in Hebrew, Greek and Latin, with criticisms and summaries of each.

To Gesner and his contemporaries the word 'fossil' included any natural object recovered from the ground, be it a mineral or the remnant of an organism. So it is hardly surprising that they struggled to make sense of these 'stony concretions', as Gesner called them. Even today fossils can be difficult to interpret: the process of fossilization sometimes not only obscures the original nature of organic remains, but also produces organic-looking objects that are actually inorganic in origin – as witness the debate about Martian microfossils.

Gesner's classical training taught him to give pride of place to naming and classifying the fossils he described. Most importantly, he was concerned with precise identification. His book was the first to present fossil illustrations (albeit fairly crude woodcuts) so 'students may more easily recognize objects that cannot be very clearly described in words'. Although some of his interpretations turned out to be wrong, many were remarkably astute. His depiction, for example, of sharks' teeth alongside 'tongue-stones', commonly thought to have fallen out of the sky, reveals that the significance of their similarity was not entirely lost on him: 'tongue stones' are indeed sharks' teeth.

Above Gesner's sketch of 'dart-like' fossils.

See also Geological strata pages 72–73, Fossil sequences pages 132–133, Prehistoric humans pages 148–149, *Archaeopteryx* pages 180–181, Burgess Shale pages 260–261, Oldest fossils pages 410–411, Martian microfossils pages 516–517

Right Gesner had difficulty interpreting the fossils of extinct aquatic molluscs known as ammonites. Some he identified as snail shells, whereas others he mistook for coiled snakes.

A new star

Tycho **Brahe** 1546–1601

Tycho Brahe, an aristocratic Dane, was a dedicated astronomical observer who, thanks to the financial support of King Frederick II, built a superb observatory on the Baltic island Hven (now Ven). Tycho saw that the planets weren't following the paths fixed by Copernicus, so he decided to dedicate his life to observing planet and star positions more precisely than anyone before.

After sunset on the clear night of 11 November 1572, Tycho noticed the flaring out of a new star in the constellation of Cassiopeia. He followed changes in its colour and brightness for 15 months and published his results in his book *Progymnasmata*. This 'nova' was not in fact a new star but an existing one that had exploded and increased enormously in brightness.

His observations of the great comet of 1577 were even more exciting. Tycho tried to measure how far it was from the Earth by comparing its position against distant stars, as seen during the night by astronomers throughout Europe. As no shift, or 'parallax', was found, Tycho concluded that the comet was beyond the Moon. Aristotle had regarded comets as meteorological objects, just above the clouds. Tycho proved that they were out there with the planets. The varying form and brightness of the great comet convinced him that it had an elliptical path, meaning that it must have crashed through the crystalline spheres supposedly carrying the heavenly bodies. Tycho suggested that these spheres didn't exist, and that the planets were unsupported in their wanderings.

Tycho's dedication to the design of large, stable and remarkably accurate naked-eye instruments transformed astronomical observation. He was also proud of being a cosmologist. This was unjustified. His 'Tychonic system' was a flawed and conservative halfway house between Ptolemy and Copernicus: the Earth became stationary again, with the Sun and Moon as orbiting satellites and the other five known planets as satellites of the Sun.

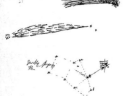

Above Tycho's drawings of the comet of 1577.

See also Earth-centred universe pages 30–31, Sun-centred universe pages 42–43, Natural magnetism pages 50–51, Halley's comet pages 84–85, A cometary reservoir pages 360–361, Supernova 1987A pages 488–489

Right Tycho in his observatory. The huge quadrant was used to measure the height of celestial objects, while the two clocks, described by Tycho as having 'the greatest possible accuracy', were used to establish the timings.

Natural magnetism

William **Gilbert** 1544–1603

The phenomenon of magnetism in which little magnetic rods align themselves in the same direction on the Earth's surface – duly pointing north and south – was well known to all ancient civilizations, and by the thirteenth century there were reports of sailors using magnetic needles floating on water as an early form of the compass. But explanations for this peculiar behaviour were rooted in old Greek models of the universe as a fixed celestial sphere from which all sorts of benign and malign influences emanated to shape our lives, including magnetic forces: magnets aligned themselves with 'poles' of the celestial sphere, as if governed by the heavens.

It was the English physician William Gilbert (or Gylberde) who put to rest these fanciful notions and laid the foundations of magnetism as a science with the publication of his book *De Magnete* in 1600. He postulated that the Earth was itself a giant lodestone. By making a small, spherical, permanent magnet out of ordinary magnetic material, he was able to show that tiny magnetic needles placed on the surface of this sphere (called a 'terrella') behave just like lodestones on the Earth's surface. Most strikingly, the experimental needles on the terrella mimicked declination, the familiar dipping of a compass out of the horizontal plane as one moves from Earth's equator towards the poles. 'It has been settled by nature', Gilbert concluded, '… that in the pole itself shall be the seat, the throne as it were, of a high and splendid power'.

Gilbert was one of the first true experimenters, relying on observation rather than philosophical speculation and preceding by several years that most famous advocate of the experimental method, Francis Bacon. Gilbert's demonstration that the Earth and not the heavens held the power went far beyond the force of magnetism and influenced all thinking about the physical world.

See also Aristotle's legacy pages 16–17, Earth-centred universe pages 30–31, A new star pages 48–49, Humboldt's voyage pages 114–115, Electromagnetism pages 134–135, Sunspot cycle pages 158–159, Maxwell's equations pages 186–187, Climate cycles pages 276–277, Eel migration pages 290–291, Geomagnetic reversals pages 304–305, Solar wind pages 388–389, Plate tectonics pages 414–415

Right Forging a magnet, from Gilbert's *De Magnete*, 1600. The experimenter hammers a hot iron bar oriented north–south with the Earth's magnetic field.

Laws of planetary motion

Johannes **Kepler** 1571–1630

Johannes Kepler was a German mathematical genius obsessed by number puzzles. He set himself the task of understanding why the planetary orbits had the shapes and sizes that they did, and how this related to the time it took a planet to complete its orbit.

Because of religious persecution, the Lutheran Kepler left Graz in 1598 and travelled to Prague to work with Tycho Brahe, succeeding him as Imperial Mathematician in 1601. Tycho was a superb observational astronomer and Kepler was set to work on interpreting Tycho's observations of Mars. After many tedious trials, rejecting time and again models that disagreed with Tycho's accurate data, Kepler finally realized that Mars had an elliptical orbit and that the Sun was at one of the 'foci'. The shackles of circularity that had dominated the analysis of planetary orbits for 2,000 years were finally broken.

Kepler's first law was published in 1609 in his book *Astronomia Nova*, together with his second law (discovered before the first). The second law described how the imaginary line from Sun to planet sweeps out equal areas in equal times, explaining why a planet moves faster when closer to the Sun. Kepler was fascinated with heavenly harmonies. His third and final law – the square of a planetary year is proportional to the cube of its average distance from the Sun – was published in 1619, hidden in his mystical *De Harmonica Mundi*.

Kepler tried to understand the forces underlying planetary movement, suggesting (incorrectly) that there was a magnetic interaction between the planets and the Sun. His *Rudolphine Tables* of 1627 were the first modern astronomical tables and used the newly invented logarithms of John Napier. They allowed astronomers to predict the positions of planets at any time in the past, present or future, and their accuracy brought Kepler much fame.

See also Astronomy before history pages 12–13, Music of the spheres pages 14–15, Earth-centred universe pages 30–31, Sun-centred universe pages 42–43, A new star pages 48–49, Logarithms pages 56–57, Newton's *Principia* pages 78–79

Right Based on the five 'Platonic solids', Kepler's 1596 model of the orbits of the planets was claimed to reveal God's scheme for the heavens.

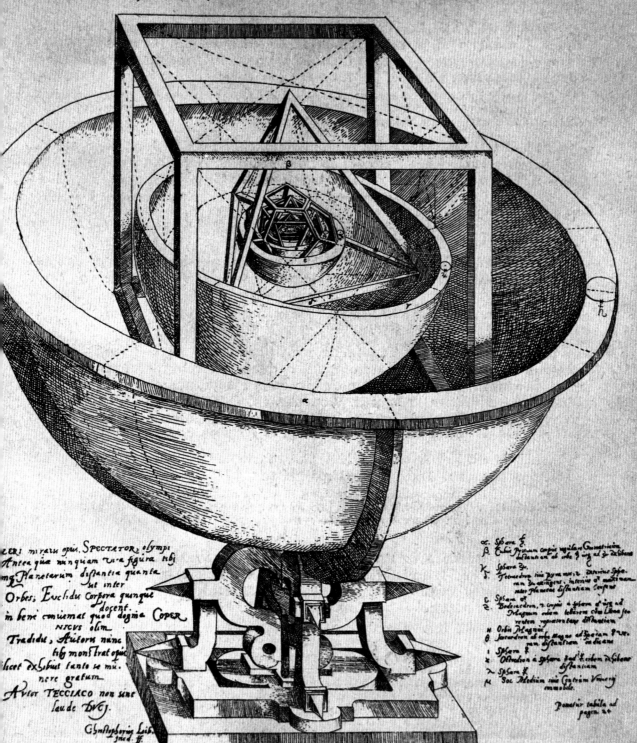

Heavens through a telescope

Galileo **Galilei** 1564–1642

In the summer of 1609, when Galileo Galilei was in Venice, he heard that two Dutchmen had combined two pieces of curved transparent glass to produce a device that made distant objects appear nearer. At that time convex lenses had been in use for about 300 years and concave ones for about 150 years. In the autumn of 1608, Hans Lippershey and Zacharias Janssen, two independent spectacle-lens-makers working in the shadow of the glass factory at Middelburg, the Netherlands, both devised telescopes. In a few months the news had spread throughout Europe. (Lippershey's apprentice is thought to have been the first to notice the effectiveness of a weak convex lens held at arm's length combined with a strong concave lens near the eye.)

Galileo was technically skilled with a gift for improvisation. By August 1609 he had made a telescope that magnified objects 8 times, and had improved this to 20 times by the end of the year. In early December 1609 he discovered that the Moon had mountains and had measured some of their heights. By mid-January 1610 he had discovered four satellites around Jupiter and named them the Medicean Stars in honour of the Grand Duke Cosimo. Looking at the Milky Way, he realized that what appeared to the naked eye as a luminous blur was in fact a myriad of faint stars. It became clear that planets, unlike stars, had discs, and Venus showed phases just like the Moon. Also the Sun, far from being the symbol of perfection proposed by Aristotle, was spotty and impure, and was rotating once every 25 days.

Galileo published his findings quickly before he was overtaken. On 13 March 1610 he sent an advanced copy of his *Sidereus Nuncius* ('Starry Messenger') to the Florentine Court. By 19 March not only had 550 copies been printed, but they had also all been sold.

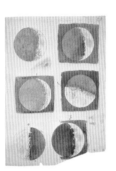

Above 'Six Phases of the Moon', Galileo's original telescopic drawings for *Sidereus Nuncius* (1610).

See also Sun-centred universe pages 42–43, Saturn's rings pages 68–69, Sunspot cycle pages 158–159, Spiral galaxies pages 160–161, 'Canals' on Mars pages 200–201, Our place in the cosmos pages 280–281, Galileo mission pages 514–515, Water on the Moon pages 522–523

Right Galileo's telescopic observations made him confident of the truth of the heliocentric Copernican system, a view he was forced to recant in 1633 under pressure from the Inquisition.

Logarithms

John Napier 1550–1617

John Napier, eighth baron of Murchiston, spent much of his time managing his Scottish estate and embroiled in religious politics. But he also made two major contributions to mathematics that saved astronomers, navigators and engineers from hours of tedious arithmetical calculations – Napier's 'bones' and logarithms.

Napier's bones were one the first modern computational aids. They were rods carved with multiplication tables that could be arranged in a lattice pattern. For long multiplications, they essentially cut out all the intermediate steps, reducing the calculation to simple additions. In this way they prefigured Napier's invention of logarithms.

At the heart of the idea of a logarithm is the relationship between an arithmetical series (such as 0, 1, 2, 3, 4, 5, 6…) and a geometrical series (1, 2, 4, 8, 16, 32, 64…). Here the base number is 2 and a multiplication such as $4 \times 16 = 64$ can be rewritten in terms of this base ($2^2 \times 2^4 = 2^6$), thereby turning it into the simple addition of the powers ($2 + 4 = 6$). As the numbers become larger, this simple device becomes ever more efficient. Also, Napier's insight was that any number could be written in terms of the power of a base. For example, 10 is approximately equal to $2^{3.32}$.

Napier's original tables, *A Description of the Marvelous Rule of Logarithms* (1614), used a slightly more complicated idea than a simple base and were aimed at simplifying trigonometric calculations for mariners. If at sea a calculation took an hour then the result would be about an hour out – Napier's logarithms cut the error down to just a few minutes. One of Napier's great admirers was Henry Briggs, the first Savilian Professor of Geometry at Oxford. They both agreed that a more practical table could be constructed using the now familiar base 10. But Napier died in 1617, and it fell to Briggs to compile it. Today the calculator has consigned logarithm tables and slide-rules to history.

See also Laws of planetary motion pages 52–53, Difference Engine pages 138–139, The computer pages 340–341, Four-colour map theorem pages 450–451, Fermat's last theorem pages 506–507

Right A popular aid to calculation, Napier's rods or 'bones' were initially made as four-sided ivory or wooden rods. In this later form the rods are mounted in a case and can be rotated.

1	2	3	4	5	6	7	8	9	10	11	12
2	3	4	5	6	7	8	9	10	11	12	13
3	4	5	6	7	8	9	10	11	12	13	14
4	5	6	7	8	9	10	11	12	13	14	15
5	6	7	8	9	10	11	12	13	14	15	16
6	7	8	9	10	11	12	13	14	15	16	17
7	8	9	10	11	12	13	14	15	16	17	18
8	9	10	11	12	13	14	15	16	17	18	19
9	10	11	12	13	14	15	16	17	18	19	20
10	11	12	13	14	15	16	17	18	19	20	21
11	12	13	14	15	16	17	18	19	20	21	22

Circulation of the blood

William Harvey 1578–1657

The English physician William Harvey was more interested in medical research than clinical practice. He was particularly fascinated by the heart, blood vessels and blood. In Harvey's day, the teachings of the Greek physician Galen were still accepted, and Harvey himself believed Aristotle's idea that blood contained 'vital spirit' that imbued animals with life. It was against this ancient philosophical backdrop that Harvey carried out one of the greatest experimental programmes of the scientific revolution and thus discovered the circulation of the blood.

Harvey's faith in the traditional view began to wane when he noticed the links between the heart and arteries and the liver and veins – two supposedly distinct systems according to accepted wisdom. Added to this information were the recent discoveries of the transmission of blood through the lungs, and of one-way valves in veins, which indicated that venous blood could travel only towards the heart. Building on these findings with his own experiments on many species of animals, Harvey established that the heart was a muscle with one-way valves and that it pushed blood out by contraction. He was troubled by his calculation that within half an hour the heart would pump out the whole mass of blood. Where did the blood go and new supplies come from? Might the same blood flow in a circle from the heart through the arteries into the veins and back again to the heart? He tested this hypothesis in a series of simple and elegant experiments, and in 1628 announced his incontrovertible conclusions in a small book of only 72 pages. Its short title is *De Motu Cordis et Sanguinis* ('On the Motion of the Heart and Blood').

Despite his careful observations, Harvey could not tell how the blood passed from the tiniest arteries to the veins. In 1661 Marcello Malpighi used a simple single-lens microscope to see in the lungs of a frog what the naked eye could not see: the passage of blood through the capillary vessels. The circle was complete.

See also The body in question pages 32–33, Human anatomy pages 44–45, Microscopic life pages 76–77, Regulating the body pages 188–189, Blood groups pages 236–237, Sickle-cell anaemia pages 358–359, Structure of haemoglobin pages 390–391, Nitric oxide pages 496–497

Right An illustration from a 1628 edition of Harvey's treatise illustrates the one-way system of valves in the veins and its role in the circulation of blood.

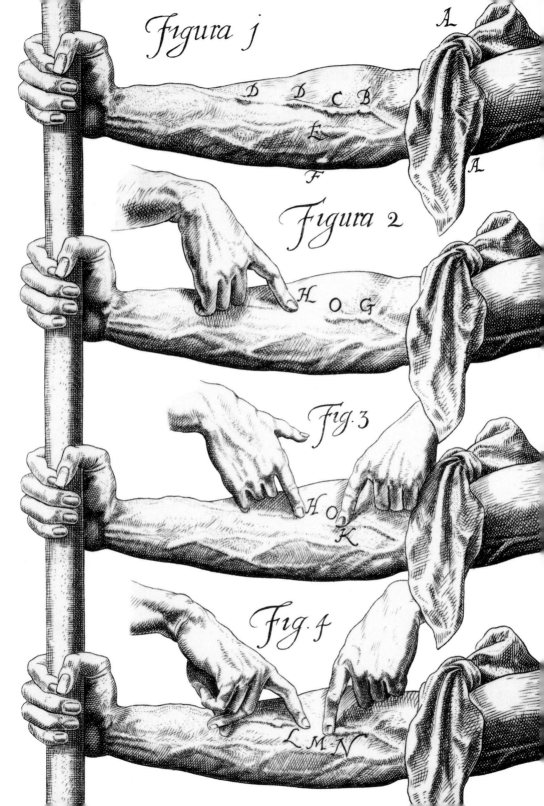

Falling objects

Galileo **Galilei** 1564–1642

According to legend, Galileo dropped two cannon balls of different weights from the Lean¡
Tower of Pisa to prove they would hit the ground simultaneously. Although the story is
apocryphal, he certainly performed experiments that radically altered our understanding of
the laws of motion. These he discussed in his *Mathematical Discourses and Demonstrations on Tw*
New Sciences, published in 1638. His 'new sciences' were all the more extraordinary because
they seemed to clash with everyday observation.

Aristotle had taught that heavy objects fall faster than light ones. This, Galileo showed,
was a conclusion erroneously drawn from the fact that air resistance slows the fall of light
objects such as feathers with relatively large surface areas. To test his claim he slowed down
the process by rolling balls down inclined planes and measuring the short time intervals by
human pulse and the quantity of water leaking from a large vessel. From these experiments
he deduced that in a vacuum all objects accelerate at the same rate, no matter what their
weight or constitution.

What's more, Galileo realized that any body moving on a horizontal plane will continue
at the same speed unless a force opposes it. Aristotle believed that to keep an object moving
a force had to be continually applied – after all, a block of wood pushed at a constant spee
across a table quickly comes to rest when we stop pushing. But Galileo showed that this
common-sense view neglected a hidden force: the frictional force between the surface and t
object. Today we call a body's tendency to continue in horizontal motion 'inertia', a concep
developed by Newton in his three laws of motion.

See also Newton's *Principia* pages 78–79, 'Weighing' the Earth pages pages 112–113

Right Imagine a heavy cannon ball attached by a string to a light musket ball. What will happen if they are released together? Using
thought experiment alone, Galileo proved that they must fall at the same speed.

Transit of Venus

Jeremiah Horrocks 1619–41

The accurate measurement of the Sun's distance from Earth was one of the main endeavours of seventeenth-century astronomers. It is an important number, providing the scale-bar of the planetary and stellar system and featuring prominently in the Newtonian gravitational calculation of the solar mass.

In 1627 Johannes Kepler's knowledge of the planetary orbits enabled him to predict when a planet would pass in front of the Sun. The English astronomer Jeremiah Horrocks improved on these calculations, and on 24 November 1639 became the first person to observe such a 'transit' of Venus. He also realized that accurate transit timings from different places on Earth would allow the Earth–planet distance and thus the Earth–Sun distance to be worked out geometrically, an idea popularized by Edmond Halley. Halley witnessed a transit of Mercury in October 1677. But as this was much quicker than a Venus transit (three hours as opposed to seven hours), it did not permit accurate distance estimates.

Unfortunately, the next Venus transits were not until 6 June 1761 and 3 June 1769. Several hundred astronomers from around the world travelled to distant places to time them, including Tahiti (on an expedition led by James Cook), Hudson Bay, Ireland and the Russian-Chinese border. The observations were bedevilled by optical irregularities, making it hard to establish when Venus entered and left the solar disc. The small black disc of Venus appeared reluctant to break free from the solar limb, and when it eventually, teardrop-like, became detached, it was seen to be well on to the disc. Astronomers standing side by side, using different instruments, recorded timings that differed by tens of seconds. Venus transits on 8 December 1874 and 6 December 1882 were equally frustrating. Still, the collective measurements did help to refine estimates of the Sun's distance from Earth, coming within 10 per cent of current radar estimates of 149,597,870 kilometres, accurate to within 2 kilometres.

See also Celestial predictions pages 26–27, Laws of planetary motion pages 52–53, Planetary distances pages 74–75, Newton's *Principia* pages 78–79, Distance to a star pages 150–151

Right Tahiti's northernmost point, Point Venus, where Captain James Cook observed the transit of Venus in 1769.

Atmospheric pressure

Blaise Pascal 1623–62

Since antiquity engineers have known that it is possible to pump water to a height of almost 10 metres but no higher. In the 1640s several scientists began to link this limit to atmospheric pressure, the weight of air above the surface of the Earth.

In 1647 and 1648 Blaise Pascal, a mathematical prodigy from Clermont-Ferrand in France, carried out a series of experiments using the newly invented barometer that led to the first clear understanding of atmospheric pressure. The barometer consisted of a tube just over a metre long, sealed at one end, filled with mercury and inverted into a dish. To the surprise of many contemporary scientists, not all the mercury flowed out of the tube, leaving a gap at the top of the tube which Pascal and others correctly interpreted as a vacuum.

This phenomenon can be explained by the idea that atmospheric pressure is the weight of the column of air above the barometer and that this weight balances the weight of mercury in the tube. At sea level the weight turns out to be roughly equal to the weight of a column of mercury 760 millimetres high and to a column of water just under 10 metres high, a fact that explained the pumping limit that engineers had wondered about for so long.

Pascal's experiments consisted of making simultaneous measurements of the pressure at various points up a mountain near his home – chronically ill, he sent his strong young brother-in-law to do the actual work. These showed that the atmospheric pressure dropped with altitude. Pascal guessed that the pressure must fall to zero at the top of the atmosphere, and this led him to the revolutionary idea that a vacuum exists above it. The international unit of pressure is the 'pascal' (Pa), equivalent to one newton per square metre. Atmospheric pressure at sea level is 101,325 Pa, often called 1 atmosphere.

See also Trade winds pages 86–87, Combustion pages 94–95, Hydrogen and water pages 98–99, Greenhouse effect pages 184–185, Weather forecasting pages 284–285, Gaia hypothesis pages 432–433, Ozone hole pages 438–439

Right Meteorological experimentation with a barometer, 1688. Pascal repeated similar experiments using red wine instead of mercury. This required a tube 10 metres long.

Rules of chance

Blaise Pascal 1623–62, Pierre de Fermat 1601–65

Gambling is as old as civilization and yet the study of games of chance in terms of probabilities did not begin until the seventeenth century. The 'game of hazard' was popular in Europe during the Middle Ages. Mentioned in Dante's *Divina Commedia*, this three-dice game involved one player throwing and the other guessing the total score. The 56 possible outcomes were enumerated in the thirteenth-century poem *De Vetula*, but what concerned mathematicians most was a problem known as 'division of stakes'. If a game is left unfinished, how should the stakes be divided fairly between the two players?

The first correct solution to this question was given in the correspondence of 1654 between the two French mathematicians Blaise Pascal and Pierre de Fermat. Fermat looked at all possible outcomes in terms of a 'probability tree', a method that soon becomes unwieldy as the number of games increases. Pascal preferred a more numerical approach, using a general mathematical formula known as the 'binomial theorem'. An easily remembered summary of the expansion of this formula is given by 'Pascal's triangle', which consists of rows of numbers, each digit the sum of the two on either side of it in the row above. Given that a game has only two outcomes – win or lose – the triangle provides a quick way of working out the possible numbers of wins and losses in a sequence of games.

Both Pascal and Fermat gave their answers in terms of ratios. The first treatment of probabilities as a measure between zero and one was provided by the Swiss mathematician Jakob Bernoulli in *Ars Conjectandi* ('Conjectural Arts'), published posthumously in 1713. He also made the important distinction between theoretical probabilities and experimental ones, thus beginning the mathematics of population sampling. In 1718 the Frenchman Abraham de Moivre described the familiar bell-shaped curve for the distribution of population characteristics and established conditions for accurate sampling. Such mathematics revolutionized financial instruments such as life policies and annuities.

See also Measuring variation pages 212–213

Right Beating probability: 'The Cheaters' by Georges de la Tour, c.1635. Despite high stakes, early gamblers played with little idea of the percentages for or against them.

Saturn's rings

Christiaan Huygens 1629–95, James Clerk Maxwell 1831–79

When Galileo Galilei turned his early telescope on Saturn, in 1610, he thought the planet had ears. The timing was unfortunate because his line of sight was nearly in the plane of Saturn's rings. When he looked again, two years later, he was amazed to see no ears at all. The rings, now viewed edge on, had disappeared.

As the seventeenth century progressed, the quality of telescopes improved. The explanation for the missing ears was first given in 1656 by the Dutch physicist Christiaan Huygens in *Systema Saturnium*, four years after he had discovered Saturn's largest satellite, Titan. His telescope would bear a magnification of 50 times. To ensure priority, even before he published his book, he presented the solution in the form of an anagram, which translates as: 'the planet is surrounded by a thin flat ring, nowhere touching, and inclined to the ecliptic'.

The rings were initially believed to be solid or liquid, but in 1675 Giovanni Domenico Cassini discovered a gap in the system, subsequently named the 'Cassini division' in his honour. A further division was found in 1837 by Johann Franz Encke, director of the Berlin Observatory.

The nature of the rings was the topic of the University of Cambridge's Adams Prize Essay in 1855. The Scottish physicist James Clerk Maxwell won by showing theoretically that only a collection of individual orbiting particles would be stable. It was only when viewed from a distance that these gave the impression of a solid ring.

Proof of Maxwell's concept came spectroscopically, and thanks to the Doppler effect. The outer edge of a solid ring would rotate faster than the inner edge; but individual orbiting particles obey Kepler's harmonic law, and their velocity decreases relative to their distance from the planet. In 1895 the American astronomer James Keeler, using a spectrometer attached to the Lick 91-centimetre refractor, proved that Maxwell's ideas were correct.

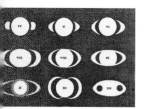

Above Huygens showed how a flattened ring around Saturn, and inclined to its orbital plane, could explain the planet's changeable 'ears'.

See also Laws of planetary motion pages 52–53, Heavens through a telescope pages 54–55, Origin of the Solar System pages 104–105, Spectral lines pages 130–131, Doppler effect pages 156–157

Right Saturn's rings proved to be made up of thousands of ringlets and narrow divisions, though their origin remains a matter of much discussion.

Boyle's *Sceptical Chymist*

Robert Boyle 1627–91

Robert Boyle is often regarded as the founder of modern chemistry. Even so, this son of an Anglo-Irish aristocrat was no scourge of the alchemists, but himself an adept who eagerly sought the Philosopher's Stone. His great work, *The Sceptical Chymist* (1661), is not so much a refutation of alchemy as a critique of those parts of it he considered ill-founded. Boyle was certainly not distinguishing 'alchemy' from 'chemistry': the transitional term 'chymistry' contains elements of both. Rather, his objective was to distinguish between the charlatans and superstitious 'ignorant puffers', who blindly followed recipes, and the informed 'chymical philosophers', who pursued the art of transmutation in a systematic 'scientific' way.

The main targets of Boyle's scepticism were the chemical theories of Aristotle, Paracelsus and Jan Baptista van Helmont. Aristotelians maintained that there were four elements from which all things were composed: earth, air, fire and water. In the sixteenth century, Paracelsus had veneered Aristotle's quartet with a system of three 'principles' from which all matter was composed: sulphur, quicksilver (mercury) and salt. This was an elaboration of the earlier alchemical sulphur–mercury theory of metals. Boyle dismissed the followers of Paracelsus and van Helmont – the so-called spagyrists – as 'vulgar chymists'.

But he refrained from offering any alternative system, stating only that there are probably more than four (perhaps even more than five) elements. His much vaunted definition of an element – 'primitive and simple, or perfectly unmingled bodies; which not being made of any other bodies, or of one another, are the ingredients of which all those called perfectly mixt bodies are immediately compounded' – did not in itself add much to Aristotle's.

Boyle expressed scepticism that such elements could in fact exist. But one of his most felicitous contributions was his insistence on experiment to learn 'what heterogeneous parts particular bodies do consist of', as exemplified by his iconic investigations into the properties of gases with the 'air pump'.

See also Aristotle's legacy pages 16–17, The body in question pages 32–33, Combustion pages 94–95, Atomic theory pages 124–125, Periodic table pages 196–197, Changes of state pages 198–199, Brownian motion pages 254–255, A new state of matter pages 510–511

Right 'Experiment with an Air Pump' by Joseph Wright of Derby, c.1730. Boyle pioneered the use of the air pump to investigate the characteristics of air and the vacuum.

Geological strata

Nicolaus Steno 1638–86

Nicolaus Steno, also known as Niels Stensen, originally left his native Copenhagen in 1660 to study medicine at Leiden in the Netherlands. He later settled in Florence, where he became a renowned anatomist and personal physician to Duke Ferdinand II. His interest in fossils was aroused when at the request of his patron he examined the head of a shark and saw that its teeth resembled fossil objects commonly known as 'tongue-stones'. He rightly concluded that they were in fact fossilized sharks' teeth.

In *Prodromus*, published in 1669, his stated aim was that 'given an object possessing a certain form, and produced by natural means, to find in the object itself evidence showing the position and manner of its production'. In question were the diverse organic and inorganic objects known at the time as 'fossils' and how they were formed. Steno showed that quartz crystals grow by precipitation just like crystals in the laboratory. By contrast, shells – fossilized or otherwise – show a pattern of growth that reflects the 'vital' accretionary processes of their former occupants, so could not have grown within the rock.

But for the organic origin of fossil shells to be plausible, Steno also had to explain why they were found inland within rocks high above sea level. Drawing on field studies in Tuscany, he argued that strata were originally precipitated on the seafloor as successive horizontal layers of sand, gravel and enclosed shells. This implied that the uplifted and tilted strata commonly found today reflect subsequent changes in the Earth's history. Steno also described two separate periods of horizontal precipitation: the first antedated life, giving rise to a lower layer that contained no fossils, whereas the second occurred after the creation of life, producing a fossil-rich upper layer. For the first time fossils could be used to reconstruct a sequence of events in the history of Earth and life.

See also Fossil objects pages 46–47, Earth cycles pages 100–101, Fossil sequences pages 132–133, Lyell's *Principles of Geology* pages 146–147, *Archaeopteryx* pages 180–181, Mountain formation pages 206–207, Burgess Shale pages 260–261, Oldest fossils pages 410–411, Martian microfossils pages 518–519

Right Traces of the past: Henry De la Beche's *Duria Antiquior*, or 'A More Ancient Dorset', with prehistoric animals obeying the 'law of nature which bids all to eat and to be eaten in their turn', *c*.1830.

Planetary distances

Giovanni Domenico Cassini 1625–1712

In 1543, when Nicolas Copernicus introduced the concept of the Sun-centred universe, it became easy to calculate the ratios between the distances of the planets to the Sun. This got even simpler in the early 1600s after Johannes Kepler noted in his harmonic law that the square of the time a planet takes to orbit the Sun is proportional to the cube of its average distance from the Sun. Yet until the time of Giovanni Domenico Cassini, the only absolute value for the scale of the Solar System was the highly erroneous one put forward in 280 BC by Aristarchus of Samos, who said that the Sun was about 20 times farther from the Earth than the Moon.

Cassini was appointed Director of the Paris Observatory by King Louis XIV. In 1671 the Sun, Earth and Mars were aligned, and the distance from Earth to Mars was at its minimum. Taking advantage of this, Cassini sent Jean Richer to Cayenne on the northeast coast of South America. Then, simultaneously, Cassini in Paris and Richer in Cayenne measured the angular position of Mars against the distant stars. Using trigonometry, and knowing that the two observing sights were 10,000 kilometres apart, Cassini worked out the distance between Earth and Mars. Using Kepler's harmonic law, he found that the distance between Earth and the Sun was 138,000,000 kilometres, only seven per cent less than the correct figure.

Through the use of trigonometry, astronomers could then calculate the sizes of the Sun and planets as they knew the angles that their discs subtended at the Earth. It transpired that the Sun was a staggering 110 times bigger than our planet.

After Isaac Newton had published *Principia Mathematica* (1687), which outlined his theory of gravity, it was realized that the Sun was about 330,000 times more massive than the Earth. Knowledge of the solar size and mass was the foundation stone of astrophysics.

Above Cassini founded a dynasty of five successive generations of astronomers who dominated French astronomy for over a century.

See also Circumference of the Earth pages 24–25, Celestial predictions pages 26–27, Sun-centred universe pages 42–43, Laws of planetary motion pages 52–53, Transit of Venus pages 62–63, Newton's *Principia* pages 78–79, Halley's comet pages 84–85, 'Weighing' the Earth pages 112–113, Distance to a star pages 150–151, Stellar evolution pages 286–287

Right Sun King: Louis XIV as 'Le Soleil' in the ballet *La Nuit*. Cassini persuaded him to redesign the Paris Observatory so that it was less ornamental and more useful.

Microscopic life

Antoni van Leeuwenhoek 1632–1723

Antoni van Leeuwenhoek was one of the great amateur scientists. His single-lens microscopes, with a magnification of up to 250 times, allowed him to see what no one had seen before. The draper from Delft in the Netherlands did not read Latin, the seventeenth-century language of science; and when he forwarded his findings to the Royal Society in London, he had to provide character references. Beginning in 1673 he wrote more than 400 communications to the Royal Society and the French Academy of Science. He described 'infusoria' (protozoa) teeming in water; human spermatozoa; the flow of blood through capillary vessels; the detailed organization of muscles, nerves, bone, teeth and hair; red blood cells and plant cells; and the fine structure of 67 species of insect (including tiny creatures parasitic on fleas). His most remarkable discovery was made in 1683 – bacteria from his mouth. Bacteria were not to be seen by other scientists for more than a century.

He also studied sexual reproduction among animals in an attempt to refute the idea that new life could arise spontaneously. The Italian physician Francesco Redi had already shown in 1668 that maggots hatch from egg-laying flies rather than forming from decaying matter, but his work was ignored. Leeuwenhoek countered the claim that sperm results from the putrefaction of semen. He went on to suggest that fertilization follows penetration of an egg by a sperm and that the egg provides the sperm with nothing but nutrients.

Leeuwenhoek began grinding simple magnifying glasses in 1671 and produced more than 400 lenses during his career. Because he could read only Dutch, he studied the illustrations of untranslated works by contemporary microscopists such as Robert Hooke (first to use the word 'cell' in describing the 'pores' of cork) and Marcello Malphigi (first to observe capillaries). As his fame grew, he became acquainted with Europe's scientific establishment and entertained royalty. The clarity and power of his lenses remained unbeaten until the nineteenth century when the technical faults of compound microscopes were overcome.

See also Heavens through a telescope pages 54–55, Circulation of the blood pages 58–59, Spontaneous generation pages 90–91, Eggs and embryos pages 140–141, Communities of cells pages 174–175, Germ theory pages 202–203, Viruses pages 232–233, Five kingdoms of life pages 426–427

Right Animalcules, including sperm, based on Leeuwenhoek's drawings. He sent 26 of his tiny microscopes to the Royal Society so that members could see his remarkable observations for themselves.

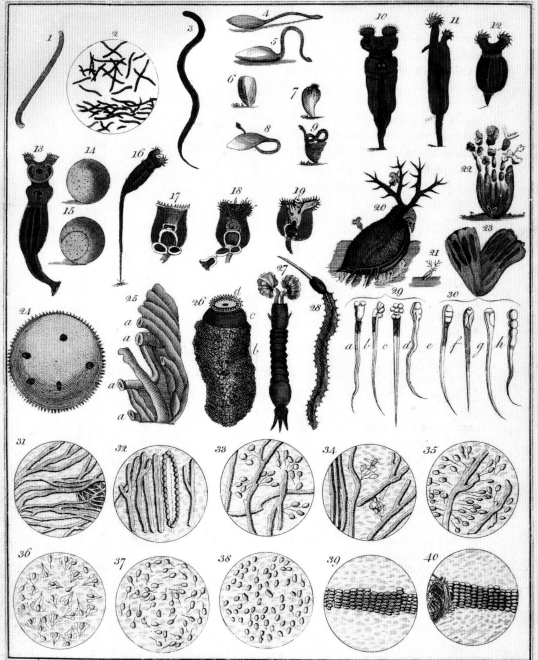

Dodd Delin.

Animalcules.

Pass Sculp.

Newton's *Principia*

Isaac Newton 1642–1727

As Lucasian Professor of Mathematics at Cambridge, Isaac Newton had secured his reputation well before the publication of his masterwork, *Principia Mathematica*, in 1687.

In 1666, five years after enrolling as a student at Trinity College, he deduced the inverse-square law, which describes how gravitational forces acting on the planets decrease with distance, and began to appreciate that it also applies to falling apples. And to express planetary motions in mathematical terms, he developed the technique of calculus; independently, in 1673–75, the German philosopher Gottfried Willhelm Leibniz devised his own version, and the subsequent battle to claim priority is notorious.

The *Principia* was the fruit of another bitter feud, in which Newton pitched his formidable intellect against that of Robert Hooke. In 1672 Hooke had delivered a tepid report to the Royal Society on Newton's paper 'Theory of Light and Colours', the forerunner of his *Opticks* (1704), which showed that white light is a 'heterogeneous mixture of differently refrangible rays'. So when in 1684 Newton learned of Hooke's casual claim to have proved the laws of planetary motion, he determined to set the record straight. From 1685 he worked feverishly, checking his calculations against the latest astronomical measurements.

The *Principia* unifies the mundane mechanics of Galileo and the celestial mechanics deduced empirically from the observations of Kepler. Newton's earlier work on the inverse-square law referred only to the centrifugal force experienced by a planet orbiting the Sun; in the *Principia* he showed how this must be balanced by an attractive force of gravity acting (at a distance) between the Sun and the planet, and demonstrated why the planet's motion must be elliptical.

Newton might never have published his extraordinary discoveries. Distrustful of revealing his work in print and sensitive to the slightest criticism, it was only with the gentle coaxing of Edmond Halley that he was persuaded to give the printers his manuscript.

Above Newton's gravity map showing the Earth and the paths of a projectile around it.

See also Unweaving the rainbow pages 36–37, Laws of planetary motion pages 52–53, Falling objects pages 60–61, Halley's comet pages 84–85, Origin of the Solar System pages 104–105, 'Weighing' the Earth pages 112–113, Wave nature of light pages 118–119, Spectral lines pages 130–131

Right The study of the house in which Newton wrote the *Principia*. A spectrum of light produced by refraction through a prism is projected onto the wall. Newton believed there must be seven 'uncompounded' colours by (spurious) analogy to the seven notes of the musical scale.

The calculus

Ian Stewart

WHEN ISAAC NEWTON MADE THE EPIC DISCOVERY that the motion of an object is described by a mathematical relation between the forces that act on the body and the acceleration it experiences, mathematicians and physicists learned quite different lessons … Acceleration is a subtle concept: it is not a fundamental quantity, such as length or mass; it is a rate of change. In fact, it is a 'second order' rate of change – that is, a rate of change of a rate of change. The velocity of a body – the speed with which it moves in a given direction – is just a rate of change: it is the rate at which the body's distance from some chosen point changes. If a car moves at a steady speed of 60 miles per hour, its distance from its starting point changes by 60 miles every hour. Acceleration is the rate of change of velocity. If the car's velocity increases from 60 miles per hour to 65 miles per hour, it has accelerated by a definite amount. That amount depends not only on the initial and final speeds, but also on how quickly the change takes place. If it takes an hour for the car to increase its speed by 5 miles per hour, the acceleration is very small; if it takes only 10 seconds, the acceleration is much greater …

So acceleration is a rate of change of a rate of change. You can work out distances with a tape measure, but it is far harder to work out a rate of change of a rate of change of distance. This is why it took humanity a long time, and the genius of a Newton, to discover the law of motion. If the pattern had been an obvious feature of distances, we would have pinned motion down a lot earlier in our history.

In order to handle questions about rates of change, Newton – and independently the German mathematician Gottfried Leibniz – invented a new branch of mathematics, the calculus. It changed the face of the Earth – literally and metaphorically. But the ideas sparked by this discovery were different for different people. The physicists went off looking for other laws of nature that could explain natural phenomena in terms of rates of change. They found them by the bucketful – heat, sound, light, fluid dynamics, elasticity, electricity, magnetism. The most esoteric modern theories of fundamental particles still use the same general kind of mathematics, though the interpretation – and to some extent the implicit worldview – is different. Be that as it may, the mathematicians found a totally different set of questions to ask. First of all, they spent a long time grappling with what 'rate of change' really means. To work out the velocity of a moving object, you must measure where it is, find out where it moves to a very short interval of time later, and divide the distance moved by the time elapsed. However, if the body is accelerating, the result depends on the interval of time you use. Both the mathematicians and the physicists had the same intuition about how to deal with this problem: the interval of time you use should be as small as possible. Everything would be wonderful if you could just use an interval of zero, but unfortunately that won't work, because both the distance travelled and the time elapsed will be zero, and the rate of change of 0/0 is meaningless. The main problem with nonzero intervals is that whichever one you choose, there is always a smaller one that you could use instead to get a more accurate answer. What you would really like is to use the smallest possible nonzero interval of time – but there is no such thing, because given any nonzero number, the number half that size is also nonzero. Everything would work out fine if the interval could be made infinitely small – 'infinitesimal'. Unfortunately, there are difficult logical paradoxes associated with the idea of an infinitesimal; in particular, if we restrict ourselves to numbers in the usual sense of the word, there is no such thing. So for about 200 years, humanity was in a very curious position as regards the calculus …

The story of calculus brings out two of the main things that mathematics is for: providing tools that let scientists calculate what nature is doing, and providing new questions for mathematicians to sort out to their own satisfaction. These are the external and the internal aspects of mathematics, often referred to as applied and pure mathematics. It might appear in this case that the physicists set the agenda: if the methods of calculus seem to be working, what does it matter *why* they work? You will hear the same sentiments expressed today by people who pride themselves on being pragmatists. I have no difficulty with the proposition that in many respects they are right. Engineers designing a bridge are entitled to use standard mathematical methods even if they don't know the detailed and often esoteric reasoning that justifies these methods. But I, for one, would feel uncomfortable driving across that bridge if I was aware that nobody knew what justified those methods. So, on a cultural level, it pays to have some people who worry about pragmatic methods and try to find out what really makes them tick. And that's one of the jobs that mathematicians do. They enjoy it, and the rest of humanity benefits from various kinds of spin-off.

Plant sex

Rudolph Jakob Camerarius 1665–1721

The idea that plants might exist in male and female forms can be traced to Theophrastus in the third century AD, but it was Rudolph Jakob Camerarius, Professor of Physic at Tübingen in Germany, who first provided experimental evidence that plants reproduce sexually.

Camerarius's surgical experiments on flowers proved conclusively that a deposit of pollen on stigmas was needed if a plant was to set seed. When he removed stamens from castor oil plants, or stigmas from maize flowers, he noticed that the plants lost the ability to set seed. Isolation of the separate male and female plants of species such as spinach and dog's mercury produced a similarly sterile result. Plants, like most animals, he concluded, reproduce sexually. Camerarius announced his findings in 1694 in his *Epistola de Sexu Plantarum*, which was addressed to the Professor of Physic at Geissen in Germany, Michael Bernard Valentini. Other contemporary botanists seem to have arrived at similar conclusions independently at about the same time, and in 1676 the English botanist and physician Nehemiah Grew read a paper to the Royal Society that identified pollen as the male part of flowers.

By the end of the century others had confirmed Camerarius's observations. When the London nurseryman Thomas Fairchild produced the first man-made inter-specific hybrid by cross-pollinating a carnation and a sweet william sometime before 1720, deliberate species hybridization and the selective breeding of plants had begun. Much later, in 1830, the Italian microscopist Giovanni Battista Amici showed that pollen grains germinate to produce a pollen tube that grows down the style and enters the ovule via a minute pore, the micropyle, to bring about sexual fusion. Thirty years later Gregor Mendel used carefully controlled cross-pollinations in peas to demonstrate the laws of heredity. Today the manipulation of sexual reproduction and Mendelian genetics are the cornerstones of modern plant breeding.

See also Birth of botany pages 18–19, Naming life pages 88–89, Mendel's laws of inheritance pages 192–193, Nitrogen fixation pages 208–209, Crop diversity pages 292–293, Green revolution pages 428–429

Right 'Cupid Inspiring the Plants with Love' by Philip Reinagle, from Robert Thornton's *The Temple of Flora*, 1804.

Halley's comet

Edmond Halley 1656–1742

In his famous *Principia Mathematica* of 1687, Isaac Newton had shown how the path of a comet could be calculated if its position is accurately measured three times during an interval of about two months. He demonstrated the procedure using the great comet of 1680. But the technique worked only if the path was assumed to be parabolic: that is, that the comet came from an infinite distance, passed the Sun and returned to an infinite distance. Newton had compiled records of a further 23 comets, but had become too busy, or too bored, to slog through the laborious calculations of their movements. The data were handed over to his friend, the Londoner and clerk to the Royal Society, Edmond Halley.

In 1696, just before his appointment as Deputy Comptroller of the Chester Mint, Halley read a paper to the Royal Society in which he gave the possible paths for the comets of 1607, 1618 and 1682. He concluded that the comets of 1607 and 1682 were apparitions of the same astronomical object.

In 1705 Halley, now Professor of Geometry at the University of Oxford, published his famous *Astronomiae Cometicae Synopsis*. Here he listed the movements of 24 comets, all assumed to be parabolic. He now saw that the comet of 1531 had a similar path to the 1607 and 1682 comets. Halley concluded that he was dealing with a single comet in a closed but very elongated orbit about the Sun and visible only when close to the Earth. The time between appearances was about 76 years. Halley wrote: 'hence I dare venture to foretell that it will return in the year 1758'. Return it did, being first sighted on Christmas Day. The comet has been called Halley's comet ever since. Its return proved that Newton's law of gravity applied at least as far as the distant borders of our planetary system.

Above Nucleus of Halley's comet seen by the Giotto spacecraft on 13/14 March 1986.

See also A new star pages 48–49, Newton's *Principia* pages 78–79, Discovery of an asteroid pages 120–121, A cometary reservoir pages 360–361, Extinction of the dinosaurs pages 458–459, Comet Shoemaker–Levy 9 pages 508–509, Water on the Moon pages 522–523

Right A detail from the 11th-century Bayeux tapestry depicts the English King Harold being informed of a comet's portentous arrival. This was an early sighting of Halley's comet.

Trade winds

George Hadley 1685–1768

In the early eighteenth century George Hadley, an English barrister with a passion for meteorology, became fascinated with the patterns of winds across the Earth's surface and their large-scale predictability. For example, sailors had long recognized that in the tropical zone north of the equator the so-called trade winds always blew from the northeast and that in the tropical zone south of the equator they blew from the southeast. What's more, near the equator itself the winds often died entirely, creating a region called the 'doldrums', much dreaded by sailors, who were often becalmed there.

In an effort to understand these effects, and as the man in charge of meteorological observations for the Royal Society in London, Hadley conducted the first serious study of the currents in the tropics. He then attempted to explain the effect in a paper entitled 'Concerning the cause of the general trade winds', which he presented to the Royal Society in 1735. Hadley first assumed that the heat from the Sun causes air to rise at the equator and to fall at the poles as it cools down. This sets up a convection cell in each hemisphere, known as a Hadley cell, in which the air near the surface moves towards the equator while air at high altitude moves towards the poles. In addition, friction between the rotating Earth and air near the surface imparts an easterly component to the wind, so explaining the direction of the trade winds in both tropics. The doldrums are the result of the uprising air where the two cells meet.

Although Hadley's work is impressive, his model of air circulation is somewhat simplistic and detailed models are needed to describe global wind patterns accurately. His work was largely ignored during his lifetime but rediscovered by John Dalton in 1793.

See also Atmospheric pressure pages 64–65, Foucault's pendulum pages 166–167, Chaos theory pages 238–239, Weather forecasting pages 284–285

Right A compass map showing winds, 1693. By the mid-17th century voyagers had brought back enough information about prevailing winds for scientists to attempt to explain them.

Naming life

Carolus **Linnaeus** 1707–78

Biologists refer to species by a Linnaean two-part name, such as *Equus caballus*, which is the formal title of the domesticated horse. The first of the two words (*Equus*) always takes an initial capital and is the name of the genus; the second (*caballus*) always takes lower case and is the name of the species. A genus may contain more than one species: the plains' zebra, for instance, is *Equus burchelli*, but the combination of both names is unique to one species.

Carolus Linnaeus, a Swedish naturalist and physician, published his system of nomenclature in *Systema Natura* in 1735. His method of referring to species works so well that it has been used, unchanged, for two-and-a-half centuries. Linnaeus introduced it because the existing Aristotelian method was collapsing beneath the weight of newly discovered species. An Aristotelian name has a generic and specific component, but the species term aimed to describe the distinguishing features of the species, enabling it to be recognized from its name alone. This worked well for small numbers of local species, but became impossible as science became global in the eighteenth century. The Aristotelian name of the tomato, for instance, had expanded into *Solanum caule inerme herbaceo, foliis pinnatis incisis, racemis simplicibus* (*Solanum* with smooth herbaceous stem, incised pinnate leaves and simple inflorescence). This name is less convenient than the Linnaean name *Solanum lycopersicum*. The species name in a Linnaean binomial makes little or no attempt to describe the species: *burchelli*, for example, simply refers to a person. Linnaeus separated the description of species from their naming.

Linnaeus was also influential because he described so many species (almost 10,000 of them), and his books became the standard reference works for anyone wanting to know the name for a living creature. He introduced many larger classificatory groups that have become familiar, including the category of mammals (formally, Mammalia), which is based on the possession of mammary glands and to which we humans belong.

See also Aristotle's legacy pages 16–17, Birth of botany pages 18–19, Darwin's *Origin of Species* pages 176–177, Neo-Darwinism pages 282–283, Five kingdoms of life pages 426–427, Diversity of life pages 468–469

Right Linnaeus classified plant species on the basis of the sexual parts of flowers, drawn here by Georg Dionysius Ehret in 1736.

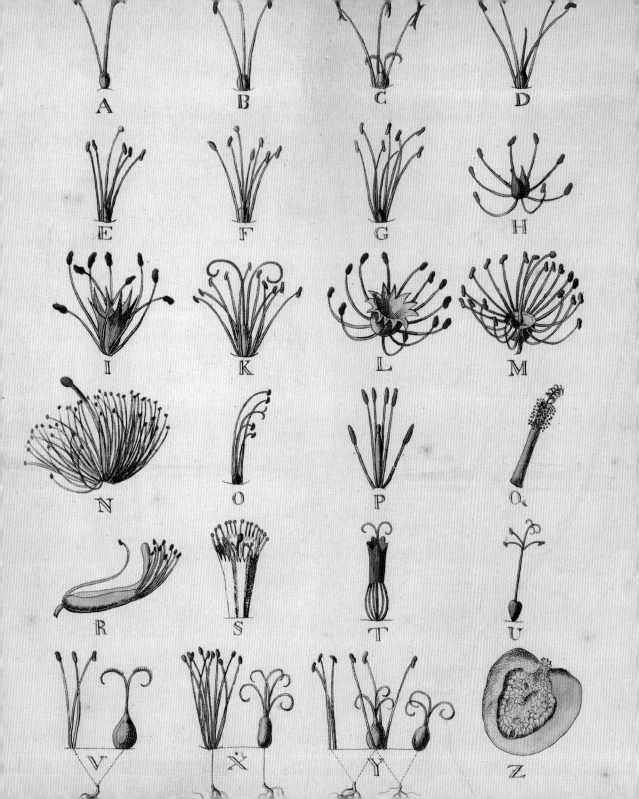

Spontaneous generation

Lazzaro Spallanzani 1729–99

Many sensible biologists believed in spontaneous generation – the idea that new life could form from decaying matter. It explained the existence of internal parasites of the human body such as tapeworms, which had no free-living counterparts, and the numerous 'animalcules' and 'infusoria' (microbes) that were revealed by the microscope but which had no clear origin. The mechanism of spontaneous generation was hotly debated and the phenomenon itself had many opponents as well as supporters. Lazzaro Spallanzani, an Italian polymath, professor of natural history and priest, provided the proof to overturn this centuries-old doctrine.

Spallanzani was familiar with Antoni van Leeuwenhoek's and Francesco Reddi's work from the previous century, which seemed to refute the idea of spontaneous generation. So he was provoked by the publication of two books in which it was claimed that the decomposition of plant and animal matter was able to generate new life forms: George Buffon's *Histoire Naturelle* (1749) and John Needham's *Nouvelles Observations Microscopiques* (1750).

Spallanzani repeated Needham's experiments, but obtained contradictory results. He tried various combinations of heated and unheated vegetable or cereal infusions; some were hermetically sealed, others left open. In flasks completely sealed and boiled for an hour, no animalcules appeared until air was allowed in again. In his essay 'Saggio di Osservazioni Microscopiche' (1765), he wrote up his conclusions and discredited spontaneous generation at the microscopic level. Nevertheless, Needham replied that Spallanzani's rough treatment – especially long boiling – had destroyed some vital principle in the specimens and invalidated the experiment. For a century the debate continued. Louis Pasteur incontrovertibly showed in 1860 that fermentation, putrefaction and infection were all due to contamination by microbes. In demonstrating how microbes could be destroyed by heat, however, Spallanzani had anticipated the French chemist's process of 'pasteurization'.

See also Microscopic life pages 76–77, Germ theory pages 202–203, Viruses pages 232–233

Right Out of a bottle? Robert Hooke was one of the most versatile of 17th-century scientists. His compound microscope allowed him to make the first detailed observations of such creatures as this 'blue fly'.

Fig: 1.

Fig: 2.

π

Johann Heinrich Lambert 1728–77

Pi is the world's most famous number – there is even a perfume named after it. Although this constant was known to the earliest civilizations, the use of the Greek letter π to denote the 'periphery' of a circle didn't appear until the 1700s. The definition of π is simple – the ratio of the circumference of a circle to its diameter – yet its precise value is much harder to pin down. It is slightly more than 3, and various approximations were used in antiquity, including 25/8 or 3.125 (Babylonian) and 256/81 or 3.16 (Egyptian). One particularly ingenious value was the square root of 10, or 3.162, which although close is totally unrelated to a circle.

The first known method for calculating π was made by Archimedes, who approximated a circle using a polygon with 96 sides. This gave a value of about 3.1418. Other people extended his method using polygons of even more sides, and by about 1600 π had been calculated to 35 decimal places. By then many other formulas for computing approximations were in use. In 1853, after 15 years of toil, an Englishman named William Shanks extended π to 707 decimal places, but it was later discovered that a mistake in his 528th place meant all subsequent places were wrong.

In 1768 Johann Lambert proved that π is 'irrational': it cannot be expressed as the ratio of two whole numbers, nor can it have a repeating pattern as a decimal. In 1882 Ferdinand Lindemann proved that π is also 'transcendental': it cannot be a solution to any algebraic equation, which means that it is impossible to square a circle with just a pair of compasses and a ruler. Still, the quest to calculate π to ever greater accuracy continued apace with the help of computers. In 1949, after 70 hours of computer time, the record stood at 2,037 decimal places. The latest calculation, declared in 1997, gives π to 51,539,600,000 decimal places. The sequence of digits still seems to be random, and yet the string of numbers '0123456789' appears six times.

See also Euclid's *Elements* pages 20–21, Circumference of the Earth pages 24–25, Celestial predictions pages 26–27

Right From fragrance to film: in the movie π, the renegade mathematician Max Cohen (Sean Gullette) seeks a number that will decode the pattern of the stockmarket.

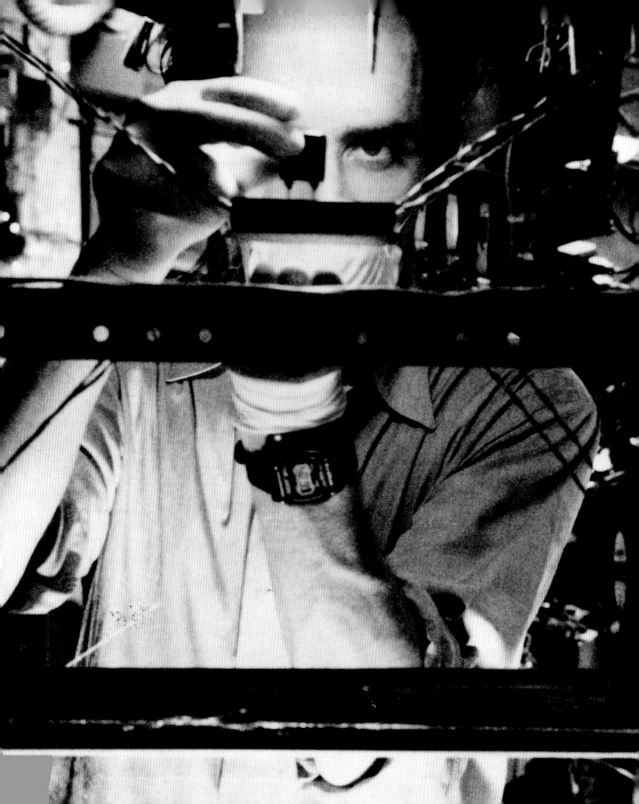

Combustion

Joseph Priestley 1733–1804, Carl Wilhelm Scheele 1742–86,
Antoine Laurent Lavoisier 1743–94

The eighteenth century was the heyday of the 'pneumatick' chemist, whose focus was air. For centuries the primary tool of chemical transformation was heat, and the central concern of the pneumatick chemists was to understand combustion.

The great pneumatick chemists were mostly avowed believers in 'phlogiston' – the 'principle' of flammability thought to be liberated from matter as it burns – and none more so than Joseph Priestley, the English political and religious nonconformist who trained as a Presbyterian minister. The fatal shortcoming of the phlogiston theory was not that it was wrong but that it was so nearly right. When Priestley observed oxygen in 1774, he interpreted it as 'dephlogisticated air': air that, for want of phlogiston, enabled substances rich in this 'combustible principle' to burn more brightly.

The following year, Priestley studied the effect of oxygen on respiration, noting how it made his breath feel 'peculiarly light and easy'. His method of making oxygen – heating mercuric oxide – was not novel. The Swedish apothecary Carl Wilhelm Scheele had performed the same experiment several years earlier, as well as having liberated oxygen from various other salts. Scheele called his gas 'fire air', and even deduced that normal air consists of one part 'fire air' to four parts 'spent air' (unreactive nitrogen).

But it is the French chemist Antoine Laurent Lavoisier who is considered oxygen's discoverer. In 1777 he gave the gas its name: *oxygène*, meaning 'acid-former', as he believed it (wrongly) to be the fundamental constituent of all acids. He went further than Priestley by showing that the gas given off by mercuric oxide is taken up in equal measure by mercury when heated in air. And he saw no need for 'dephlogisticated air'; this 'pure air' was a substance in its own right. Lavoisier's oxygen theory of combustion, promoted enthusiastically in France but initially resisted in England, gave chemistry the unifying principle it needed.

See also Boyle's *Sceptical Chymist* pages 70–71, Hydrogen and water pages 98–99, New elements pages 122–123, Atomic theory pages 124–125, Periodic table pages 196–197, Citric-acid cycle pages 320–321

Right Marie-Anne Pierrette Paulze married Lavoisier at 14, became his personal assistant, translated vital papers by Priestley and Cavendish, and illustrated chemical apparatus in Lavoisier's textbook.

Discovery of Uranus

William Herschel 1738–1822

In antiquity, seven moving objects were seen in the sky: the Moon, the Sun, Mercury, Venus, Mars, Jupiter and Saturn. On 13 March 1781, Uranus was added to this list, and the known size of the planetary system doubled.

William Herschel, the son of a Hanoverian Guards' bandmaster, had moved to Bath in England as organist at the Octagon Chapel. In his spare time he started making reflecting telescope mirrors out of speculum metal, an alloy of copper and tin. Tuesday evening, 13 March 1781, found him in the garden of 19 New King Street with a wooden altazimuth telescope he had made himself. The mirror was 6.2 inches (15.7 centimetres) in diameter and the focal length was 7 feet (2.13 metres). Herschel was surveying a region of the constellation Gemini and counting the stars of different brightness in an attempt to map our galaxy. He was also recording any unusual objects in the field of view. In the region of the star 1 Geminorum, he came across a 'star' that was both brighter and larger than usual and he suspected at first that it was a comet. As he increased the magnifying power of his telescope, he saw the image become larger and larger.

Herschel wrote to many astronomers telling them of his discovery. Turning their telescopes to the sky, they noticed that, unlike a comet, the object was moving very slowly and steadily. It also lacked a head or tail. Anders Lexel found that the orbit was nearly circular, with a period of about 83 years. Clearly, the object was a planet.

In 1782 Herschel proposed that the new planet be called Georgium Sidus ('George's Star'). King George III was so pleased with the discovery that he appointed him as his court astronomer at a salary of £200 a year. By 1850, following the suggestion of the German astronomer Johann Bode, the planet was being referred to as 'Uranus', the mythological father of Saturn.

See also Transit of Venus pages 62–63, Planetary distances pages 74–75, Discovery of an asteroid pages 120–121, Discovery of Neptune pages 162–163

Right 'Georgium Sidus' is eclipsed by the Duke of Wellington, the British prime minister who in 1829 carried through the Catholic Emancipation Act. King George IV had unsuccessfully tried to marry a Roman Catholic.

Hydrogen and water

Henry Cavendish 1731–1810

The idea that water is an element has such ancient roots that one can understand the discomfort many scientists felt when they realized it was moribund. Yet this was inevitable once water's elemental constituents, hydrogen and oxygen, had been isolated. As 'inflammable air', hydrogen had been known to British chemist Robert Boyle in the seventeenth century, and surely others before him; Carl Wilhelm Scheele made it in 1770, and suspected it might be pure 'phlogiston'; and Henry Cavendish reached the same (mistaken) conclusion four years earlier.

But the behaviour of this 'inflammable air' held some surprises. In 1774 John Warltire, a friend of Joseph Priestley, ignited the gas mixed with common air in a sealed copper flask, and found 'dew' on the walls. In France, Pierre Joseph Macquer, a colleague of Priestley's rival Antoine Laurent Lavoisier, burnt inflammable air and observed that a porcelain plate held above the flame 'was wetted by drops of a liquid like water'. In England, James Watt and Priestley subsequently repeated these experiments.

Strange, then, that the synthesis of water from its elements is traditionally ascribed to Henry Cavendish, who, in 1781, repeated a process already conducted four times earlier. What's more, as an avowed 'phlogistonist' rather than a disciple of Lavoisier's oxygen theory, Cavendish had little chance of correctly interpreting his findings. Yet no one previously had so thoroughly investigated the process as Cavendish. He went on to show that Priestley's 'dephlogisticated air' (oxygen) would unite with twice its volume of 'inflammable air' (hydrogen), revealing the composition (H_2O) of the substance so formed.

Cavendish spent three years refining his experiments before announcing his discovery to the Royal Society in 1784, in the meantime allowing Lavoisier to repeat them and stake his own claim. Lavoisier also showed that water could be split back into hydrogen and oxygen when passed through a red-hot gun barrel. The oxygen reacted immediately with the iron to form rust. So Lavoisier named the 'inflammable air' *hydrogène*: water-former.

See also Boyle's *Sceptical Chymist* pages 70–71, Combustion pages 94–95, New elements pages 122–123, Periodic table pages 196–197, Alien intelligence pages 398–399, Water on the Moon pages 522–523

Right Cavendish used electrical sparks to ignite a mixture of common air and 'inflammable air' in sealed glass vessels. Exploding in 'a proper proportion', they condensed into dew that was 'plain water'.

James Hutton is hailed as the founder of modern geology and perhaps the first student of the Earth who may properly be called a geologist. A product of the eighteenth-century Scottish Enlightenment, he studied in Paris, France, and Leiden in the Netherlands, made money manufacturing sal ammoniac (ammonium chloride) and was interested in steam engines, canals and agricultural improvements; but unlike many of his contemporaries he never held a university position.

In 1785 Hutton published his *Theory of the Earth* (expanded with 'Proofs and Illustrations' in 1795), a theoretical and archaic Newtonian philosophy of endless terrestrial cycles – and criticized at the time for being so. Still, his theory, or rather the 1802 reworking of it by his friend the mathematician John Playfair, became hugely influential, especially through the work of fellow Scot Charles Lyell, and later Charles Darwin.

Hutton held that natural Earth processes, although destructive, are necessary for the formation of fertile soil to support plant, animal and human life. But if these processes worked in one direction only, all land would end up in the sea and life would come to an end. Indubitably, a wise Deity would design a compensatory mechanism to regenerate land and preserve life. So a search of Nature should reveal the processes that continually reconstitute the land.

Fossiliferous strata are formed on the seabed and yet the detritus of which they are composed comes from land, so some continuing process turns loose detritus into solid rock. For Hutton this process was heat and pressure, and it suggested that the Earth had a central fire, accounting also for earthquakes, volcanoes and mineral veins. The overall process is a steady-state 'uniformitarian' cycle, with new continents continually being created out of the debris of former ones. So the Earth wheels on forever with 'no vestige of a beginning, no prospect of an end'.

Vaccination

Edward Jenner 1749–1823

'But why think? Why not try the experiment?' wrote the Scottish surgeon John Hunter to his pupil Edward Jenner in 1775. And so, on 14 May 1796, Jenner inoculated eight-year-old James Phipps with pus from a cowpox sore of a milkmaid, Sarah Nelmes. In Jenner's experience, milkmaids who had contracted cowpox, a mild disease prevalent in cattle, were spared the grossly disfiguring and often-fatal outbreaks of smallpox. Could this immunity be produced by arm-to-arm inoculation directly from a cowpox pustule, he wondered?

Phipps developed a mild fever and some blistering from which he recovered fully. Six weeks later Jenner inoculated him with smallpox. Phipps remained healthy. In 1798 Jenner announced the successful 'vaccination' of 23 patients against smallpox (*vacca* is Latin for cow). His method was instantly adopted throughout Europe, being much safer than inoculation with smallpox matter, a traditional Asian practice promoted by Lady Mary Wortley Montague, wife of the British consul in Constantinople, on her return to England early in the eighteenth century.

Meanwhile doctors debated whether smallpox and cowpox were two distinct diseases or the same disease with differing virulence. We now know they are distinct, but the idea that a mild version of a serious disease might confer immunity was to inspire Louis Pasteur half a century later while he was developing his germ theory of disease. He found that the microbes that cause chicken cholera could be weakened in culture, and that birds injected with these 'attenuated' bacteria were protected from the disease. He went on to produce attenuated anthrax bacilli and rabies virus, demonstrating their success as vaccines in a series of dramatic animal experiments in the early 1880s. The attenuation of infective disease agents remains a goal of vaccine development today.

Following a 14-year mass vaccination campaign, the World Health Organization announced in 1980 that smallpox had been eradicated once and for all.

See also Cellular immunity pages 204–205, Antitoxins pages 214–215, A magic bullet pages 262–263, Biological self-recognition pages 430–431, Monoclonal antibodies pages 448–449, Prion proteins pages 466–467

Right A 19th-century painting recording a doctor vaccinating a child against smallpox.

Origin of the Solar System

Pierre-Simon Laplace 1749–1827

The importance of spin in the formation of the planets was illuminated by the French mathematician and astronomer Pierre-Simon (Marquis de) Laplace in his popular book *Exposition du Système du Monde*, published in 1796. It helped to explain well-known regularities such as why planets revolve about the Sun in nearly circular orbits, in the same direction and in roughly the same plane.

Laplace suggested that the Sun originated as a giant nebula or cloud of gas that was in rotation. As the gas contracted, the rotation of the cloud would accelerate and an outer rim of gas would be left behind owing to centrifugal force. Turbulence would produce a series of equatorial rings. Material in each ring would then slowly condense into a planet. As the outer ring edge would be spinning faster than the inner, the planet would rotate about its axis in the same direction as the original nebular rotation. Finally, the core of the nebula would condense into the present-day Sun.

One of the main drawbacks of the hypothesis is that the primordial Sun would be spinning so fast that it hovers on the brink of equilibrium. Today's Sun plods around every 25 days and is far from this condition. By suggesting that the solar wind dampens the spinning of the ageing Sun, Laplace's nebular hypothesis has survived the test of time, and has even been applied to the formation of satellites around giant planets.

In the same 1796 book, Laplace calculated how big the Sun would have to be for all the light emitted from its surface to be eventually pulled back by gravity. This 'black hole' calculation was dropped from the 1808 edition. The discovery that nothing could move faster than light was a century away.

See also Saturn's rings pages 68–69, Discovery of an asteroid pages 120–121, General relativity pages 278–279, A cometary reservoir pages 360–361, Apollo mission pages 424–425, Black-hole evaporation pages 440–441, Galileo mission pages 514–515, Water on the Moon pages 522–523

Right René Descartes's conception of the universe had each star surrounded by a whirlpool of matter, or 'vortex', in which its planets circulate. This illustration from his *Principles of Philosophy* (1644) shows a comet travelling from one vortex to the next.

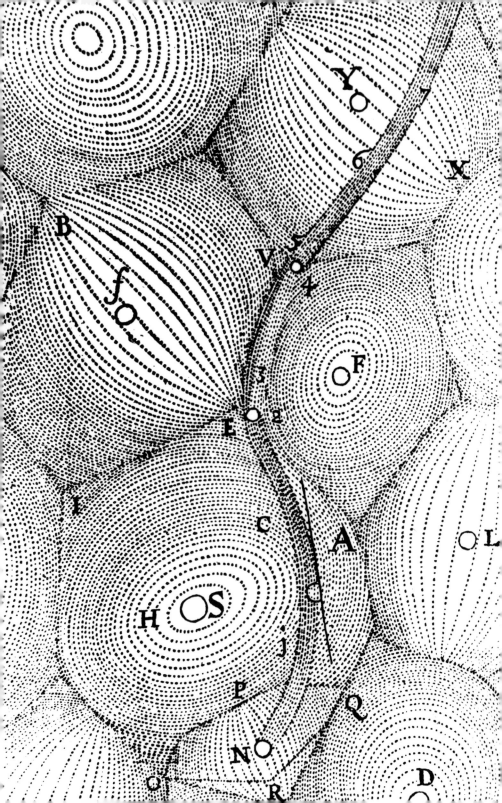

Comparative anatomy

Georges Cuvier 1769–1832

Jean Léopold Nicholas Frédéric, later also Dagobert, the Baron Cuvier, otherwise know
more simply as Georges Cuvier, survived revolution to become France's Minister of th
Interior in 1832. One of the greatest naturalists of the time, Cuvier, like so many of hi
contemporaries, was not bound by scientific discipline. He made fundamental contribu
geology, palaeontology and zoology. With Alexandre Brongniart he independently disc
the principle of stratigraphic geology at about the same time as William Smith, the En
surveyor. But he is best remembered as one of the founders of comparative anatomy.

The science of comparative anatomy, independently founded by Cuvier and the Sco
surgeon John Hunter (1728–93), allowed the reassembly of fragmentary, incomplete and
mixed fossil bones of extinct creatures into anatomically viable forms. The basis for such
reconstruction was the realization that all vertebrate animals have a common skeletal pl

Cuvier made innovative and startling reconstructions of fossil mammals, contributi
the belief that extinction occurred. From a lithograph of its skeleton, Cuvier demonstr
1796 that the extinct *Megatherium* of South America was a giant ground sloth. He
reconstructed some of the earliest known primitive mammals of the Tertiary age from
strata of the Paris basin, including an extinct marsupial mammal (1804) and the tapir-
Palaeotherium (1804). Despite his knowledge of the underlying common connection betw
the vertebrates, it did not immediately lead him to an acceptance of evolutionary theo

Cuvier was an implacable anti-evolutionist and fundamentally opposed to the ideas
developed by his French contemporaries such as Jean-Baptiste Lamarck. Cuvier espous
catastrophism. As he said in his 'Preliminary Discourse' of 1812: 'life on earth has ofte
disturbed by terrible events: calamities which initially perhaps shook the entire crust of
earth to a great depth'.

See also Fossil objects pages 46–47, Geological strata pages 72–73, Earth cycles pages 100–101, Catastrophist geology page
108–110, Acquired characteristics pages 128–129, Fossil sequences pages 132–133, Lyell's *Principles of Geology* pages 14(
Invention of the dinosaur pages 154–155, *Archaeopteryx* pages 180–181, Extinction of the dinosaurs pages 458–459

Right Cuvier's reconstruction of the skeleton of *Megatherium*. Although as large as a modern elephant, this giant ground sloth
up on its hind legs, supported by its thick tail.

1ᵃ

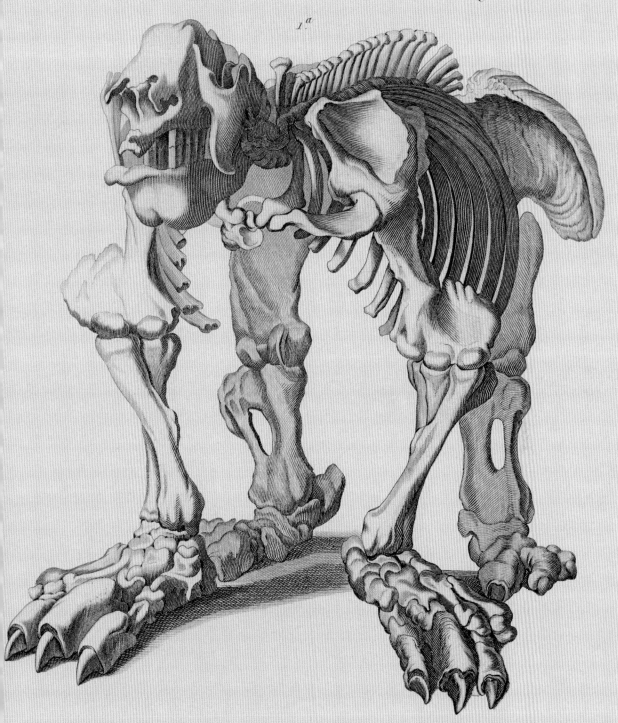

Catastrophist geology

James Hall 1761–1832

Sir James Hall was a hero in one of the lesser-known conflicts of the late eighteenth century: the battle between the Neptunists and the Plutonists. These opposing groups, with their science-fiction names, were serious scholars, pioneers in the young science of geology. The Neptunists said Earth's land surface had first been laid down under the sea, then exposed and weathered to its present shape; the Plutonists believed it had been created by a dynamic process that included earthquakes and the formation of rock by heat and pressure.

For years Hall argued against the claims of his friend and fellow Scotsman, the arch-Plutonist James Hutton. Hall's objections began to ease, however, after he heard that a sample of bottle glass, after it spilled accidentally and cooled slowly, had 'lost every character of glass, and had completely assumed [a] stony structure'. In 1797 Hall decided to put Hutton's ideas to the test. He managed to replicate the process – crystallization or devitrification – under controlled conditions. Then he took a sample of hornblende, a dark silky-looking igneous rock, from a quarry near Edinburgh, and subjected it to intense heat in an iron foundry. 'I removed the crucible and allowed it to cool rapidly,' he wrote. 'The result was a black glass, with a tolerably clean fracture.' Then he heated hornblende over an open fire and let it cool slowly. This time the result was 'a substance, differing in all respects from glass… rough, stony and crystalline'.

After artificially producing different rock types in the laboratory, Hall became known as the founder of experimental geology. In his later years he became increasingly interested in volcanism, 'and the truly Plutonic operations by which our continents have been elevated'. In some respects he anticipated the theory of continental drift, advanced by Germany's Alfred Wegener in 1915.

See also Earth cycles pages 100–101, Comparative anatomy pages 106–107, Lyell's *Principles of Geology* pages 146–147, Inside the Earth pages 250–251, Rocks of ages pages 252–253, Continental drift pages 270–271, Plate tectonics pages 414–415, Eruption of Mount St Helens pages 462–463, Galileo mission pages 514–515

Right 'Plutonists' believed that the main mechanism for shaping rocks was heat, a force observed in eruptions of Vesuvius. Sir William Hamilton, one of the 18th century's greatest vulcanologists, climbed the volcano more than 60 times. For him, it was like a 'body full of bad humours'.

Population pressure

Thomas Robert Malthus 1766–1834

The English Enlightenment drew to a close not with revolution as in France, but with an increasing concern about the growing number of poor people. The optimism of the political philosophers William Godwin and the Marquis de Condorcet, champions of the perfectibility of mankind, was arrested by the work of a retiring clergyman, Thomas Malthus. His 1798 *Essay on the Principle of Population* foretold that struggle and strife were the ultimate lot of all species of plants and animals including the human race. Essentially a text for social scientists and politicians, Malthus's work was one of the first to attempt a systematic study of human society.

Malthus relied on two central postulates – that food is essential for human life and that sexual desire is a constant – and drew a series of chilling conclusions. Because population outstrips food supply (population increases geometrically, whereas subsistence increases arithmetically), mass starvation will eventually result unless human numbers are kept in check. In such a harsh scheme it is not surprising that the four checks he saw operating were unpleasant: infant mortality, epidemics, famine and prostitution.

Malthus's book, published anonymously, met with both avid admirers and harsh critics. In subsequent editions, bearing his name, he provided further supporting evidence for his early assertions – he had travelled to several countries to collect more data. He also developed his argument, admitting that moral restraint (delayed marriage and sexual continence) might be a better way of countering population growth. In 1805 he was appointed England's first professor of political economy, at the newly established East India College at Haileybury. Politicians used his ideas to shape the 1834 Poor Law Amendment Act, which revoked 'outdoor relief' (doles) from public funds to the unemployed, the old and the sick. His essay also influenced Charles Darwin and Alfred Russel Wallace. In Malthus's competitive struggle for life they saw a mechanism for evolution: natural selection.

See also Darwin's *Origin of Species* pages 176–177, Crop diversity pages 292–293, Green revolution pages 428–429

Right Cartoon by Thomas Hood commenting on the barbarous practice of separating husbands and wives and placing them in separate wards in the workhouse, 1832.

'Weighing' the Earth

Henry Cavendish 1731–1810

'The richest of all learned men, and very likely the most learned of all the rich' was how a French contemporary once described the shy English scientist Henry Cavendish, the first man to 'weigh' the Earth. Cavendish inherited his wealth at the age of 40 from his father but lived modestly, spending his money only on books and scientific apparatus. Throughout his life he made major contributions to chemistry and the understanding of electricity. And in 1798, at the age of 70, he carried out an experiment of such precision that scientists were later to herald it as a new era in the measurement of small forces.

The experiment measured the force of gravity between iron spheres. Knowing the mass of the spheres and the distance between them allowed Cavendish to work out a value for the gravitational constant, G, according to Newton's famous equation $F = Gm_1m_2/r^2$, where m_1 and m_2 are the masses of the spheres and F and r are respectively the force and distance between them. Cavendish used a torsion balance that measures the twisting force on a wire supporting a horizontal rod with iron balls the size of tomatoes at either end. When a pair of larger iron balls the size of melons were placed close to the balance, it rotates very slightly owing to the gravitational attraction between the spheres.

Using his results, Cavendish calculated that the density of the Earth was 5.45 times that of water, a result that is only 1.3 per cent lower than the value accepted today. Curiously, many scientists have pointed out that Cavendish made a simple mathematical error in the paper outlining his results to the Royal Society. He gives the value of the Earth's density as 5.48 when his data clearly indicate that it should be 5.45. Even the greatest minds are capable of making mistakes.

See also Circumference of the Earth pages 24–25, Falling objects pages 60–61, Newton's *Principia* pages 78–79, Inside the Earth pages 250–251, Rocks of ages pages 252–253

Right Hercules supporting the world, from the 'Camerino', a fresco by Annibale Carracci, 1596.

Humboldt's voyage

Friedrich Wilhelm Heinrich Alexander von Humboldt 1769–1859

In 1799 the German naturalist Alexander von Humboldt, accompanied by the French botanist Aimé Bonplant, set out on a journey to the Americas from which they would return – despite intermittent newspaper reports of their deaths – five years later. Entering a largely unexplored continent, they travelled up the Orinoco in Venezuela. At that time in Europe it was widely assumed that two major water systems could not be interconnected. Crossing the watershed into the Amazon basin, Humboldt and Bonplant returned to the Orinoco via the still disputed Casiquiare Canal, so proving the mysterious connection between the Orinoco and the Amazon.

Humboldt continued his scientific quest through South America, Mexico and finally the United States. Along the way, he collected an abundance of botanical and geological specimens, examined the volcanoes of Colombia and Ecuador, observed a meteor shower, measured the decline in magnetic intensity between the poles and the equator, and even climbed the volcano Chimborazo in Ecuador, which was then believed to be the highest mountain on Earth. This feat was accomplished without mountaineering equipment – he passed out, and his lips and gums bled – and set a world altitude record unsurpassed until Joseph Louis Gay-Lussac's 1804 ascent in a hot-air balloon.

Unprecedented in scale, Humboldt's journey established South America as a field for scientific exploration and greatly influenced Charles Darwin, who admitted 'my whole course of life is due to having read and re-read as a youth his *Personal Narrative*'. He was the first person to establish a link between a specific ecological environment and the plants and animals adapted to it. Anticipating modern ecology by some 200 years, he insisted that nature should be seen from a global perspective – hence his invention of isotherms and isobars to mark equal temperature and pressure levels on the world map. Such was the extent of his influence on the sciences that many of his other individual contributions are today taken for granted.

See also Natural magnetism pages 50–51, Atmospheric pressure pages 64–65, Trade winds pages 86–87, Mountain formation pages 206–207, Weather forecasting pages 284–285, Geomagnetic reversals pages 304–305, Gaia hypothesis pages 432–433, Eruption of Mount St Helens pages 462–463

Right Humboldt's research spawned entire new fields of scientific enquiry. Celebrated as an explorer as well as a scientist, he was second in fame only to Napoleon during his lifetime.

Electric battery

Alessandro Volta 1745–1827

Until the 1790s the only form of electricity that could be experimented with was static electric charge – the kind produced when you rub glass or amber with a cloth. Although this could be stored in a device known as a Leyden jar, the quickness of the discharge imposed severe limitations on what could be studied – if not demonstrated. In 1746 the French physicist and abbé Jean-Antoine Nollet discharged a Leyden jar in front of King Louis XV by sending a current through a chain of 180 royal guards.

A convenient steady source awaited the invention of the battery in 1799 by the Italian physicist Alessandro Volta. The discovery owes much to Volta's compatriot, the anatomist Luigi Galvani. In 1791 Galvani had noticed he could cause the legs of dead frogs to twitch by touching them with a metal probe, and concluded that the metal was transporting a fluid – 'animal electricity' – from the nerve to the muscle. Volta recalled how when he had once put a coin on his tongue, a coin of a different metal under his tongue and connected them with a wire, the coins tasted salty. So he proceeded to recreate Galvani's experiments and rapidly convinced himself that a current flowed when two different, physically connected metals are applied to any moist body – in other words it was the presence of the metals, and not the frogs' muscle tissue, that was creating the electricity.

Around 1799 he constructed stacks of alternating discs of silver and zinc, separated by layers of cardboard soaked in brine. This 'voltaic pile' provided the first controlled source of continuous electric current, opening the way to several practical applications such as the telegraph and electroplating, and setting the stage for major advances in the theories of electric current and electrochemistry.

See also New elements pages 122–123, Electromagnetism pages 134–135, Maxwell's equations pages 186–187, Power from the nucleus pages 330–331, Nerve impulses pages 366–369

Right The first electric battery. Volta realized that the continual production of electric current required an energy source, but failed to identify that source as chemical changes in the pile.

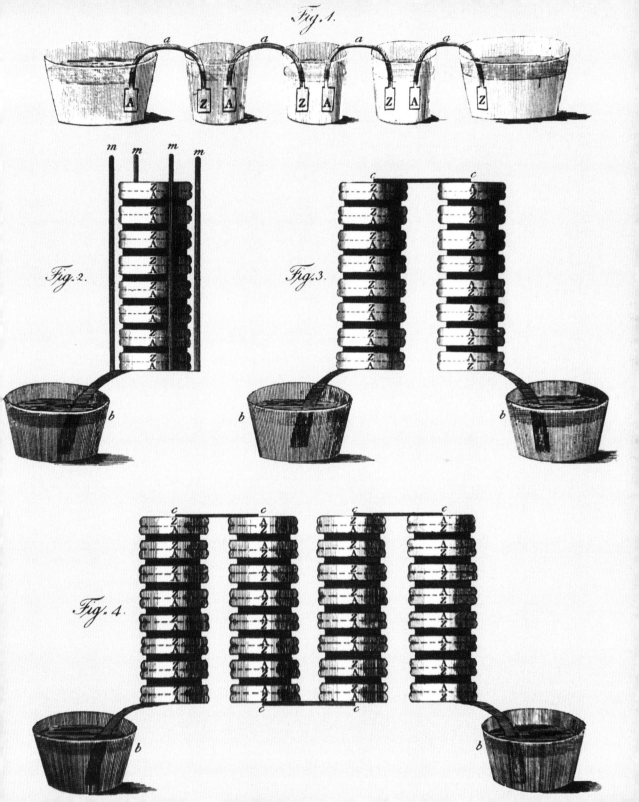

Fig. 1.

Fig. 2.

Fig. 3.

Fig. 4.

Wave nature of light

Thomas Young 1773–1829

The nature of light was a topic that had occupied natural philosophers for centuries. In 1675 Isaac Newton proposed in a lecture to the Royal Society that light was a stream of tiny particles. His rival, Christiaan Huygens, challenged this 'corpuscular' theory, arguing for a view of light based on waves travelling through an all-pervading medium called the ether. He wrote his *Treatise on Light* in 1678 but his procrastination meant that it was not published until 1690. Meanwhile it was the corpuscular theory that became dominant, mainly because of Newton's prestige.

Around 1800 the English polymath Thomas Young – famous also for deciphering the Rosetta stone – began a series of experiments that revitalized Huygens's wave theory. Young shone light through two thin slits in a piece of card and onto a screen. On reaching the screen, the light made a banded pattern of light and dark which he attributed to interference: light regions appear where the crests of the waves spreading out from the two slits enhance each other; dark regions appear where the crests of one wave are cancelled out by the troughs of the other. As this pattern could not easily be explained by Newton's corpuscles, it proved that light acts like a wave.

Support for the wave description grew in the early 1800s, and it eventually became part of James Clerk Maxwell's theory of electromagnetic radiation. Unfortunately, the quantum revolution of the early twentieth century showed that it told only half the story: Albert Einstein's 1905 explanation of the photoelectric effect had light behaving as a stream of particles, or photons. What's more, electrons, previously regarded as particles, sometimes behave like waves. We seem to need both models. As Sir William Bragg quipped in the 1920s, 'On Mondays, Wednesdays and Fridays light behaves like waves, on Tuesdays, Thursdays and Saturdays like particles, and like nothing at all on Sundays.'

See also Maxwell's equations pages 186–187, The quantum pages 234–235, Wave–particle duality pages 300–301

Right 'Wave Interference Pattern' by Berenice Abbott, c.1958. Some 150 years after Young demonstrated light interference, Abbott illustrated the phenomenon with a short exposure of two sets of waves.

Discovery of an asteroid

Guiseppe Piazzi 1746–1826, Carl Friedrich Gauss 1777–1855

Copernicus and Kepler both thought there was a 'hole' in the Solar System. Something was missing between the orbits of Mars and Jupiter. This suspicion was reinforced by the Titius–Bode law (1764), which described a peculiar arithmetical relationship of the planetary distances to the Sun and, astonishingly, fitted Uranus, which was discovered in 1781. Twenty-four European astronomers divided up the zodiac and started looking.

Although the Italian astronomer Guiseppe Piazzi was one of these 'celestial police', he was spending most of his time industriously compiling his new catalogue of 6,748 bright stars. Fortunately, his observatory in Palermo, Sicily, was Europe's southernmost. He had clear skies, an excellent climate and a superb Jesse Ramsden theodolite-type telescope built in London in 1789.

On the evening of 1 January 1801, Piazzi noted a 'new' faint star in the constellation of Taurus. The star had moved when he checked his observations the next night. This movement continued on the third and fourth nights. After measuring the star's position on 24 nights between 1 January and 11 February, he wrote to various astronomers to announce his discovery. At first he thought the object was a comet, but in late February his calculations showed that its orbit was nearly circular, leaving 'no doubt that this new star is a true planet'.

Piazzi named his new planet 'Ceres' in honour of the patron goddess of Sicily. But he worried about its being very dim and small. It was soon clear that this was a new species of celestial body – a fragment of a full-sized planet, or 'asteroid'.

The German mathematician Carl Gauss devised a new method to calculate the orbit of Ceres from the meagre 41 days' worth of data collected by Piazzi. Ceres had moved only three degrees in that time, and then been lost in the solar glare. Gauss's forecast allowed Franz von Zach to relocate Ceres at the end of 1801.

Above For decades asteroids appeared in lists of planets and on diagrams of the Solar System. From Asa Smith's *Ilustrated Astronomy*, drawn in 1889.

See also Discovery of Uranus pages 96–97, Origin of the Solar System pages 104–105, Discovery of Neptune pages 162–163

Right Ceres, the largest of the asteroids, is some 940 kilometres in diameter. Despite the impression given by this computer illustration, recording surface detail is far from easy.

New elements

Humphry Davy 1778–1829

Gravity is universal, but chemical affinity is elective: some substances react together and others don't. In 1800 chemistry had recently acquired a new vocabulary and a theory of burning from Antoine Lavoisier, but it awaited its Newton, who would explain affinity in terms of simple forces. Then Alessandro Volta announced that when two different metals are separated by wet cardboard, an electric current flows between them. Humphry Davy, a young Cornishman working in Bristol, England, believed that this could not be due to mere contact – a chemical reaction must be generating electricity.

Appointed lecturer at the Royal Institution, London, in 1801, Davy attracted paying audiences large enough to maintain a research laboratory there. He was at first made to work on tanning and agriculture, but by 1806 he was doing 'pure' research. On dipping wires connected to a big version of Volta's battery into water, he noticed that oxygen and about twice its volume of hydrogen bubbled around them. Davy was sure that the ratio should be exact, as it was in the formation of water, and that there shouldn't be any by-products. Using apparatus of silver, gold and agate, he confirmed this hunch, concluding that electricity and chemical affinity were manifestations of one power.

Researching in autumnal bursts, the following year he tried using electric currents to break down other substances, notably caustic potash and soda. With molten potash and sparks flying, he obtained globules of a light and highly reactive 'potagen' like the alchemists' long-sought alkahest. He danced about the laboratory in ecstatic delight. Experiments on the soft material, which floated on water, bursting into flames, convinced him that it was a metal, which he renamed 'potassium'. From soda, he obtained the analogous sodium, afterwards isolating calcium and other metals too. More systematic chemists such as Jöns Jacob Berzelius and Davy's assistant Michael Faraday brought new order into chemistry by developing Davy's Newtonian insight that affinity was electrical.

See also Combustion pages 94–95, Hydrogen and water pages 98–99, Electric battery pages 116–117, Atomic theory pages 124–125, Electromagnestism pages 134–135, Periodic table pages 196–197

Right Davy's lectures at the Royal Institution became hugely popular, his eloquence and the novelty of his experiments attracting large audiences. As a result he gained sufficient funding for research into electrochemistry.

Atomic theory

John Dalton 1766–1844

The concept of matter as a collection of small indivisible particles called 'atoms' probably originated with Democritus in Greece in the fifth century BC, but it was not generally accepted until the nineteenth century. Even as late as 1900 there were eminent scientists who scornfully disputed the existence of these invisible entities: 'Who has ever seen one?' was constantly iterated by the German physicist Ernst Mach.

John Dalton didn't have to see an atom to deduce its existence and instead asked some very simple questions. Why does water always contain the same ratio of hydrogen to oxygen? Why when carbon dioxide is made are the proportions of oxygen and carbon always the same? His answer to these questions came in 1808 when he published the first volume of *A New System of Chemical Philosophy*. Atoms (the elements – carbon, hydrogen, oxygen and so on) are proposed to be tiny (invisible) spherical bodies of fixed mass; each different chemical element has its own distinct kind of atom; and atoms combine in definite proportions to form molecules (which Dalton called compound atoms). This was the conceptual revolution that established the model chemists have used ever since.

Dalton's world was centred on the industrial town of Manchester, where he taught mathematics and natural philosophy in what was to become the University of Manchester. He was very much a man of the provinces, and commented on London during one of his rare visits that 'it is a surprising place and well worth one's while to see once, but the most disagreeable place on earth for one of a contemplative turn to reside in constantly'. Yet his ideas about the atomic nature of matter reached across national boundaries and set the stage for the great discoveries of the twentieth century.

See also Boyle's *Sceptical Chymist* pages 70–71, Combustion pages 94–95, Hydrogen and water pages 98–99, Benzene ring pages 190–191, Periodic table pages 196–197, Changes of state pages 198–199, The electron pages 228–229, The quantum pages 234–235, Model of the atom pages 272–273, Wave–particle duality pages 300–301, Buckminsterfullerene pages 486–487, A new state of matter pages 510–511

Right Dalton's list of atomic weights. Included in the list are compounds as well as pure elements. The chemical symbols he devised never became widely established.

ELEMENTS.

	Element	W.t		Element	W.t
☉	Hydrogen.	1	⊕	Strontian	46
⊖	Azote	5	⊛	Barytes	68
●	Carbon	54	Ⓘ	Iron	50
◖◗	Oxygen	7	Ⓩ	Zinc	56
⊻	Phosphorus	9	Ⓒ	Copper	56
⊕	Sulphur	13	Ⓛ	Lead	90
◉	Magnesia	20	Ⓢ	Silver	190
⊗	Lime	24	Ⓖ	Gold	190
⫶	Soda	28	Ⓟ	Platina	190
⫶	Potash	42	⊛	Mercury	167

Mapping the elements
Peter Atkins

WELCOME TO THE PERIODIC KINGDOM. This is a land of the imagination, but it is closer to reality than it appears to be. This is the kingdom of the chemical elements, the substances from which everything tangible is made. It is not an extensive country, for it consists of only a hundred or so regions (or elements), yet it accounts for everything material in our actual world. From the hundred elements that are at the centre of our story, all planets, rocks, vegetation and animals are made. These elements are the basis of the air, the oceans and the Earth itself. We stand on the elements, we eat the elements, we *are* the elements. Because our brains are made up of elements, even our opinions are, in a sense, properties of the elements and hence inhabitants of the kingdom.

The kingdom is not an amorphous jumble of regions, but a closely organized state in which the character of one region is close to that of its neighbour. There are few sharp boundaries. Rather, the landscape is largely characterized by transitions: savannah blends into gentle valleys, which gradually deepen into almost fathomless gorges; hills gradually rise from plains to become towering mountains. These are the images, the analogies, to keep in mind as we travel through the kingdom. The principle to keep in mind is that not only is the material world built from a hundred or so elements but these elements also form a pattern …

The real world is a jumble of awesome complexity and immeasurable charm. Even the inanimate, inorganic world of rocks and stones, rivers and ocean, air and wind, is a boundless wonder. Add to that the ingredient of life, and the wonder is multiplied almost beyond imagination. Yet all this wonder springs from about one hundred components that are strung together, mixed, compacted and linked, as letters are linked to form literature. It was a great achievement of the early chemists – with the crude experimental techniques available but also with the ever-astonishing power of human reason (as potent then as now) – to discover this reduction of the world to its components, the chemical elements. Such reduction does not destroy its charm but adds understanding to sensation, and this understanding only deepens our delight.

Then there came a grander achievement. Although these elements are matter, and few thought that one might be related to another, chemists saw through the superficialities of appearance and identified a kingdom of relationships, of family ties, of alliances, of affinities. Through their experimentation and reflection, a land rose from the sea, and then the elements were seen to form a landscape. Just as important, the landscape, when examined in a wide variety of lights, was seen to be structured, not merely a random collection of gorges and pinnacles; in particular, it was found to be periodic in variation. That was the most astonishing discovery of all, for why should matter display any sort of periodicity?

As so often in the development of science, comprehension springs from simple concepts that operate just below the surface of actuality, and constitute the true actuality. Once atoms were known – and their constitution elucidated in terms of that great invention of the mind, quantum mechanics – the foundation of the kingdom was exposed. Simple principles – the enigmatic exclusion principle, in particular – showed that the periodicity of the kingdom was a representation of the periodicity of the electronic structure of atoms.

The structure, layout and probable extensions of the kingdom are now fully understood. There are deeper currents flowing than presented here, but those currents are known. Yet despite our comprehension of the kingdom, it remains a mysterious place. The properties of the regions are rationalizable, and within certain ranges we can predict with confidence the chemical and physical properties of an element and the types of compounds it forms. The kingdom – the periodic table – is the single most important unifying principle of chemistry: it hangs on walls throughout the world, and the best way of mastering chemistry and inspiring new lines of chemical research lies in an understanding and use of its layout. But it is a land of conflict, as we have seen. The properties of a particular region are the outcome of competition, of influences that pull in different directions, sometimes several influences. These influences are usually finely balanced, and it is difficult, even from the heights of experience, to be absolutely sure that an element will not have a particular quirk of personality that will defeat a prediction or open up a new and exciting avenue of investigation.

Just as the letters of the alphabet have the potential for surprise and enchantment, so, too, do the elements of the kingdom. Unlike an alphabet, which has little infrastructure, the kingdom has sufficient structure to make it an intellectually satisfying aggregation of entities. And because these entities are balanced, living personalities, with quirks of character and not always evident dispositions, the kingdom is always a land of infinite delight.

Acquired characteristics

Jean-Baptiste Pierre Antoine de Monet Lamarck 1744–1829

Like Darwin, Malthus and Newton, the Frenchman Jean-Baptiste de Lamarck has given his name to the English language. But in contrast to the adjectives derived from these other pioneers, 'Lamarckian' refers to only a minor portion of his thought: the inheritance of acquired characteristics. Lamarck believed in this doctrine (now called 'soft heredity'), but it was merely an incidental feature of his philosophy of nature, and he shared it with virtually all of his contemporaries. From Hippocratic times to the end of the nineteenth century, naturalists assumed that characteristics acquired by parents could influence the fundamental nature of the offspring.

Unlike most of his peers, however, Lamarck used this notion of soft heredity in the service of a theory of evolutionary change. He was 65 years old when, in 1809, his *Philosophie Zoologique* put all the elements of his long and systematic study of the cosmos into a final form. His earlier monographs on chemistry, meteorology, geology and invertebrate animals had hinted at what was to come. Lamarck's mature statement of a lifetime's study offered a grand vision of how the world had developed. Animals, Lamarck argued, possess a quality he called *besoin*, variously translated as 'need' or 'want'. His most famous example was that of the giraffe's long neck, gradually created by this species' 'need' to graze on the leaves at the tops of trees. Generations had produced the characteristic form of the giraffe, able to fill an ecological niche.

Lamarck's notions of organic change over time challenged geologists and biologists (he himself coined the term 'biology') until Darwin offered another version. Lamarck's theories were sufficiently powerful that many naturalists who rejected the notion of organic species as the product of time, rather than special creation, tried to counter his ideas. Although natural selection offered a more compelling mechanism, even Darwin drew on notions of soft heredity, and so was 'Lamarckian' in that sense.

See also Darwin's *Origin of Species* pages 176–177, Mendel's laws of inheritance pages 192–193, Crop diversity pages 292–293, Random molecular evolution pages 422–423, Directed mutation pages 494–495

Right In 1546 the French naturalist and traveller Pierre Belon saw captive giraffes in Egypt – 'a very beautiful beast and of the sweetest nature there could be'.

Spectral lines

Joseph von Fraunhofer 1787–1826

Isaac Newton had used a prism to split light into its constituent colours. But it was the Bavarian glass- and lens-maker Joseph von Fraunhofer who quantified the dark narrow spectral lines that revolutionized chemistry and astronomy.

Von Fraunhofer was famous for his 'achromatic doublet' telescope lenses. In 1814 he used them to devise a spectroscope for measuring the bending of the different colours in a sunlight beam passed through a prism. Instead of peering at the light with the naked eye, as Newton had done, von Fraunhofer viewed it through a small telescope mounted on a circular scale. The English chemist William Hyde Wollaston had noted seeing seven dark spectral lines in 1802; von Fraunhofer counted 574, mapping the wavelengths of 324. He lettered the darkest and most prominent of the lines A through to K, a system still used today.

Von Fraunhofer had no idea how the lines were produced. By 1859, however, the chemist Robert Bunsen and the physicist Gustav Kirchoff, both German, had shown that each chemical element absorbs and emits its own characteristic combination of wavelengths, producing a unique 'fingerprint' of spectral lines. Because gaseous chemical elements in the solar atmosphere would therefore remove light of specific wavelengths from the direct light beam, they could work out the chemical composition of the Sun. Spectral analysis also allowed scientists to work out the chemical composition of laboratory samples, and eventually led to the discovery of caesium, rubidium, neon and argon, elements previously undetectable on Earth.

These spectroscopic endeavours were furthered by William Huggins, a Londoner who became an amateur astrophysicist after selling his family drapery business in 1854. He analysed lines in solar, lunar, planetary, cometary, stellar and nebular light. These and other studies revealed great similarities, proving the universe is made of the same elements as Earth. But there also seemed to be some notable exceptions – this being how helium was initially found in the Sun. The actual proportions of the stellar elements were not worked out until the 1920s.

Above A spectroscope being used to compare the flame of a Bunsen burner with that of a candle (1873).

See also Unweaving the rainbow pages 36–37, Doppler effect pages 156–157, Periodic table pages 196–197, Stellar evolution pages 286–287, Expanding universe pages 306–307, Planetary worlds pages 512–513

Right Spectra from various light sources, including the Sun, stars and different elements. From a lithograph published in Paris in 1872.

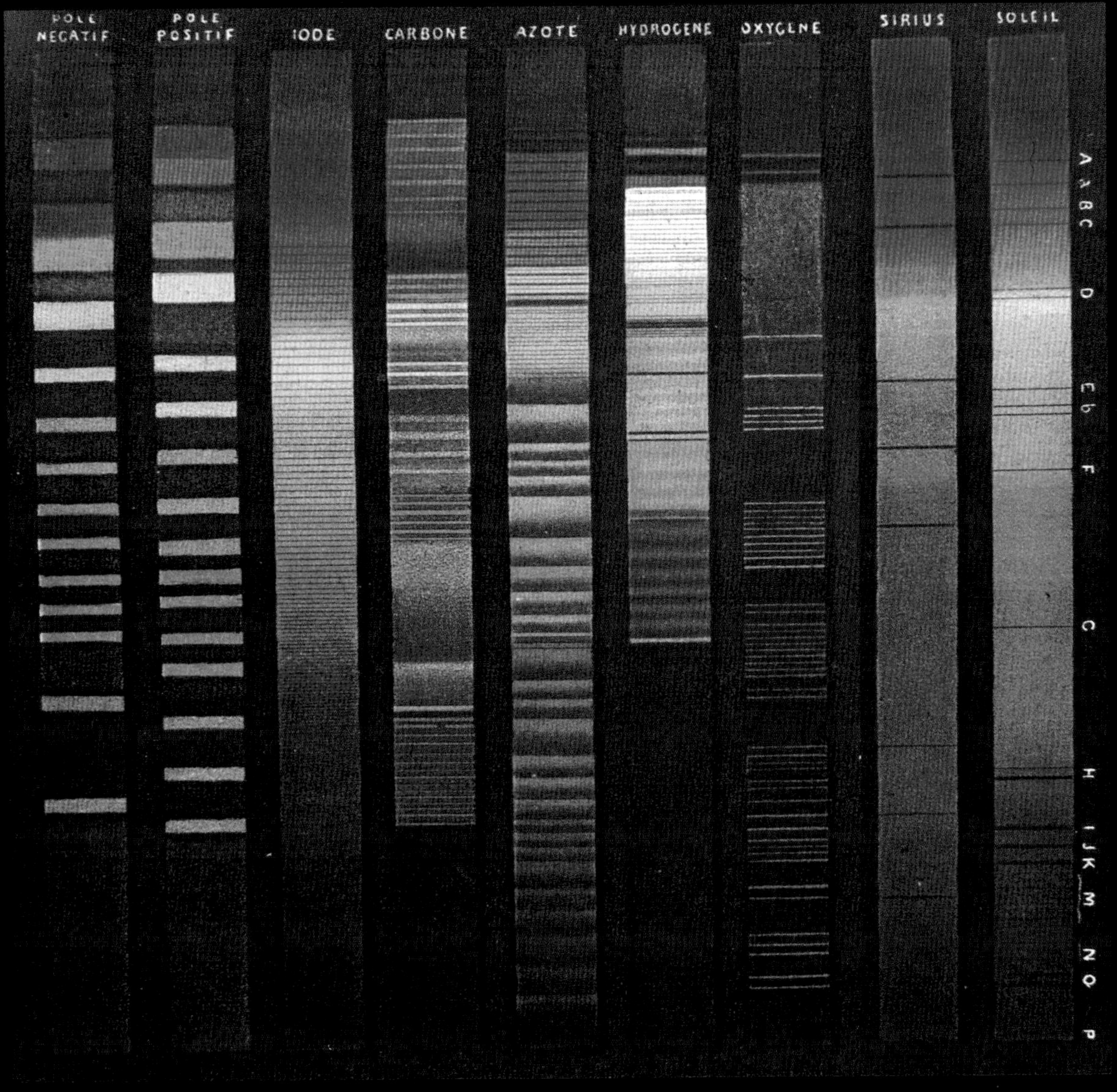

POLE NEGATIF | POLE POSITIF | IODE | CARBONE | AZOTE | HYDROGENE | OXYGENE | SIRIUS | SOLEIL

A B C D E b F G H I J K M N O P

Fossil sequences

William Smith 1769–1839

William Smith, a mostly self-taught English surveyor and canal engineer, independently discovered the stratigraphic method of geological mapping and pioneered its use. Like his French contemporaries Georges Cuvier and Alexandre Brongniart, Smith first illustrated the succession of fossils and strata and the way in which fossils could be used for stratigraphic correlation. Fossils were shown to have a fundamental importance for the relative dating of strata. This had enormous practical potential for the Industrial Revolution. Building stone, iron ore and coal were in great demand, and landowners wanted to know if such valuable natural resources lay hidden beneath their fields and pastures.

As a practising canal engineer, Smith was caught up in the rapid expansion of the canal network. The accurate prediction of the rock substrate for any canal-building proposal was economically vital. By identifying characteristic fossils in strata, measuring their inclination, drawing vertical sections, identifying rock outcrops at the surface and delimiting their geographical distribution, Smith was able to map out the actual and predicted outcrop of strata over large areas of countryside.

Over several years, his geological mapping extended across the country, and in 1815 Smith published *The Geological Map of England and Wales*, one of the earliest detailed geological maps on a large scale in the world. For a largely individual effort, it was a remarkable achievement.

Coming from someone with a relatively humble background, Smith's work was only belatedly recognized. The predominantly middle-class Geological Society of London awarded Smith the first Wollaston medal in 1831, while his nephew, a professor of geology at Trinity College, Dublin, secured him an honorary doctorate in 1835. As a final act of recognition, King William IV procured him a pension in 1832.

See also Fossil objects pages 46–47, Geological strata pages 72–73, Earth cycles pages 100–101, Lyell's *Principles of Geology* pages 146–147, Mountain formation pages 206–207, Extinction of the dinosaurs pages 458–459

Right Smith's survey work during canal construction introduced him to a variety of rock sequences of different ages.

Strata		Uses
London Clay, forming Highgate, Harrow, Shooters, and other detached hills	}	Septarium, from whi... No building Stone of materials wh... island.
Clay or Brick-earth, with interspersions of Sand and Gravel	}	These strata con... different purpos...
Sand, or light Loam, upon a sandy or absorbent Substratum	}	
Chalk { Upper Part, soft, contains Flints		Flints, the best road
Under Part, hard, none...............		Good Lime for wate...
Green Sand, parallel to edge of Chalk		Firestone, and other
Blue Marl, so kindly for the growth of oak as to be called in some places the oak-tree soil.		
Purbeck Stone, Kentish Rag, and Limestone of the Vale of Pickering.		
Iron Sand and Carstone, which, in Surry and Bedfordshire, contains Fuller's-earth, and, in some places, Yellow Ochre and Glass Sand	}	Some Lime used on
Dark Blue Shale produces a strong clay soil, chiefly in pasture, in North Wilts and Vale of Bedford.		
Cornbrash, a thin Rock of Limestone, chiefly arable		Makes tolerable road
Forest Marble Rock, thin beds, used for rough Paving and Slate.		
Great Oolyte, Rock, which produces the Bath Freestone	}	The finest buildin... architecture wh...
Under Oolyte, of the vicinity of Bath and the midland counties	}	
Blue Marl, under the best pastures of the midland counties.		
Blue Lias Limestone, makes excellent Lime for water cements.		
White Lias, now used for printing from MS. written on the stone.		
Red Marl and Gypsum, soft Sandstone and Salt Rocks, and Springs.		
{ Magnesian Limestone... — soft Sandstone }		Small quantities of C
Coal districts, and the Rocks and Clays which accompany the Coal	}	Grind-stones, Mill-... clay from the C...
—— Generally a Sandstone beneath.		
Derbyshire Limestone...............		Lead, Copper, and I
Red and Dun-stone, of the southern and northern parts, with interspersions of Limestone, marked blue...........	}	Some good building S
Various.		
Killas, or Slate, and other strata, of the mountains on the western side of the island, with interspersions of Limestone, marked blue	}	The Limestone polish... Tin, Copper, Lead,

Part on which Lime is rarely used as a Manure.   *Part on which Lime is generally used.*

Electromagnetism

Hans Christian Oersted 1777–1851, André Marie Ampère 1775–1836,
Michael Faraday 1791–1867

On 21 July 1820 the Danish physicist Hans Christian Oersted published a six-page paper in Latin announcing his discovery of electromagnetism. While lecturing to a class of students, he had noticed that a compass needle is deflected when brought close to a wire carrying an electric current. This was the first unification of the basic forces of nature – a prime goal of nineteenth-century natural philosophy.

The paper was rapidly translated into several European languages, and almost immediately sent investigators off in new directions. In Paris André-Marie Ampère proposed that all electromagnetic phenomena could be explained in terms of short-range electric forces acting in straight lines, in accordance with Newtonian concepts. In London, the chemical assistant at the Royal Institution, Michael Faraday, demonstrated that a current-bearing wire could rotate around a magnet (or vice versa). Thus was created the first electric motor. What's more, he argued that the circular motion could not be explained by Ampère's theory.

He also made a note in his diary in 1822: 'Convert magnetism into electricity!' But it was not until 29 August 1831 that Faraday, now director of the laboratory, did so. He wound two coils of wire on opposite sides of a soft iron ring. When a current passed through one coil, the ring became magnetized and momentarily induced a current in the other coil. This was in effect the first electric transformer. Within six weeks, Faraday had also invented the dynamo. Here a permanent magnet is pushed and pulled through a coil to induce an electric current in the wire. All generation of electricity to this day, no matter what the primary source of energy, is based on this principle.

See also Natural magnetism pages 50–51, Electric battery pages 116–117, Maxwell's equations pages 186–187, Quantum electrodynamics pages 352–353, Unified forces pages 416–417.

Right Faraday in his laboratory at the Royal Institution. Without formal education and ignorant of mathematics, he laid the foundations of electromagnetism.

Deciphering hieroglyphics

Jean François **Champollion** 1790–1832

In 1799 soldiers in Napoleon's army discovered an inscribed slab at Rosetta in Egypt that destroyed the classical myth that Egyptian hieroglyphs were not writing but arcane symbols encapsulating Egyptian wisdom. The ancient Egyptians used three writing systems: the pictorial hieroglyphs; a simplified version, hieratic; and its derivative, demotic. By 600 BC only the demotic script was in common use and this had died out by the fifth century AD. These scripts were used to write ancient Egyptian, the ancestor of the Coptic language that survived into the seventeenth century – just long enough for European scholars to record it. One vital tool for deciphering an unknown script is knowledge of the language it was used to write – and Coptic provided this.

By 1799 there was a burgeoning interest in things Egyptian, and serious attempts were now made to read the demotic script. The Rosetta stone provided the vital key. Its inscription was written in three scripts: Greek, demotic and hieroglyphic. The Greek text gave scholars the meaning of the demotic inscription. They could now recognize demotic as a largely phonetic script in which many signs represented single letters. Royal names provided the starting point for matching the signs, and soon the script was successfully deciphered.

The English scholar Thomas Young went several steps further, recognizing some underlying similarities between the demotic and hieroglyphic signs, and deducing that hieroglyphs were used phonetically in names. But it was a young French linguistic genius, Jean François Champollion, who in 1822 made the key breakthrough. He correctly matched the Greek and hieroglyphic signs in the name Ptolemy, allowing him to read the names of known rulers in other inscriptions. He could now reject the traditional view of hieroglyphs as purely symbolic and see that most of the signs in the Rosetta text also had phonetic values. By 1824 he had built up a sufficient corpus of identified signs to publish a convincing decipherment.

See also Origins of counting pages 10–11, The double helix pages 374–375, Public-key cryptography pages 454–455

Right The Rosetta stone with its text in three scripts, the keystone to the decipherment of the ancient Egyptian language.

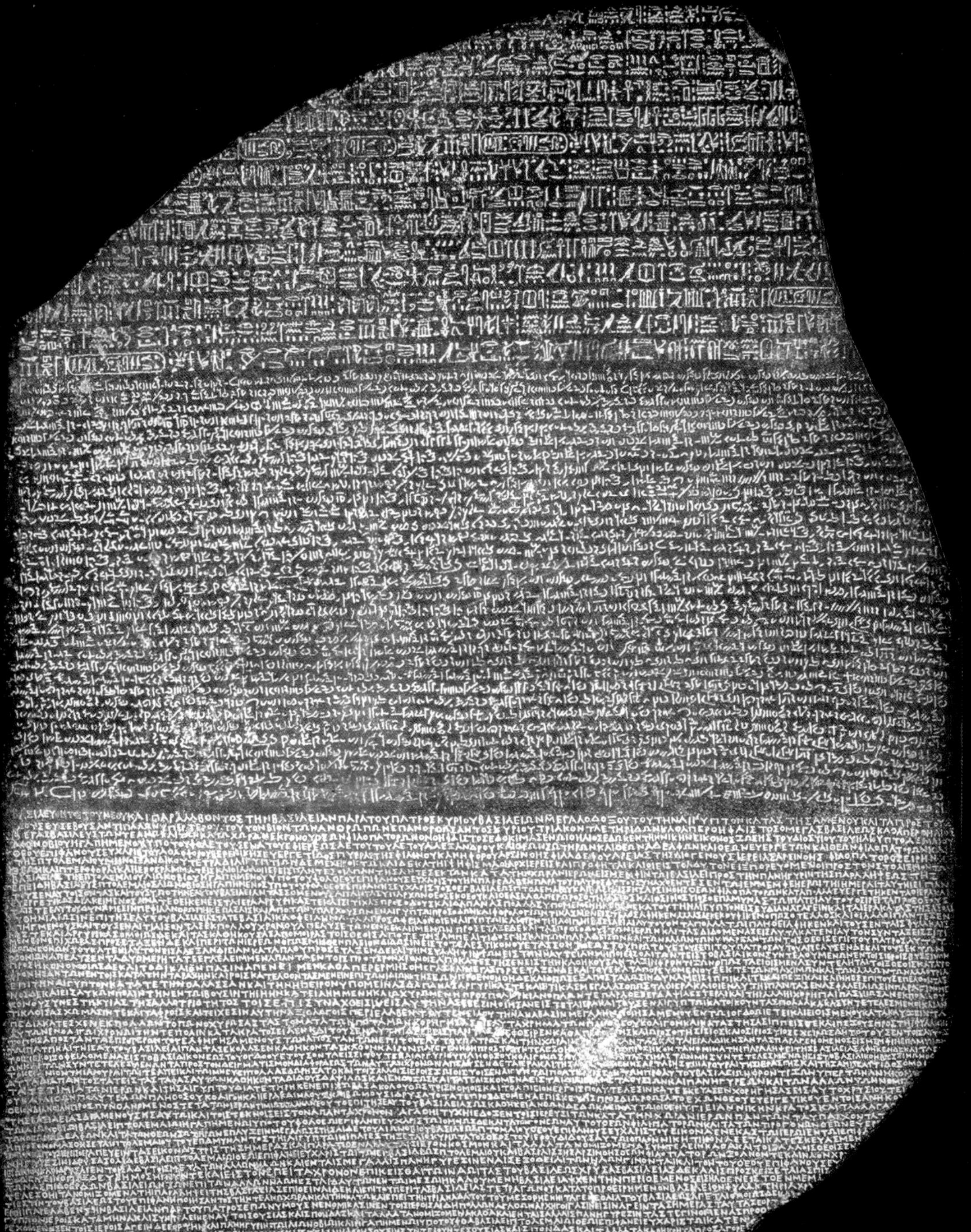

Difference Engine

Charles Babbage 1791–1871

Numerical tables of trigonometry and logarithms were still calculated by hand and then set by a printer, and so were generally riddled with inaccuracies. As these tables were used in navigation and finance, this had serious practical repercussions. In 1819 the English mathematician Charles Babbage produced the first design of his Difference Engine No. 1, with which he hoped to automate the production of such tables by repeated addition performed by trains of gear wheels. His inspiration was the Jacquard loom, which automated weaving with the use of punch-cards. A prototype Difference Engine was built by 1822, and the British government gave its backing for the construction of a fully working machine. By 1834, however, the project was over-budget and behind schedule. In the late 1840s Babbage designed Difference Engine No. 2, which computed to 31 figures of accuracy. But by then government funding had dried up. (The engine was eventually built by the Science Museum in London in 1991.)

Meanwhile Babbage had turned his attention to the real precursor of the modern computer, his Analytical Engine. This general-purpose device was designed to handle not just a single mathematical function, but many different computations. The input data and control device were encoded on punch-cards, little different from those used by IBM a century later, and there was a separate mill, which did the arithmetical calculations, and a store, which held numbers during calculations. Like previous engines, the whole process was fully automated by steam power and had a printed output. This engine was also never built. Much of Babbage's work is remembered in the writings of Ada Augusta, Countess of Lovelace, the only daughter of Lord Byron and his wife Annabella. A gifted mathematician, she was fascinated by Babbage's work but her life was tragically cut short at the age of 36. She wrote that 'We may say most aptly that the Analytical Engine weaves algebraic patterns.' These patterns were never seen.

See also Logarithms pages 56–57, The computer pages 340–341

Right Babbage's Difference Engine, as built by the Science Museum, London. One of the reasons why Babbage failed to build his computational devices was that parts could not be machined precisely.

Eggs and embryos

Karl Ernst von Baer 1792–1876

The early stages of reproduction are obscure in mammals because the offspring are not born until they are well developed – after nine months of gestation in humans, for example. Aristotle had supposed that fluid (semen) provided by the male somehow worked inside the female to form a new embryo. William Harvey challenged this view in the seventeenth century with his doctrine of *ex ovo omnia* – every living organism comes out of an egg. After Harvey, biologists showed that many creatures do indeed begin life with a single-celled egg. But the initial egg is more difficult to observe in mammals than in birds, fish or insects, which lay their eggs externally. Mammals remained a possible exception.

The mammalian egg was finally discovered in 1826 by the German biologist Karl Ernst von Baer. He found the first egg in the dog owned by his head of department, and later confirmed his observations in other mammal species. This finally laid to rest the idea that life begins with some kind of formative fluid. Every animal type begins its development with an egg cell. The discovery was the cornerstone of the cell theory that life is built from cells and that cells are formed only from other cells. It also led to much of our modern understanding of mammalian, including human, reproduction.

In addition, von Baer laid the foundations of modern embryology. He described the embryonic development of several vertebrate species from the egg stage to birth or hatching. He described the main kinds of cells (called germ layers) that gave rise to later adult organs. He also argued that development is 'epigenetic' not 'preformationist' – that is, development proceeds from the homogeneous to the heterogeneous, rather than being a process of growth from a miniaturized adult form within the embryo. Preformationism did not survive, in biology, after von Baer's time.

Above Nicolaas Hartsoeker's 1694 woodcut of a spermatozoon containing a homunculus.

See also Aristotle's legacy pages 16–17, Microscopic life pages 76–77, Communities of cells pages 174–175, Genetics of animal design pages 460–461

Right The German zoologist Ernst Haeckel believed that embryo development is a speeded-up replay of the evolution of the species. From left to right, pig, cow, rabbit and man (1891).

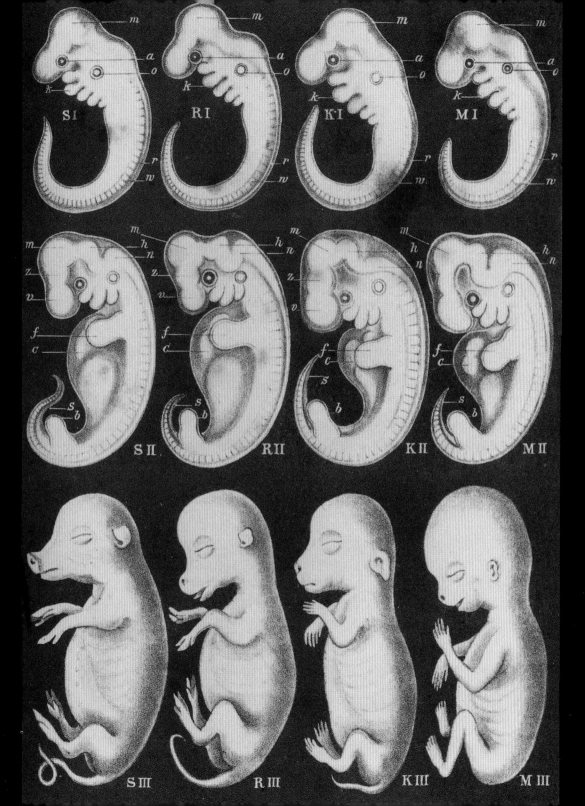

Synthesis of urea

Friedrich Wöhler 1800–82

It was widely held that living or 'organic' matter was fundamentally distinct from inorganic matter and that some 'vital force' animated the living world. Although the idea of vitalism rested more on religious conviction than on scientific evidence, it seemed impossible to make organic matter from inorganic reagents. That changed in 1828 when the German chemist Friedrich Wöhler told Jöns Jacob Berzelius, 'I can prepare urea without requiring kidneys or animal, either man or dog'. Urea was of biological origin, but Wöhler had made it from ammonia and cyanic acid, which, combined in ammonium cyanate, can be transformed to a compound identical to the natural product.

Wöhler himself was cautious about the philosophical implications of what others called his 'epoch-making' experiment, saying 'one must have for the production of cyanic acid (and also of ammonia) always initially after all, an organic substance, and a natural philosopher would say that the vital aspect has not yet disappeared from either the animal carbon or the cyanic compounds derived therefrom'. Meanwhile, Berzelius maintained that urea should be regarded as a substance on the borderline between organic and inorganic.

Later studies of digestion and fermentation by Justus von Liebig and Louis Pasteur revealed ever more common ground between the chemical principles of the organic and inorganic worlds. Liebig argued that fermentation could be regarded as a purely chemical transformation, while Pasteur believed life could be created artificially, proposing that chirality – the existence of molecules in mirror-image forms – was fundamental to life's chemistry.

Proof that the chemical processes of life required no vital force came in 1897, when Eduard Buchner demonstrated fermentation in the absence of cells. He ground and pressed yeast cells to obtain a cell-free juice (an 'unorganized ferment') that converted sugar to alcohol, by virtue, it was presumed, of the chemical substances named by the German physiologist Willy Kühne in 1876: 'enzymes'.

See also Spontaneous generation pages 90–91, Germ theory pages 202–203, Enzyme action pages 218–219, Synthesis of ammonia pages 256–257, Origin of life pages 370–371

Right Polarized light micrograph of a crystal of urea. The synthesis of urea was the beginning of the end for the idea that some 'vital force' animated the living world.

Non-Euclidean geometry

Nicolai Ivanovich **Lobachevsky** 1793–1856, **Janos Bolyai** 1802–1860,
Georg Friedrich Bernhard **Riemann** 1826–1866

For 2,000 years Euclidean geometry was considered the most logical of mathematical systems and a true description of our three-dimensional world. But this citadel had a small weakness, an itch mathematicians just had to scratch. Euclid included as one of his postulates the statement that if two lines are not parallel then they will meet at one point. This seems fairly uncontroversial, but mathematicians felt it was too complicated to be self-evident and sought to prove it from simpler axioms. But it was not until the nineteenth century that two independent mathematicians made the crucial breakthrough.

The Russian mathematician Nicolai Lobachevsky was a professor at the University of Kazan. Amid his additional duties as museum curator, chief librarian and rector, he found time in 1829 to publish *On the Principles of Geometry*. In this he supposed that Euclid's parallel postulate was false, and on that basis managed to construct a seemingly bizarre and counter-intuitive geometry that was nevertheless mathematically perfectly consistent. As early as 1823 a Hungarian mathematician, Janos Bolyai, had worked out the same non-Euclidean geometry, but his publication on the 'absolute science of space' was delayed until 1932. Even then it was nearly lost to posterity, for it was hidden in the appendix of a singularly unsuccessful textbook written by his father.

By modifying Euclid's starting points Lobachevsky and Bolyai had produced a new kind of geometry. In 1854 Bernhard Riemann generalized their findings by showing that various non-Euclidean geometries are possible, given the appropriate number of dimensions, coordinate system and method of establishing distances. In the next decade the study of configurations of space and their relationships to each other ('topology') also yielded a menagerie of exotic objects, such as the Möbius band which has only one face and one edge. Such discoveries led to renewed interest in the true geometry of space and time.

See also Euclid's *Elements* pages 20–21, Perspective pages 40–41, General relativity pages 278–279, Fractals pages 446–447

Right 'Möbius Strip II' by M.C. Escher. Two Möbius strips 'zipped' together form a Klein bottle, with only one surface and no boundary.

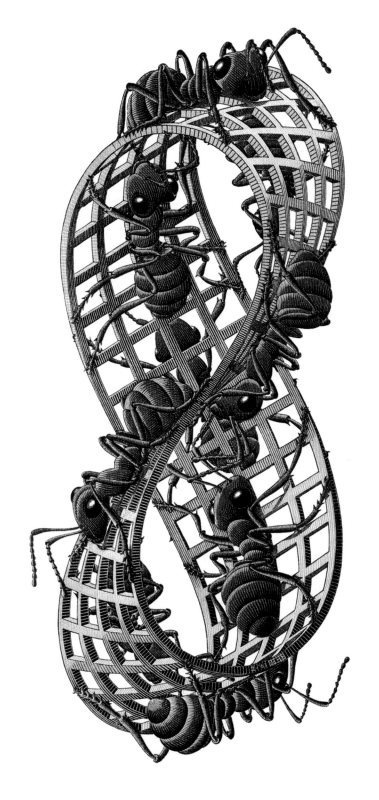

Lyell's *Principles of Geology*

Charles Lyell 1797–1875

Lyell's name is probably the best known in the whole of Earth science. An Oxford-educated barrister from a Scottish background, he was geology's Darwin, yet his fame rests not with some major theory but with a book.

Principles of Geology, first published as three volumes between 1830 and 1833, sold more than 15,000 copies and eventually ran to 11 editions (the last in 1872). Clearly and attractively written, the first volume famously served as a 'Beginner's Guide to Geology' for Darwin when he set sail in the *Beagle* on 27 December 1831.

Today Lyell is associated with the dictum that 'the present is a key to the past', otherwise known as the principle of 'uniformitarianism'. He was opposed to Georges Cuvier's catastrophist views and followed James Hutton's Newtonian approach, which stated that natural phenomena can be reasonably explained only by agencies that are observably effective: that is, modern or 'actual' causes. In Lyell's own words, *Principles* was 'an attempt to explain the former changes at the Earth's surface, by reference to causes now in operation'.

Originally, Lyell's approach even extended to the fossil record. He expected to find fossil representatives of all kinds of organism even in the oldest rocks, and to begin with it looked as if he might be right. Consequently, Lyell was initially unsympathetic to Darwin's theory of evolution. By mid-century, however, the fossil record had revealed some development of life through geological history, and Lyell was persuaded by T. H. Huxley's advocacy of Darwinism. In fact Lyell carried Darwin's views into the most sensitive field of all: the evolution and antiquity of man.

See also Geological strata pages 72–73, Earth cycles pages 100–101, Catastrophist geology pages 108–109, Fossil sequences pages 132–133, Ice ages pages 152–153, Darwin's *Origin of Species* pages 176–177, Mountain formation pages 206–207, Rocks of ages pages 252–253, Burgess Shale pages 260–261, Extinction of the dinosaurs pages 458–459

Right Reading the rocks: Lyell's work was thoroughly grounded in the first-hand experience he gained on his travels through France and Italy. His catholic interests ranged from rock dating to coal formation, glaciation to volcanism.

Prehistoric humans

Edouard Lartet 1801–71, Louis Lartet 1840–99

Edouard Lartet was a French landowner whose enthusiasm for fossils was aroused by Georges Cuvier's lectures on anatomy. In 1834 Lartet was shown a large tooth dug from Sansan Hill in Gascony; he recognized it as that of a mastodon, a kind of extinct elephant. Lartet's further excavation produced countless fossil mammal bones, including a distinctly ape-like jawbone.

Named *Pliopithecus antiquus* in 1836, it was the first fossil anthropoid to be found, and made possible the conjecture that perhaps apes and humans had a prehistory. Edouard and his son Louis subsequently made some of the most important early finds that did indeed demonstrate such a prehistory. In 1852 a human-like bone was pulled out of a rabbit hole in a hillside near Aurignac in the south of France. A trench dug into the hill revealed a cave entrance blocked by a limestone slab behind which lay seventeen skeletons. They looked so modern that they were reburied in the local cemetery. Lartet heard of the find in 1860 and excavated the cave floor to find more isolated human bones and the bones of extinct animals. In 1863 he claimed this as evidence for the antiquity of humans and their coexistence with extinct animals, but most contemporary scholars were unconvinced. Then, in 1868, Louis made a breakthrough.

This time, human-like remains seemed to have been purposefully buried along with personal ornaments. They lay beneath a natural rock shelter near Les Eyzies in the Dordogne, an area known locally as Cro-Magnon. There were at least five people, including a baby, a young woman, two young men and an older man, along with perforated shells and animal teeth, stone tools and the scattered bones of lion, reindeer and mammoth. In 1874 these anatomically modern humans were named 'Cro-Magnons'.

See also Comparative anatomy pages 106–107, Neanderthal man pages 170–171, Iceman pages 504–505

Right Paintings on the walls of the Lascaux cave in France are so fine that their modern re-discoverers found it hard to believe they were the work of 'primitive' Stone Age hunters.

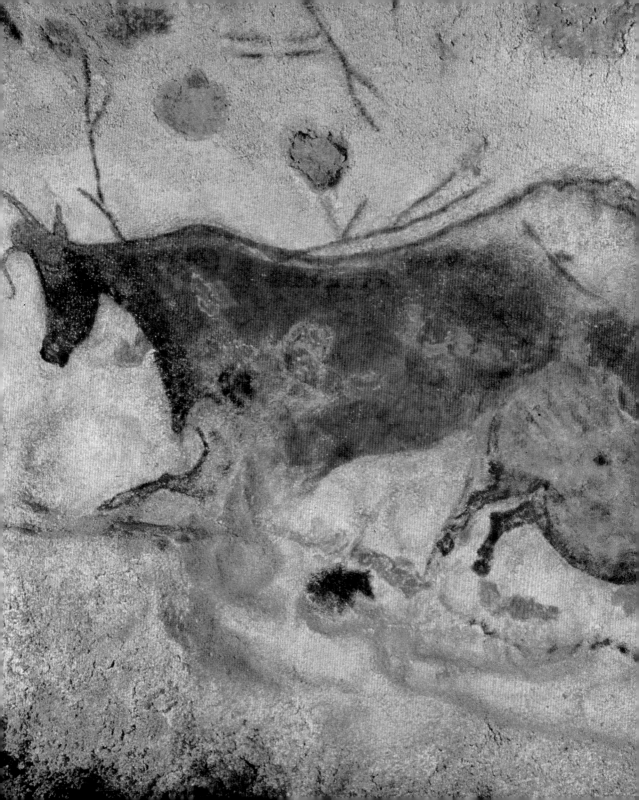

Distance to a star

Friedrich Wilhelm Bessel 1784–1846

Above Medieval astrolabe used to determine the position of celestial bodies.

Ever since Copernicus, in 1543, had the Earth going around the Sun and not vice versa, astronomers had realized that the orbiting Earth would change position by 300 million kilometres every six months. Nearby stars therefore appear to move with respect to the more distant stars, a movement known as 'parallax'. Unfortunately, as the average distance to the bright stars visible to the naked eye is now known to be about 20 million times the average distance of the Earth from the Sun, the parallax movement is very small.

By 1700 astronomers had convinced themselves of the immensity of the distance between stars, simply by comparing the relative brightness of the Sun, stars and planets. And by the 1830s they realized that bright, rapidly moving stars were likely to be the closest. Star 61 Cygni was a prime candidate. It was racing across the sky at more than 0.14 degrees per century. By the 1830s the engineering stability of a telescope's equatorial mounting had become first-class. Excellent bisected objective lenses were also being made. These 'heliometers' were specifically designed for making accurate measurements of the angular separation of close stars.

As often happens in scientific research, many people were trying to be the first to make the breakthrough. The German astronomer Friedrich Bessel was observing from Königsberg, near Heidelberg, Germany, using a 16-centimetre heliometer that had been made by Joseph Fraunhofer. At the end of 1838 he announced that 61 Cygni had a parallax of one third of a second of arc, putting it about 10 light-years away. After five years of work at the Dorpat Observatory in Estonia, the Russian astronomer Wilhelm Struve announced in 1840 that the star Vega was 13 light-years away. At nearly the same time the Scottish astronomer Thomas Henderson, royal astronomer at the Cape of Good Hope, found that Alpha Centauri was only 4 light-years away. Practical astronomy had finally established the scale of the universe.

See also Celestial predictions pages 26–27, Earth-centred universe pages 30–31, Transit of Venus pages 62–63, Our place in the cosmos pages 280–281, Planetary worlds pages 512–513

Right The 10th-century Persian astronomer al-Sufi's *Book of the Fixed Stars* located more than 1,000 stars, with many constellations shown as seen from the Earth and on a globe.

صورة برساوش

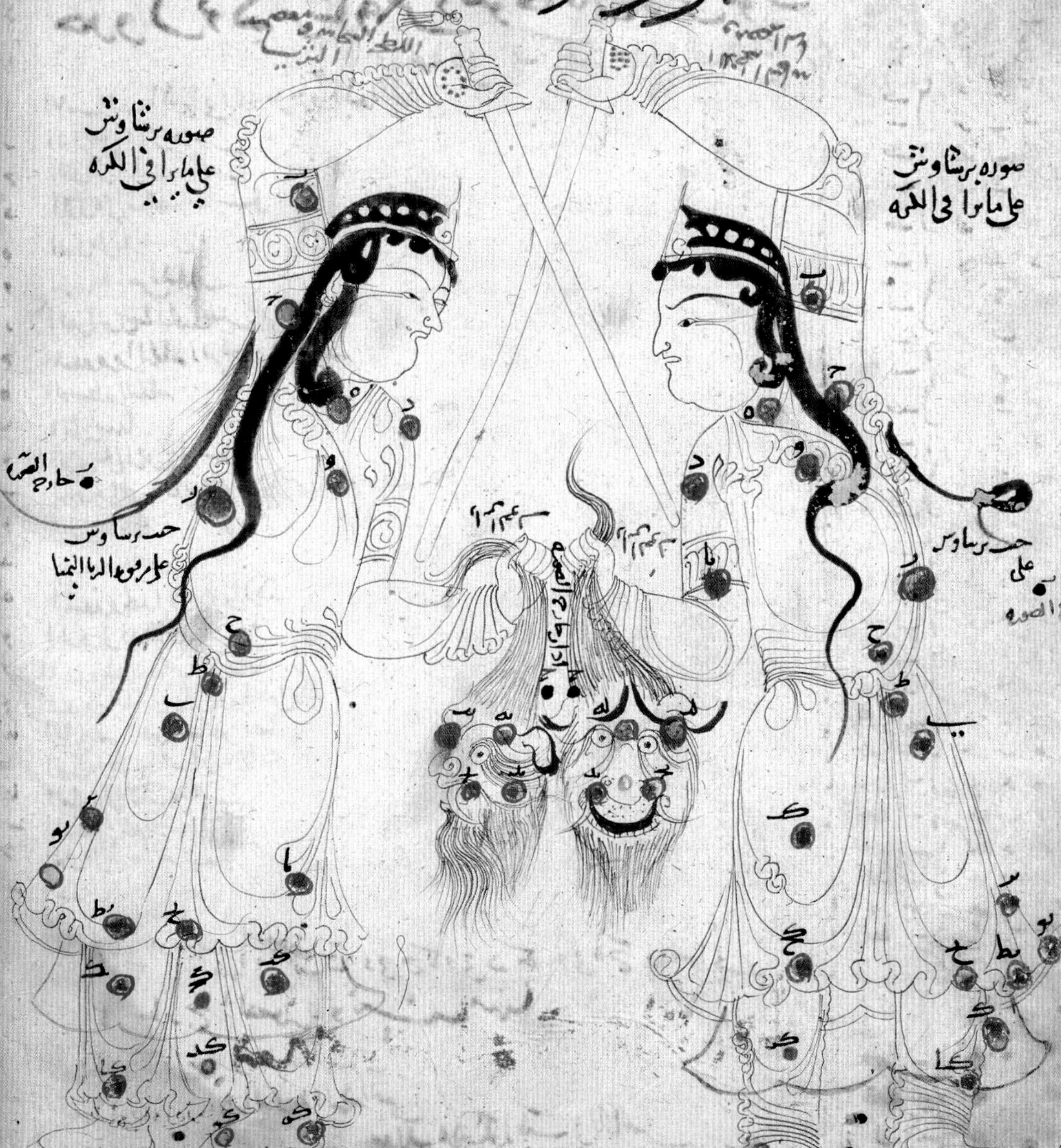

كوكبة برشوش وهو حامل راس الغول

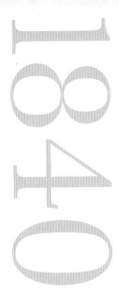

Ice ages

Jean Louis Rodolphe Agassiz 1807–73

That high latitudes were once inundated by glaciers was one of the great scientific discoveries of the nineteenth century. Many individuals contributed to its development but one name in particular is linked with the theory of the ice age: Louis Agassiz.

The Swiss-born son of a protestant pastor, Agassiz obtained a medical degree before working on fossil fishes (ultimately describing over 1,700 new species). Meanwhile he also became interested in glaciers – there was speculation at that time that the boulders dotting the Northern European plains had been brought there by glacial action. Agassiz soon concluded that an ice sheet, formed before the elevation of the Alps, had once extended well beyond the mountains. His collaborator, the German botanist Karl Schimper, protested that Agassiz had not recognized his contribution to this idea, and their friendship fell apart.

After careful observations in the Mont Blanc region, Agassiz published the first general discussion of Alpine glacial phenomena along with his own theory of the ice age. Never one for understatement, he claimed that ice had extended from the North Pole southwards across the Alps to the Atlas Mountains and over northern Asia and North America. While promoting his theory at the 1840 British Association meeting in Glasgow, Agassiz visited the Scottish Highlands with the influential English geologist William Buckland. Pointing out widespread glacial phenomena such as the accumulation of rock debris and scratched rock surfaces, Agassiz convinced Buckland that Scotland had been glaciated. Buckland in turn converted Charles Lyell and Roderick Murchison to the idea.

Agassiz went on to study the nature of glaciers with the Edinburgh physicist James Forbes, who first demonstrated that the surface of a glacier flows faster than the ice beneath it. Their results were published by Agassiz in 1847, who moderated his views on the extent of the great ice sheet and its timing. We now know that there have been several ice ages in Earth's history.

See also Greenhouse effect pages 184–185, Mountain formation pages 206–207, Climate cycles pages 276–277, Ozone hole pages 438–439

Right Glacier in the Alps, Switzerland, c.1815. Agassiz proposed that an ice sheet had once covered northern Europe and had acted like a vast and complex glacier. This conflicted with the orthodox idea that the Earth was gradually cooling.

Invention of the dinosaur

Richard Owen 1804–92

In 1842 the British anatomist Richard Owen coined the term 'dinosaur' ('terrible lizard') and its scientific category 'Dinosauria' to distinguish recently discovered fossils of the giant reptiles *Iguanodon* and *Megalosaurus* from known living reptiles. Little did he appreciate what sort of beast he was in fact unleashing. His dinosaurs have become universal icons, eclipsing dragons and all other mythical creatures. Owen's contemporaries thought that he himself was something of a monster: Charles Darwin was warned that Owen was 'not only ambitious, very envious and arrogant, but untruthful and dishonest'.

With his taxonomic trick, Owen stole the initiative from Gideon Mantell and William Buckland, who had first discovered and described the giant reptile fossils. In doing so, Owen achieved the greatest scientific coup of the mid-nineteenth century. He redefined this extinct group of creatures, transforming Mantell's 'lowly creeping' serpent-like animals into something much more ponderous and imperious, befitting Victorian values and Owen's rabid anti-evolutionary views. Owen reckoned his dinosaurs to be up to six times the size of an elephant – this was long before the discovery of the really big sauropod fossils.

Owen was also an adept publicist. When the Great Exhibition of 1851 in London closed, Joseph Paxton's spectacular steel and glass building was relocated to Sydenham, where it became the 'Crystal Palace'. This gave Owen a golden opportunity to promote his Dinosauria concept. He joined forces with the artist Benjamin Waterhouse Hawkins to supervise the construction of life-size dinosaur statues of concrete, stone and iron to be shown as part of the world's first theme park. The grand opening by Queen Victoria drew crowds of thousands and was heralded by a dinner for leading scientists inside the *Iguanodon* on New Year's Eve, 1853. The models can still be seen in the park, although the palace is long gone.

Above Anti-evolutionist: Owen thought that Darwin's work 'would be forgotten in ten years'.

See also Comparative anatomy pages 106–107, Darwin's *Origin of Species* pages 176–177, *Archaeopteryx* pages 180–181, Extinction of the dinosaurs pages 458–459

Right Benjamin Waterhouse Hawkins's 'Extinct Animals Model-Room', at the Crystal Palace, Sydenham. The animals, like the scientists, seemed generally to have been engaged in combat.

Doppler effect

Christian Johann Doppler 1803–53

A steam engine pulling a truck-load of playing trumpeters convinced the Austrian physicist Christian Johann Doppler that the drop in pitch of the sound he heard as they passed by was directly proportional to their velocity – a principle he first raised in 1842.

The effect is familiar in everyday life. When an ambulance speeds towards us with its siren wailing, sound waves are squashed together by its motion (higher frequency), whereas when it rushes away the sound waves are stretched out (lower frequency). The principle also applies to light and other electromagnetic waves. The rotation of the Sun stretches light waves from its receding western limb to longer wavelengths, producing a 'redshift' in its spectrum, and squeezes together light waves from its approaching eastern limb, producing a 'blueshift'. Comparisons of Doppler effects show that stars more massive than the Sun typically spin about 100 times faster.

In 1868 William Huggins used his stellar spectroscope to measure the 'radial' velocity of stars along the line of sight, towards or away from us. By 1887 stellar observations could be used to measure the velocity of the Earth around the Sun. The star's radial velocity could also be combined with its velocity perpendicular to the line of sight (obtained from decades of observations of its movement across the sky) to give its actual velocity through space. These measurements revealed that the Sun orbits around the galactic nucleus, its period being some 200 million years. The variation of stellar orbital velocities with distance from the galactic centre shows that the galaxy has a massive spherical halo.

The redshift in light from distant galaxies led Hubble to deduce in 1929 that the universe was expanding – a secondary Doppler effect caused by space itself stretching. Today's search for planets outside our Solar System is also built around the Doppler principle, which reveals their mass and orbital radius. Over 100 Jupiter-sized planets have been found to date.

See also Spectral lines pages 130–131, Expanding universe pages 306–307, Planetary worlds pages 512–513

Right Seeing the unseen: high-speed photograph of a bullet passing through the hot air above a candle flame, showing the shock waves caused by its passage and its turbulent wake.

Sunspot cycle

Heinrich Samuel Schwabe 1789–1875

Telescopic observations by Galileo Galilei and Johannes Fabricius around 1610 had shown that sunspots were solar surface phenomena and not satellites in low orbits around the Sun or intervening cloud in the Earth's atmosphere. We had to wait until 1843 before it was discovered that the spotted appearance of the Sun changed with time.

Heinrich Schwabe, an apothecary in Dessau, Germany, was fascinated by astronomy. Wishing to pursue astronomical research as well as professional pharmacy, he decided to concentrate on a branch of astronomy that would occupy him in the daytime. His first thought was that by observing the Sun he might find a new planet inside the orbit of Mercury, during one of its transits of the solar disc.

Schwabe could not help but notice sunspots through his small 5-centimetre telescope. Soon he was making daily counts. From 1825 onwards he watched the Sun assiduously. After carefully collating his results, he announced, in 1843, that the number of spots on the solar disc waxed and waned with a periodicity of 10 years. By 1851 the Swiss astronomer Rudolf Wolf had put together a larger data set and obtained a more accurate period of 11.1 years.

It was soon found that sunspot periodicity was echoed in the periodicity of Earth's magnetic storms and aurorae. Some astronomers even reckoned they could detect the periodicity in weather conditions and animal and plant growth rates. By 1858 the wealthy English amateur astronomer R. C. Carrington was reporting that the latitude of the spots changed throughout the cycle, starting at about 40 degrees and then slowly moving towards the solar equator. Sunspots also showed that equatorial regions of the Sun were spinning faster than polar ones. This led to the proposal in 1961 that the Sun's equatorial magnetic field lines are dragged along so quickly that they form magnetic 'pipes' which float upwards, break through the solar surface and produce pairs of spots.

Above Sunspots drawn in 1875 by Etienne Trouvelot, one of the most skilled scientific artists of the pre-photographic era.

Right As shown in this photograph, sunspots each have a dark core or umbra, surrounded by a lighter penumbra. Their behaviour indicates that they are depressions rather than lumps.

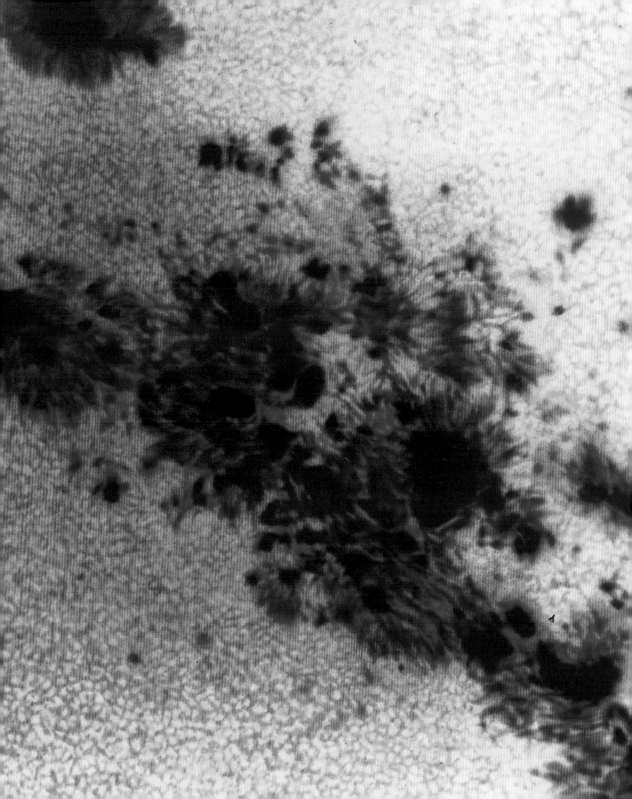

Spiral galaxies

William Parsons, third Earl of Rosse 1800–67

Scattered among the stars are hazy patches of light called 'nebulae'. Ptolemy recorded seven in the second century AD. Charles Messier noted 103 through his telescope in the early 1770s, mainly as objects to be avoided in his search for comets. The greatest cataloguer of nebulae was William Herschel, who by 1802 had listed 2,500. But astronomers were still unsure of their nature. Some were undoubtedly clouds of gas and dust, whereas others gave the impression of being made up of stars, some in our galaxy and some way beyond.

In 1845 William Parsons, third Earl of Rosse, built an enormous telescope in the grounds of his castle at Parsonstown (now Birr) in Ireland. Called 'Leviathan', it had a speculum metal mirror 72 inches (1.83 metres) across that allowed him to examine the nebulae as never before. Many of them revealed complex structures, which Parsons recorded in detailed pencil drawings. In particular he was the first to discover that some nebulae are spiral in shape.

In 1864 William Huggins found that the spectra of bright nebulae, such as that of Orion, were typical of a luminous gas (the emission type), whereas others, such as the Andromeda nebula (M31), had spectra typical of star light (the absorption type). But any stars that were there were so far away that no one could see them individually. Then, in 1885, a star flared up in M31, with a further four fainter novae appearing in 1917.

The true nature of spiral nebulae was finally revealed by Edwin Hubble. In 1924 he used the Mount Wilson 100-inch (2.54-metre) Hooker telescope to photograph M31, and managed to make out some giant stars, including Cepheid variables. These 'lighthouse' stars allowed him to determine that the Andromeda nebula was nearly a million light-years away, eight times the distance of the farthest star in our galaxy, and large enough to be a galaxy of its own. It was soon clear that the universe consisted of numerous galaxies, perhaps as many as a hundred billion.

Above Rosse's drawing of the magnificent spiral nebula – later identified as galaxy M51 in the constellation Canes Venatici, 1850.

See also Spectral lines pages 130–131, Our place in the cosmos pages 280–281, Expanding universe pages 306–307, Afterglow of creation pages 412–413, Great Attractor pages 498–499

Right A spiral galaxy thought to lie 10–20 million light-years away. Hubble produced a comprehensive classification system for galaxies, dividing them into normal spirals, barred spirals, elliptical and irregular types.

Discovery of Neptune

John Couch **Adams** 1819–92, Urbain Jean Joseph **Le Verrier** 1811–77,
Johann Gottfried **Galle** 1812–1910

By the early 1800s it was clear that the newly discovered planet Uranus was not behaving itself. By 1832 it was fully half a minute of arc from the position in which orbital calculations indicated it should be. Several astronomers suspected that an unknown planet beyond Uranus was exerting a gravitational influence. But how to find it?

There were three obvious clues. First of all, Uranus had until 1822 been accelerating, but had then become retarded. Knowledge of its position at that time gave the zodiacal constellation of the postulated planet. Also, the Titius–Bode law indicated that the planet might be about 38 times farther from the Sun than Earth. The final clue came from the acceleration of Uranus, which indicated the probable mass and brightness of the new planet.

John Couch Adams, a young Cambridge astronomer, had calculated a rough position for the new planet by 1845. He presented his results to the Astronomer Royal, George Airy, who ignored them until stirred into action by publication of a similar prediction from Urbain Jean Joseph Le Verrier working from the Paris Observatory. As there was no suitable telescope at Greenwich, Airy asked James Challis to carry out a search using the 29.8-centimetre refractor at the Cambridge Observatory. On 29 July the plodding Challis rather painstakingly started looking, unfortunately assuming the search object was 20 times fainter than it turned out to be.

Le Verrier, who used a different method of calculation, contacted Johann Gottfried Galle at the Berlin Observatory and asked him to look at a certain spot in the sky for the new planet. Fortunately, Galle had to hand a new sky map of the area. Almost as soon as he started looking, on 23 September 1846, a previously unrecorded 'star' was found; and what's more, it showed a slight disc. England had been beaten, even though Challis later found that he had actually seen Neptune three times without recognizing it.

See also Laws of planetary motion pages 52–53, Discovery of Uranus pages 96–97, Origin of the Solar System pages 104–105, Discovery of an asteroid pages 120–121, Planetary worlds pages 512–513

Right Planetary pride: French cartoons showing 'M. Adams looking for M. Le Verrier's planet' (top) and 'M. Adams discovering the new planet in the account by M. Le Verrier' (bottom).

Laws of thermodynamics

Benjamin Thompson, Count Rumford 1753–1814, Sadi Carnot 1796–1832,
James Prescott Joule 1818–89, Rudolf Clausius 1822–88

Above Joule's paddle-wheel apparatus converted mechanical work directly into heat by stirring water vigorously.

The first law of thermodynamics says that work and heat are both ways of transferring energy from one place to another. No matter how it is transferred, the total amount of energy never changes. Early in the nineteenth century, heat was thought to be a fluid, called 'caloric', that could flow from a hot body to a cooler one without being created or destroyed. Many observers doubted the caloric theory, including Count Rumford, who noted that the boring of cannon barrels produced an immense amount of heat. But the British natural philosopher James Prescott Joule, who in 1847 made careful measurements of how much heat is produced by a known amount of work, gets the credit for discovering the first law.

Long before Joule, the second law of thermodynamics was discovered by a young French military engineer, Sadi Carnot. By analogy to the turning of water wheels, he reasoned that caloric could make a steam engine go by running through it from high temperature to low. So what matters is not only the quantity of heat but also its temperature. Rudolf Clausius, who, in a famous paper of 1850, rescued Carnot's work from obscurity, coined the term 'entropy' for the quantity of heat divided by its absolute temperature. Heat at high temperature has low entropy. Once it runs downhill to lower temperature its entropy has increased whether it has done work along the way or not.

Clausius summed up the first and second laws this way: the energy of the universe is constant, and the entropy of the universe tends to a maximum. Today we associate entropy with disorder. So when fuel is burned, energy is converted from a highly organized chemical form in the fuel into heat at high temperature, which invariably winds up as the same amount of heat at ambient temperature. This gives it the highest entropy it can attain. Once that happens, the entropy of the universe has been increased forever.

See also Information theory pages 354–355, Afterglow of creation pages 412–413, Black-hole evaporation pages 440–441

Right Count Rumford, a Massachusetts-born American (and former British spy), sought ways of applying scientific ideas to everyday life. Here he is seen warming himself in front of a Rumford domestic stove.

J.ˢ Gillray des. & f.ˢᵗ ad vivum.

Foucault's pendulum

Jean Bernard Léon Foucault 1819–68

In 1851 the French physicist Jean Bernard Léon Foucault conducted an unusual experiment in the Pantheon, a monument to great minds in Paris. He hung from the building's ceiling an iron ball the size of a pumpkin attached to 67 metres of steel wire and set the contraption swinging back and forth, like a pendulum. He then carefully noted the plane in which this motion occurred. During the course of the day, this plane of motion slowly changed, rotating clockwise at a rate of 11 degrees per hour. Foucault then invited scientists to come and witness the effect, explaining that his strange experiment proved that the Earth actually rotates about its own axis.

To understand why, imagine Foucault's pendulum set up at the North Pole. Once set in motion, the pendulum's movement is entirely independent of the Earth's and the planet simply rotates beneath it. An observer at the pole would see the plane of the pendulum's motion turn clockwise through 360 degrees once every 24 hours. But this rate depends on the pendulum's latitude: in Paris the pendulum rotates through 360 degrees every 32 hours, at the equator the plane of motion doesn't change at all, and in the southern hemisphere it rotates in an anticlockwise rotation.

Scientists sometimes explain the motion of the pendulum by imagining that a pseudo-force, known as the Coriolis force, is acting on it. The Coriolis force is not a real force, as the thought experiment at the North Pole shows, but appears real to people moving in the same frame of reference as the Earth. The Coriolis force also explains why weather patterns in the northern hemisphere tend to turn clockwise while those in the southern hemisphere rotate anticlockwise.

Many science museums around the world have set up a copy of Foucault's pendulum to demonstrate the effect to their visitors.

See also Circumference of the Earth pages 24–25, Trade winds pages 86–87, 'Weighing' the Earth pages 112–113, Weather forecasting pages 284–285

Right Foucault's pendulum in the Pantheon, Paris, 1851. To observers the pendulum appeared to turn through 360 degrees in 32 hours, but in fact this was due to the Earth's rotation beneath it.

Cholera and the pump

John Snow 1813–58

Asiatic cholera wreaked havoc in Britain in 1832, and then died away. Many thought it was a 'filth' disease caused by noxious vapours from putrefying vegetable and animal matter. Others argued that it was contagious, transmissible by person-to-person contact and avoidable by quarantine. While debate about its cause and spread continued, the return of the disease provided the circumstances for the anaesthetist John Snow to establish the modern science of epidemiology.

On the basis of his experience of cholera during an outbreak in 1848, which killed 7,000 Londoners in one month alone, Snow wrote a pamphlet in which he argued that the disease could not be spread by a poison in the air as it affected the intestines, not the lungs. It was more likely to result from infected sewage seeping into wells or running into rivers from which drinking water was taken. A further outbreak in London's Soho in 1854 provided him with dramatic proof. Living nearby, he immediately suspected the affected area's water supply – the pump in Broad Street (now Broadwick Street). When he plotted the cholera deaths, he found they were all clustered around an area no more than 450 metres across, at the centre of which was Broad Street. He was convinced that removal of the handle of the pump would end the epidemic. It did.

Following interviews of the Broad Street residents, it was clear that almost all the victims had drunk water from the pump. What's more, visitors who had used the pump had also succumbed, there were far fewer deaths in a nearby prison with its own well, and brewery workers who drank free beer rather than water had remained healthy. Snow then conducted a London-wide survey, using simple statistical methods to confirm his theory that cholera was a specific, water-borne disease. But recognition of his achievement had to await Robert Koch's identification of the cholera microbe in 1884.

See also Vaccination pages 102–103, Population pressure pages 110–111, Germ theory pages 202–203, Antitoxins pages 214–215, A magic bullet pages 262–263, Prion proteins pages 466–467, AIDS virus pages 472–473

Right Cholera strikes the Turkish army during the Balkan War, 1912. The disease killed up to 100 soldiers a day.

Neanderthal man

Hermann Schaaffhausen 1816–93

Although the discovery in 1856 of human-like bones in a cave above the Neander Valley near Düsseldorf, Germany, did not at first cause much of a stir, it eventually changed the way we humans see ourselves. To a Western world still firmly adhering to the Judeo-Christian biblical version of the creation story, the scientific discoveries of the early nineteenth century were increasingly disconcerting. Human fossil remains like those of the Neanderthaler were found with worked stone tools and the bones of extinct Ice Age animals such as mammoths, giant deer and woolly rhinos. The antiquity of humans appeared far greater than previously thought, and our ancestry was clearly different from the biblical version.

The first description of the Neanderthal remains by the German anatomist Hermann Schaaffhausen explained away the distinctive eyebrow ridges and thick-walled curved bones as belonging to some 'barbarous and savage race' of pre-Roman times. Not until 1863 were the fossils recognized for what they truly are: the bones of an extinct species of human. Called *Homo neanderthalensis* by William King, a professor of geology in Galway, Ireland, this was the first acknowledgement that *Homo sapiens* had extinct yet still human relatives.

We now know that the Neanderthals occupied much of Europe from Wales south to Gibraltar and eastwards to the Caucasus. Having evolved from an even older human relative, *Homo heidelbergensis*, more than 200,000 years ago, they survived the dramatically fluctuating climate changes of the final part of the Quaternary Ice Age until around 28,000 years ago when they became extinct. During their last 12,000 years they were encroached on by modern humans, *Homo sapiens*, who first moved out of Africa around 150,000 years ago. Despite the overlap, recent analysis of Neanderthal DNA suggests that the two peoples did not interbreed.

See also Prehistoric humans pages 148–149, Ice ages pages 152–153, Java man pages 216–217, Taung child pages 298–299, Olduvai Gorge pages 392–393, Ancient DNA pages 476–477, Nariokotome boy pages 480–481, Out of Africa pages 492–493, Iceman pages 504–505

Right Neanderthal man, from H. G. Wells's *Outline of History* (1920). The Neanderthals' heavy brow ridge and sloping forehead remain to pose unresolved questions about who these people were and what became of them.

Mauve dye

William Henry **Perkin** 1838–1907

In 1856 William Henry Perkin, an 18-year-old student at the Royal College of Chemistry in London, was beginning his career under the German chemist August Wilhelm Hofmann. Hofmann was the world's leading expert on coal tar, the sticky black residue produced at gas works. When distilled, coal tar releases hydrocarbons – particularly the pungent 'aromatics' such as benzene, toluene, naphthalene and anthracene. The synthesis of phenol, widely used as a disinfectant, from benzene, and of the yellow dye picric acid from phenol, showed in the 1840s that coal-tar products had commercial potential.

A still more precious goal was to make the antimalarial drug quinine, traditionally extracted from the bark of the Peruvian cinchona tree. Hofmann suspected it could be more cheaply synthesized from coal-tar extracts. Working in his home laboratory in 1856, Perkin tried to prepare quinine from an anthracene derivative. The result was a brown sludge that most chemists would have discarded. But Perkin was intrigued, and tried the same reaction using aniline, a benzene derivative. He obtained a black solid that dissolved in methylated spirit to give a beautiful purple solution. Perkin found that silk would take up this 'aniline purple', so he took it to the renowned dyers of Scotland for testing. They had mixed feelings, as the real demand was for cotton dyes. Nevertheless, Perkin persuaded his father and brother to set up a business manufacturing the dye, and from 1857 their factory began to supply this 'mauveine' or 'mauve' (French for mallow, a purple-pink flower).

Aniline dyes began to proliferate – fuchsine or magenta, aniline blue, aniline violet, blacks and greens – and the demands of fashion drove the discovery of new classes of synthetic dyes such as azos. In 1868 two German chemists synthesized alizarin, the natural colourant of the popular madder red dye. By the end of the century, dye companies had expanded and diversified into the major chemicals companies of today: Hoechst, BASF, Bayer, Agfa, Ciba and Geigy.

Above Original mauve dye, prepared by Perkin. By 1873, at the age of 35, Perkin had ensured that his factory and patents guaranteed his 'retirement'.

See also Benzene ring pages 190–191, Dynamite pages 194–195, Synthesis of ammonia pages 256–257, A magic bullet pages 262–263, Nylon pages 316–317

Right An 1862 silk dress, dyed with Perkin's mauve – a colour 'which has been very much wanted in all classes of goods and could not be obtained fast on Silks'.

Communities of cells

Rudolf Virchow 1821–1902

Scientists were seeking ever-smaller units of biological analysis. Giovanni Battista Morgagni showed that diseases were located in specific organs of the body (1761), whereas Marie François Xavier Bichat highlighted the centrality of the tissues (1799). But even though cells had been identified, there was still no recognition of their importance. This changed in 1839 when Theodor Schwann proposed his 'cell theory'. Inspired by the German botanist Matthias Jakob Schleiden, he argued that cells are the structural and functional building blocks of all plants and animals. Schwann also rightly suggested that eggs are cells and that all life starts as a single cell. But he was mistaken in his belief that in embryological development and certain pathological conditions such as pus formation, new cells could crystallize from the fluid surrounding cells (the 'cytoblastema') – it would be several decades before cell division was fairly well understood.

Cell theory was enthusiastically promoted by the German pathologist Rudolf Virchow. In his classic work *Cellularpathologie*, published in 1858, he emphasized his credo *omnis cellula e cellula* (every cell from a cell). The idea of cellular continuity became a tenet of later nineteenth-century biology. Using a political analogy, he saw cells as living in a 'cellular democracy' or 'republic of the cells'. Further, he claimed that all disease arises from disturbances of the normal life processes within cells.

Although this dynamic view of pathology ignored diseases caused by external factors – Virchow was sceptical of Pasteur's germ theory – it was brilliantly effective at explaining cancers. Tumours were seen as abnormal cells that revolt against the organism by continual division. Virchow also explained the spread of cancer cells to distant organs, drawing on his work on the formation of blood clots and their dislodgement to form 'emboli' (a term he coined). And his unrestrained cancer cells have found a modern interpretation in terms of molecular biology. Many cancers are clones of one renegade cell in which the genetic mechanism that controls cell division has failed.

See also Microscopic life pages 76–77, Eggs and embryos pages 140–141, Regulating the body pages 188–189, Germ theory pages 202–203, Hayflick limit pages 394–395, Human cancer genes pages 456–457

Right Cell division in the root-tip cells of garlic *Allium sativum*. In Virchow's 'cellular democracy', cells are social classes and the organs and tissues are their territory.

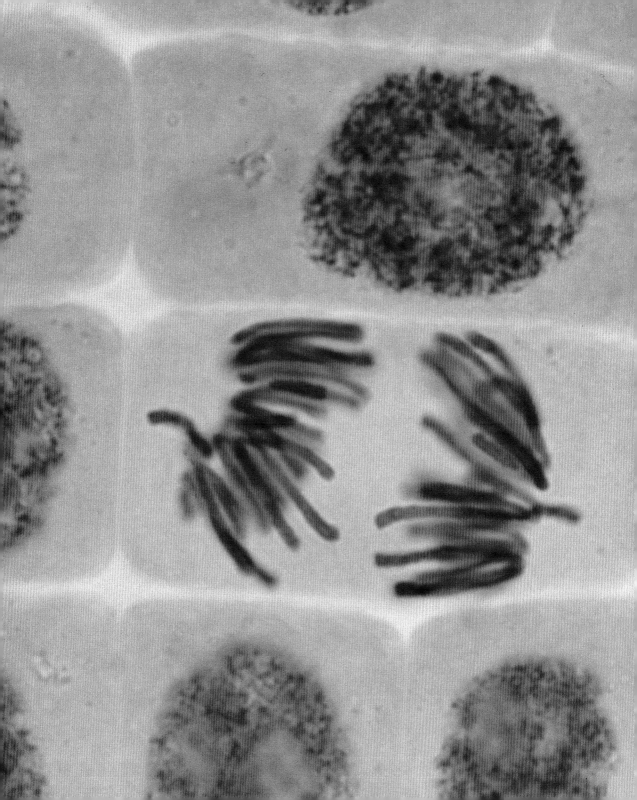

Darwin's *Origin of Species*

Charles Robert Darwin 1809–82

Darwin's *Origin of Species*, one of the greatest books ever written, has two main theories. One is that all species of life on Earth have arisen by evolution from other pre-existing species. This contradicted Christian dogma, according to which each species has a separate origin and remains fixed in form. The second is that the process that drives evolution is natural selection: some individuals in a population have more offspring than others; offspring tend to inherit the attributes of their parents; so later generations contain more of the sort of individual that, in previous generations, left more offspring.

The individuals who leave the most offspring tend to be those best adapted to the local conditions. So natural selection causes living creatures to evolve to be well-adapted for life. This conclusion also upset a religious belief, because adaptation (or 'design') in life had previously been explained supernaturally, by the action of God. In natural selection, adaptation has a natural explanation.

Some attributes, such as a peacock's tail or the antlers of a stag, appear not to help living creatures survive the rigours of the environment but to help them compete to attract members of the opposite sex. Darwin published a later book explaining these attributes by the theory of sexual selection – a special case of natural selection. Individuals (usually males) compete against each other for a limited number of mates, rather than for limited environmental resources.

Darwin conceived natural selection in the 1830s, but waited 20 years before publishing it – and then because another British naturalist, Alfred Russel Wallace, had independently developed much the same theory and presented it to Darwin. Darwin and Wallace jointly published the theory in 1858, but evolution and natural selection were little noticed until Darwin's *Origin of Species* came out a year later. Only 1,250 copies were printed, and every one was snapped up on the first day of publication.

Above Darwin said of the *Origin*, 'I see no good reason why the views given should shock the religious feelings of anyone.'

Right Spectacular squid: costume sketch for a New Orleans Mardi Gras parade in 1873, on the theme of 'Missing Links, or Darwin's *Origin of Species*'.

The first humans
Richard Leakey

ANTHROPOLOGISTS HAVE LONG been enthralled by the special qualities of *Homo sapiens*, such as language, high technological skills and the ability to make ethical judgements. But one of the most significant shifts in anthropology in recent years has been the recognition that despite these qualities, our connection with the African apes is extremely close indeed …

In 1859, in his *Origin of Species*, Darwin avoided extrapolating the implications of evolution to humans. A guarded sentence was added in later editions: 'Light will be thrown on the origin of man and his history.' He elaborated on this short sentence in a subsequent book, *The Descent of Man*, published in 1871. Addressing what was still a sensitive subject, he effectively erected two pillars in the theoretical structure of anthropology. The first had to do with where humans first evolved (few believed him initially, but he was correct), and the second concerned the manner or form of that evolution. Darwin's version of the manner of our evolution dominated the science of anthropology up until a few years ago, and it turned out to be wrong.

The cradle of humankind, said Darwin, was Africa. His reasoning was simple:

In each great region of the world, the living mammals are closely related to the evolved species of the same region. It is, therefore, probable that Africa was formerly inhabited by extinct apes closely allied to the gorilla and chimpanzee: as these two species are now man's nearest allies, it is somewhat more probable that our early progenitors lived on the African continent than elsewhere.

We have to remember that when Darwin wrote these words, no early human fossils had been found anywhere; his conclusion was based entirely on theory. In his time, the only known human fossils were of Neanderthals, from Europe, and these represent a relatively late stage in the human career.

Anthropologists disliked Darwin's suggestion intensely, not least because tropical Africa was regarded with colonial disdain: the Dark Continent was not viewed as a fit place for the origin of so noble a creature as *Homo sapiens*. When additional human fossils began to be discovered in Europe and in Asia at the turn of the century, yet more scorn was heaped on the idea of an African origin. This attitude prevailed for decades. In 1931, when my father told his intellectual mentors at Cambridge University that he planned to search for human origins in East Africa, he came under great pressure to concentrate his attention on Asia instead. Louis Leakey's conviction was based partly on Darwin's argument and partly, no doubt, on the fact that he was born and raised in Kenya. He ignored the advice of the Cambridge scholars and went on to establish East Africa as a vital region in the history of our early evolution. The vehemence of anthropologists' anti-Africa sentiment now seems quaint to us, given the vast numbers of early human fossils that have been recovered in that continent in recent years. The episode is also a reminder that scientists are often guided as much by emotion as by reason.

Darwin's second major conclusion in *The Descent of Man* was that the important distinguishing features of humans – bipedalism, technology and an enlarged brain – evolved in concert. He wrote:

If it be an advantage to man to have his hands and arms free and to stand firmly on his feet … then I can see no reason why it should not have been more advantageous to the progenitors of man to have become more and more erect or bipedal. The hands and arms could hardly have become perfect enough to have manufactured weapons, or to have hurled stones and spears with true aim, as long as they were habitually used for supporting the whole weight of the body … or so long as they were especially fitted for climbing trees.

Here, Darwin was arguing that the evolution of our unusual mode of locomotion was directly linked to the manufacture of stone weapons. He went further and linked these evolutionary changes to the origin of the canine teeth in humans, which are unusually small compared to the daggerlike canines of apes. 'The early forebears of man were … probably furnished with great canine teeth,' he wrote in *The Descent of Man*; 'but as they gradually acquired the habit of using stones, clubs, or other weapons for fighting with their enemies or their rivals, they would use their jaws and teeth less and less. In this case, the jaws, together with the teeth, would become reduced in size.'

These weapon-wielding, bipedal creatures developed a more intense social interaction, which demanded more intellect, argued Darwin. And the more intelligent our ancestors became, the greater was their technological and social sophistication, which in turn demanded an ever-larger intellect. And so on, as the evolution of each feature fed on the others. This hypothesis of linked evolution was a very clear scenario of human origins, and it became central to the development of the science of anthropology.

According to this scenario, the original human species was more than merely a bipedal ape: it already possessed some features we value in *Homo sapiens*. The image was so powerful and plausible that anthropologists were able to weave persuasive hypotheses around it for a very long time. But the scenario went beyond science: if the evolutionary differentiation of humans from apes was both abrupt and ancient, a considerable distance was inserted between us and the rest of nature. For those with a conviction that *Homo sapiens* is a fundamentally different kind of creature, this viewpoint offered comfort.

Archaeopteryx

Richard Owen 1804–92, Thomas Henry Huxley 1825–95

The discovery in 1860 of a single fossil feather in the Solnhofen limestones of Bavaria, southern Germany, helped to transform our understanding of evolution by providing the first convincing 'missing link' between two major groups of animals: reptiles and birds.

Bavaria's fine-grained Jurassic limestones were quarried for high-quality lithographic printing stone, but often the split stone also revealed exquisitely preserved fossils which the quarrymen sold to collectors. At that time possession of feathers was considered a uniquely bird-like attribute. So where there was a feather, there should also be a bird. Six months later, in 1861, a virtually complete bird skeleton, with imprints of asymmetrical flight feathers around its wing bones, was duly found.

The specimen was bought in 1862 by Richard Owen, superintendent of the natural history collections at the British Museum in London. Owen was one of the great anatomists of the day but also a severe critic of Darwin's evolutionary ideas. Nevertheless, Owen's masterly description of the 170-million-year-old fossil showed that, while it was 'unequivocally a bird', it also had features he thought were found only in embryos of living birds.

But the English biologist Thomas Henry Huxley realized that this *Archaeopteryx* ('ancient wing'), with its mixture of reptilian and bird characteristics, provided an excellent example of Darwinian evolution. His conclusion was reinforced by the discovery in the same deposits of a small two-footed dinosaur (*Compsognathus*) resembling *Archaeopteryx*. For Huxley there was no inherent problem in linking separate classes of animals despite their seemingly different anatomy and physiology. Once a common ancestor was found, other gaps could be bridged, and Owen's embryological evidence would support the theory – embryonic development was at that time believed to be a speeded-up replay of the evolution of species.

Feathered dinosaurs have now been found in Liaoning, China, and most experts regard birds as a group of raptor-like dinosaurs.

See also Comparative anatomy pages 106–107, Invention of the dinosaur pages 154–155, A living fossil pages 324–325, Extinction of the dinosaurs pages 458–459

Right One of the world's most famous fossils, *Archaeopteryx* has many features of a small dinosaur, with a long bony tail and toothed jaws. But its feathers suggest it was a bird and could fly.

Mapping speech

Pierre Paul Broca 1824–80

The idea that different regions of the brain are involved with different psychological functions came to prominence in the early nineteenth century with the work of Franz Josef Gall. He suggested that mental faculties might be reflected in the shape of the brain, and hence of the skull – a notion vulgarized by the 'phrenologists' who claimed to be able to deduce a person's character by feeling the bumps on his or her head. Phrenology was soon abandoned by orthodox medical practitioners and became the business of quacks.

But the concept of 'cerebral localization' was rehabilitated in 1861 by Paul Broca, a Parisian surgeon, pathologist and anatomist. Clinical experience with aphasic patients and subsequent postmortem investigations allowed him to associate the loss of ability to speak with damage to specific regions of brain cortex. In 1861 he proposed that the centre for articulate speech was in the third frontal convolution of the left cerebral hemisphere ('Broca's area').

Unfortunately, the precision with which Broca isolated his speech centre did not stand the test of time. Nevertheless, his work was the first clear-cut demonstration that specific functions can be mapped to particular areas in the brain, and it was soon supported by John Hughlings Jackson's research with epileptic patients. These and other mapping studies proved useful in neurosurgery. In 1884, for example, John Rickman Godlee operated on a patient to remove a walnut-sized tumour from between the frontal and parietal lobes; it was found exactly as predicted from the patient's symptoms. In the twentieth century neuroscientists focused on identifying brain regions associated with various behaviours and emotions. This inevitably led to surgical intervention in psychiatric conditions, the most notable practitioner being Antonio Egaz Moniz. In 1949 he won a Nobel Prize for pioneering prefrontal lobotomy – the severance of the connections of the frontal lobes from the rest of the brain – in the control of schizophrenia and other mental disorders. 'Psychosurgery' has now fallen into disrepute.

See also Language instinct pages 386–387, Right brain, left brain pages 400–401, Images of the mind pages 478–479

Right A place for everything and everything in its place: although phrenology lost its scientific credibility by the 1850s, it remained a popular art, as this 1923 chart shows.

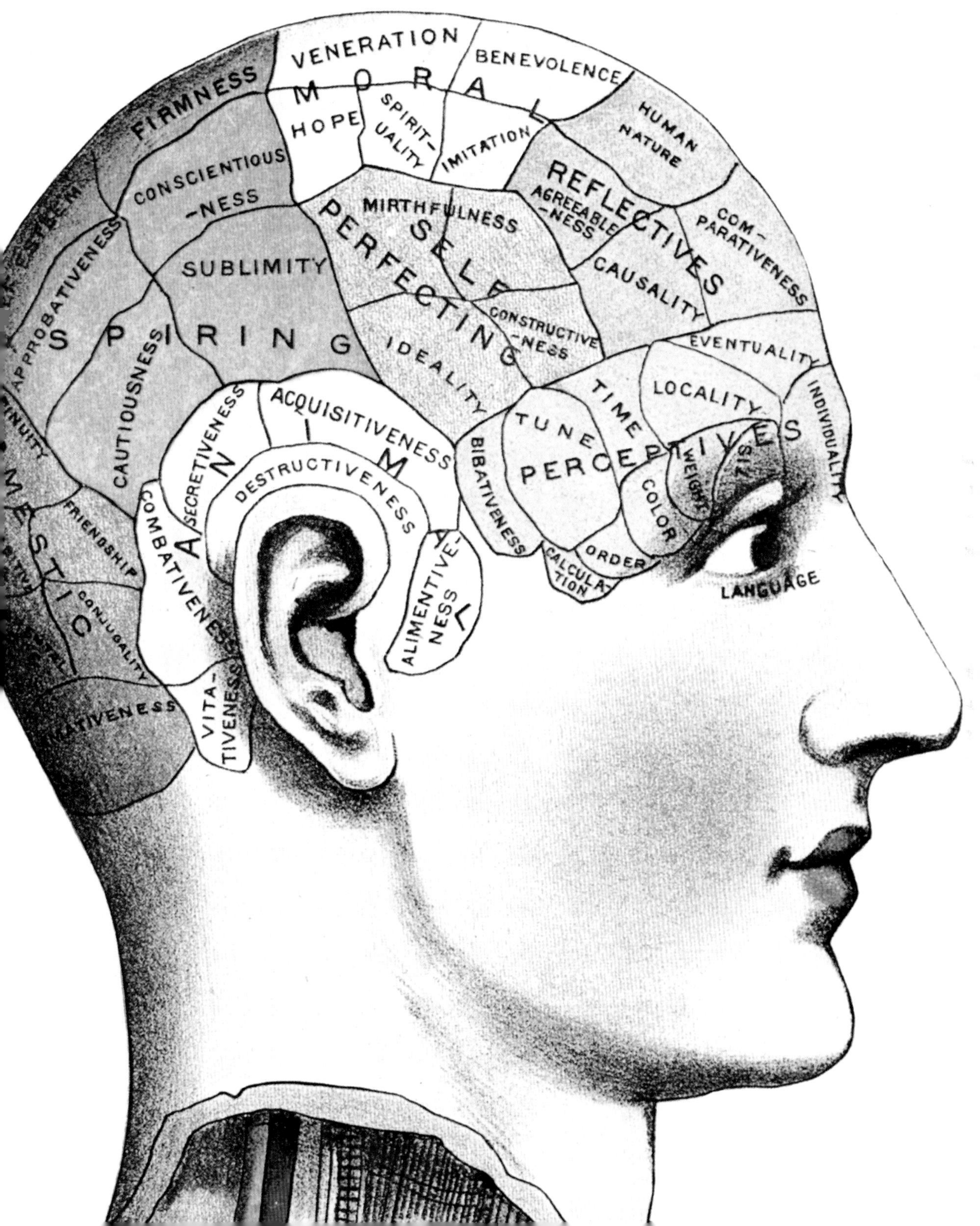

VENERATION
BENEVOLENCE
MORAL
FIRMNESS
HUMAN NATURE
HOPE
SPIRIT-UALITY
IMITATION
REFLECTIVES
CONSCIENTIOUS-NESS
AGREEABLE-NESS
COM-PARATIVENESS
ESTEEM
APPROBATIVENESS
MIRTHFULNESS
SELF
SUBLIMITY
PERFECTING
CAUSALITY
SPIRING
CONSTRUCTIVE-NESS
EVENTUALITY
CONTINUITY
CAUTIOUSNESS
IDEALITY
LOCALITY
INDIVIDUALITY
MESTIC
ACQUISITIVENESS
TIME
SECRETIVENESS
TUNE
SIZE
FRIENDSHIP
COMBATIVENESS
DESTRUCTIVENESS
BIBATIVENESS
PERCEPTIVES
WEIGHT
CONJUGALITY
ALIMENTIVE-NESS
COLOR
MATIVENESS
VITA-TIVENESS
ORDER
CALCULA-TION
LANGUAGE

Greenhouse effect

John Tyndall 1820–93

The first man to climb the Weisshorn in the Alps and almost the first to conquer the Matterhorn – his guides refused to follow him up the final slopes – was also the first to propose the greenhouse effect. In 1863 John Tyndall, the British physicist, draftsman, geologist and mountaineer, reported experiments on the radiative properties of gases, such as oxygen, nitrogen and carbon dioxide (or 'carbonic acid' as Tyndall called it). He discovered that there were huge differences in the abilities of these gases to absorb and transmit heat, noting that oxygen and nitrogen were almost transparent to heat, while water vapour, carbon dioxide and ozone were very nearly opaque.

This puzzling behaviour of otherwise colourless and invisible gases led him to a startling conclusion: water vapour was so common in the Earth's atmosphere and such an efficient absorber of heat that it must have an important role in regulating the temperature at the surface. Without it the Earth would be 'held fast in the iron grip of frost'. Tyndall went on to describe how variations in the levels of water vapour and carbon dioxide could lead to climate change, the now famous greenhouse effect. (In 1896 the Swedish chemist Svante Arrhenius also pointed out that atmospheric carbon dioxide was a 'heat trap', speculating that a slight fall in levels might set off an ice age.)

Tyndall also explained why the sky is blue: large molecules in the atmosphere scatter blue light from the Sun's rays more strongly than other colours. The same reasoning explains why the Sun is red when it sets. Close to the horizon, light from the Sun has to travel farther through the atmosphere to reach an observer's eye. During this journey, blue light and other colours are scattered, leaving only the red light. This is known as the 'Tyndall effect'.

See also Unweaving the rainbow pages 36–37, Ice ages pages 152–153, Nitrogen fixation pages 208–209, Climate cycles pages 276–277, Power from the nucleus pages 330–331, Gaia hypothesis pages 432–433, Ozone hole pages 438–439

Right Smokestacks from a factory in Pittsburgh, Pennsylvania, belch black smoke into the atmosphere in the 1890s. Carbon dioxide emissions from the burning of fossil fuels contribute greatly to the greenhouse effect.

Maxwell's equations

James Clerk Maxwell 1831–79

The Scottish physicist James Clerk Maxwell mixed electricity and magnetism, and out came light. Electric and magnetic forces were already known to be somehow connected. In the early nineteenth century Hans Christian Oersted had seen that electric currents could deflect a compass needle, and Michael Faraday found the opposite effect – a moving magnet induces an electric current in a loop of wire. Faraday thought that all of this could be explained by magnetic and electric fields of force that extend out from magnets and electrical charges. During the 1860s Maxwell used this idea to devise a system of equations that completely described the two forces, unifying them into a single force field: electromagnetism.

One solution of his equations, he found, is a wave. The wave is made of undulating electromagnetic fields and travels in empty space at the colossal pace of 300 million metres per second. This was a give-away. Back in 1676, the Danish astronomer Olaus Roemer had been the first to measure the speed of light. He noticed how Jupiter's moon Io seemed to be slightly ahead of its calculated orbit when the Earth was close to Jupiter, and slightly lagging when we were farther away. This could be explained if the light took some time to reach us. Roemer worked out a speed somewhat over 200 million metres per second, and this was revised by later measurements to about 300 million. To Maxwell, the conclusion was obvious: light was an electromagnetic wave. His equations even showed why light slows down in transparent materials such as water and glass.

Other scientists found it all harder to accept. But then, in 1888, Heinrich Hertz discovered electromagnetic waves of much longer wavelength than light, which were predicted by Maxwell's theory and which we now call radio waves. Radio waves, microwaves, millimetre waves, infrared, visible light, ultraviolet, X-rays and gamma rays make up the full spectrum, all of them spawned by Maxwell's unified electromagnetism.

See also Unweaving the rainbow pages 36–37, Electric battery pages 116–117, Wave nature of light pages 118–119, X-rays pages 220–221, Solar wind pages 388–389, Afterglow of creation pages 412–413, Unified forces pages 416–417, Gamma-ray bursts pages 434–435

Right Maxwell's depiction of a uniform magnetic field disturbed by an electric current. He gave mathematical form to Faraday's intuitive discovery that electricity and magnetism create fields of force.

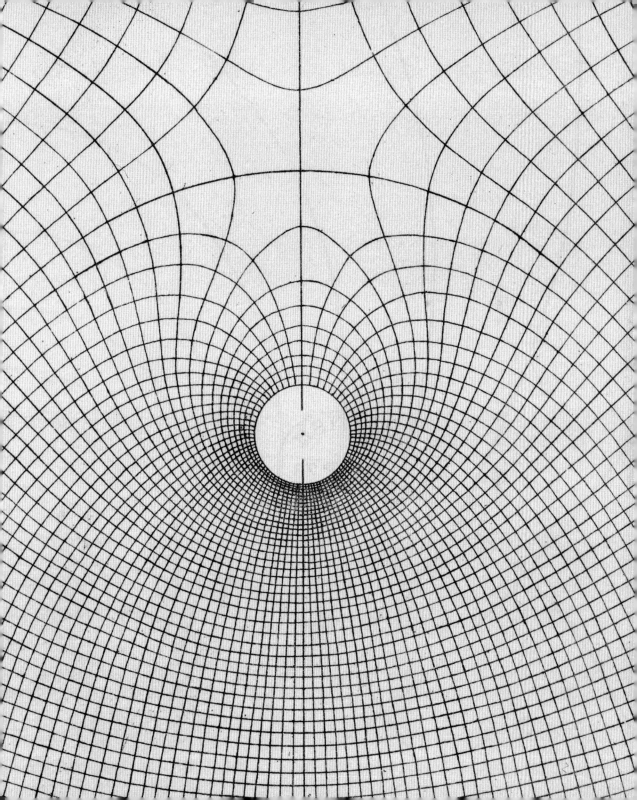

Regulating the body

Claude **Bernard** 1813–78

The French physiologist Claude Bernard abandoned hospital medicine for the laboratory. He dismissed research at the bedside as a passive exercise where observation not experiment predominated. What's more, it was not conducive to precise scientific understanding as it invariably dealt with disease at its end point. In his view the best place for the study of disease processes was the controlled environment of the laboratory.

He outlined his manifesto for scientific medicine in *An Introduction to the Study of Experimental Medicine* (1865). No mere polemic, it was grounded in the brilliant and wide-ranging physiological experiments that he carried out in the 1840s and 1850s. A bold vivisector, Bernard worked out the digestive function of secretions of the pancreas; the liver's role in synthesizing glucose with the aid of a substance he called 'glycogen' (it was previously thought that animals could only break down fats, sugars and proteins, not make them); the control of blood vessels (and so blood flow) by nerves; the carriage of oxygen by red blood cells and the abolition of this role by carbon monoxide; and the nature of curare poisoning and its link with the nervous control of muscles.

All this led him to propose that the body creates its own unchanging internal environment (*milieu intérieur*) for its community of living cells. Complex physiological mechanisms are directed to keeping the blood and tissue fluids stable in the face of external change – to keeping constant the amount of water, temperature, oxygen supply, pressure and chemical composition. In 1902 the Harvard physiologist Walter Cannon coined the term 'homeostasis' for these balancing mechanisms. His research on shock among soldiers in the First World War showed the importance of helping the body to maintain its own equilibrium, for instance by replacing lost fluids to stabilize blood pressure. These principles remain the crux of surgery and emergency care.

See also Circulation of the blood pages 58–59, Communities of cells pages 174–175, Enzyme action pages 218–219, Conditioned reflexes pages 242–243, Citric-acid cycle pages 320–321, Nitric oxide pages 496–497

Right Bernard vivisecting a rabbit. In 1870 he separated from his wife – a supporter of animal welfare.

In 1890 Friedrich Kekulé told the story of how, 25 years previously, he had discovered the structure of the benzene molecule. At the time he had been Professor of Chemistry at Ghent University and, one evening in 1865, he fell asleep in his chair by the fire and had a dream in which a snake formed a circle by biting its own tail. On waking, Kekulé realized that he had found the answer to the formula of benzene, which is C_6H_6. At the time chemists could not understand how this fitted with the fact that an atom of carbon usually bonds four hydrogen atoms. But if the molecule consisted of a ring of six carbon atoms, each with a hydrogen attached, and with alternate double bonds around the ring, then the problem was solved.

Kekulé's formula for benzene also solved another puzzle: why were all the carbon atoms in the molecule the same as one another? Replacing one of benzene's hydrogen atoms with another kind of atom always gave exactly the same product. Clearly, if the molecule was a perfectly symmetrical ring, then all carbons were equal, all products the same.

In the years that followed his discovery Kekulé deduced the structure of other compounds in this area of organic chemistry, opening up the study of 'aromatic' compounds (so called because of their characteristic odours). For this reason Kekulé's name will be forever linked to the structure of benzene and similar ring-structure compounds. But did he really come up with the answer in a dream? Or had he read a remarkable booklet by an Austrian chemist, Johann Josef Loschmidt, published a few years earlier? In it, Loschmidt proposed exactly the same kind of molecular arrangement for benzene. Kekulé was seemingly aware of Loschmidt's ideas on chemical bonding, although he never admitted these as the source of his inspiration.

See also Atomic theory pages 124–125, Mauve dye pages 172–173, Model of the atom pages 272–273

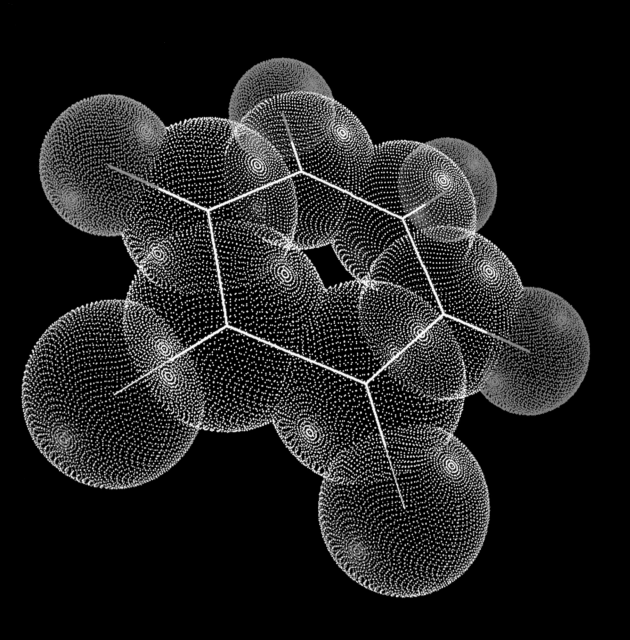

Mendel's laws of inheritance

Gregor Mendel 1822–84

Offspring clearly resemble their parents, and this fact alone means that there must be some biological mechanism of inheritance. The starting point for our modern understanding of that mechanism is provided by some experimental crosses between strains of the garden pea. The experiments were conducted by an Austrian monk, Gregor Mendel, working in his monastery at Brünn (now Brno, in the Czech Republic).

Mendel began with two pea strains, differing in an observable characteristic such as flower colour (one strain had purple, the other white, flowers). He crossed them, and the offspring were all purple. He then crossed the offspring among themselves, and found that the second generation had purple and white flowers in a 3:1 ratio. He explained the result by suggesting that coloration is controlled by two kinds of 'factor'; a pea plant inherited one factor from each parent. The purple factor was 'dominant' to the white factor, and all the first-generation plants were purple. But a quarter of the second-generation plants inherited two white factors and were white. Mendel's findings were published in an obscure journal in 1865. We now know that much of inheritance is controlled by pairs of genes inherited like Mendelian factors.

Why did Mendel succeed in making sense of inheritance where others had failed? One reason is that he worked quantitatively, using probability theory (for instance, to explain the 3:1 ratio). He also concentrated on discrete characteristics such as purple and white flowers. Others had studied attributes such as height, which does not have discrete states. Offspring tend to have heights intermediate between their parents, and it is much more difficult to discern the laws of inheritance with these continuously varying traits.

Mendel's theory was little appreciated for 35 years, because it was thought to be a rule about a few properties of peas rather than a general theory of inheritance. In 1900 three biologists – Hugo de Vries, Karl Correns and Erich Tschermak von Seysenegg – independently worked out Mendel's laws, although each later unselfishly credited Mendel as the discoverer.

Above Biologists since Mendel have delighted in picking over his pea-breeding results, which often seem to fit his theories too well.

See also Acquired characteristics pages 128–129, Measuring variation pages 212–213, Inborn metabolic errors pages 258–259, Genes in inheritance pages 264–265, Neo-Darwinism pages 282–283, Sickle-cell anaemia pages 358–359, The double helix pages 374–375

Right After his election as abbot in 1868, Mendel became an administrator first, and an experimental gardener only second. Only long after his death was he recognized as a true scientist.

Dynamite

Alfred Bernhard Nobel 1833–96

Alfred Nobel came from a family of Swedish scientists and inventors, and his father ran an explosives factory in St Petersburg, Russia. In 1863 the two men began to investigate nitrogylcerine, an explosive oil first made in 1847 by the Italian chemist Ascanio Sobrero. He had added glycerol (glycerine) to a mixture of concentrated nitric and sulphuric acids. Under such conditions the glycerol molecule becomes nitrated – attached to each of its three carbon atoms is a nitro group (NO_2). These have the potential to oxidize the molecule, releasing enough energy and gas to cause the whole sample suddenly to decompose explosively.

Although nitroglycerine was a powerful explosive, it was liable to detonate without warning: a factory that Nobel set up in Sweden in 1864 to manufacture it blew up killing five people, including his younger brother Emil. The Swedish government refused to allow him to rebuild the factory. He then began to experiment with ways to make nitroglycerine safe to handle, carrying out his research on a barge. He discovered that if the oil was absorbed by a porous and finely ground rock known as 'kieselguhr', then it became safe to handle and could be exploded only by a detonator. (Nobel had patented just such a device, based on mercury fulminate, in 1863.) His new explosive, which could be moulded into sticks and wrapped in greaseproof paper, was patented in 1867, and he called it 'dynamite'. Nobel also invented gelignite, which was also made from nitrogylcerine but included nitrocellulose and sodium nitrate, which had the advantage of being more powerful and yet safer to store.

Both dynamite and gelignite were ideal for blasting quarries, railway cuttings and tunnels, and as a result Nobel became extremely rich. He left his wealth to found the eponymous prizes.

See also Mauve dye pages 172–173, Synthesis of ammonia pages 256–257, Nylon pages 316–317

Right Although Nobel came to be viewed as a mad scientist viciously manufacturing destruction, he actually thought that his explosives would outlaw war by making it too horrible.

Periodic table

Dmitri Ivanovich Mendeleyev 1834–1907, Julius Lothar Meyer 1830–95

In 1858 the Italian chemist Stanislao Cannizzaro published the first reliable list of atomic weights. Soon other chemists were using this information to arrange the elements in order of increasing atomic weight and noted a periodic repetition of similar properties at regular intervals. (The 27-year-old English chemist John Newlands devised a crude periodic table in 1865, but his contemporaries did not take it seriously.)

The credit for drawing up the first widely accepted periodic table goes to the Russian chemist Dmitri Mendeleyev, Professor of Chemistry at St Petersburg University, who understood its real significance. In 1869 he was writing a chemistry textbook and wondered how best to discuss the 65 elements then known. He had written their names, atomic weights and some of their properties on 65 cards and, one cold winter's day, he stayed indoors and began to arrange them in rows and columns, rather like a game of patience.

Suddenly he saw that there was an underlying pattern to his arrangement and, more importantly, he realized that there were gaps in the layout that would one day be occupied by elements yet unknown. Mendeleyev was so certain that he was right that he predicted the properties of several of the missing elements. (When these were discovered in the years that followed, they were just as he had postulated.) Mendeleyev quickly published his periodic table with his predictions. About the same time, the German chemist Julius Lothar Meyer made a discovery that also revealed the pattern of the elements. He had plotted a graph of atomic weight against atomic volume, which showed the same periodic relationship between the elements, but a reviewer delayed his paper on the subject, and Mendeleyev's work was published first.

Although there are now 115 known elements, and a modern periodic table has many more columns and rows than Mendeleyev's table, the pattern that he devised is still a recognizable part of it.

Above Mendeleyev was described by the Scottish chemist Sir William Ramsay as 'a peculiar foreigner, every hair of whose head acted in independence of every other'.

See also Boyle's *Sceptical Chymist* pages 70–71, Atomic theory pages 124–125, Spectral lines pages 130–131, Radioactivity pages 224–225, Model of the atom pages 272–273, The neutron pages 312–313, Power from the nucleus pages 330–331

Right Mendeleyev's first published periodic table. In his own words, 'I saw in a dream a table where all the elements fell into place as required. Awakening, I immediately wrote it down'.

но въ ней, мнѣ кажется, уже ясно выражается примѣнимость вы
ставляемаго мною начала ко всей совокупности элементовъ, пай
которыхъ извѣстенъ съ достовѣрностію. На этотъ разъ я и желалъ
преимущественно найдти общую систему элементовъ. Вотъ этотъ
опытъ:

			Ti=50	Zr=90	?=180.
			V=51	Nb=94	Ta=182.
			Cr=52	Mo=96	W=186.
			Mn=55	Rh=104,4	Pt=197,4
			Fe=56	Ru=104,4	Ir=198.
		Ni=Co=59		Pl=106,6	Os=199.
H=1			Cu=63,4	Ag=108	Hg=200.
	Be=9,4	Mg=24	Zn=65,2	Cd=112	
	B=11	Al=27,4	?=68	Ur=116	Au=197?
	C=12	Si=28	?=70	Sn=118	
	N=14	P=31	As=75	Sb=122	Bi=210
	O=16	S=32	Se=79,4	Te=128?	
	F=19	Cl=35,5	Br=80	I=127	
Li=7 Na=23		K=39	Rb=85,4	Cs=133	Tl=204
		Ca=40	Sr=87,6	Ba=137	Pb=207.
		?=45	Ce=92		
		?Er=56	La=94		
		?Yt=60	Di=95		
		?In=75,6	Th=118?		

а потому приходится въ разныхъ рядахъ имѣть различное измѣненіе разностей,
чего нѣтъ въ главныхъ числахъ предлагаемой таблицы. Или же придется предпо-
лагать при составленіи системы очень много недостающихъ членовъ. То и
другое мало выгодно. Мнѣ кажется притомъ, наиболѣе естественнымъ составить
кубическую систему (предлагаемая есть плоскостная), но и попытки для ея образо-
ванія не повели къ надлежащимъ результатамъ. Слѣдующія двѣ попытки могутъ по-
казать то разнообразіе сопоставленій, какое возможно при допущеніи основнаго
начала, высказаннаго въ этой статьѣ.

Li	Na	K	Cu	Rb	Ag	Cs	—	Tl
7	23	39	63,4	85,4	108	133		204
Be	Mg	Ca	Zn	Sr	Cd	Ba	—	Pb
B	Al	—	—	—	Ur	—	—	Bi?
C	Si	Ti	—	Zr	Sn	—	—	—
N	P	V	As	Nb	Sb	—	Ta	—
O	S	—	Se	—	Te	—	W	—
F	Cl	—	Br	—	J	—	—	—
19	35,5	58	80	190	127	160	190	220.

Changes of state

Johannes Diederik **van der Waals** 1837–1923

The description of the behaviour of many particles – atoms or molecules in gases, liquids and solids; electrons in a metal – is a matter of statistics. It stems from nineteenth-century work that tried to connect a 'microscopic' view, in which gas particles move according to Newton's laws, to a 'macroscopic' view, in which the relationship between the pressure, temperature and volume of a gas is described by empirical 'gas laws'.

Temperature is a measure of the kinetic energy – the energy of motion – of the gas particles. Pressure results from collisions of the particles with the walls of a confining vessel. In the 1860s James Clerk Maxwell deduced the probability that any randomly chosen particle in a gas at a certain temperature has a particular velocity. This establishes the underlying character of a gas, from which the other properties emerge. In 1872 Ludwig Boltzmann showed how this 'probability distribution' *must* result from particles moving at random.

The kinetic theory assumes that the gas particles are infinitely small and that they remain so far apart as to never feel one another's presence. This works well for rarefied gases, but less so for denser gases. In his doctoral thesis, completed in 1873, the Dutch scientist Johannes Diederik van der Waals set himself the task of modifying the theory to account for the deviations. Accepting that gas particles have a small but finite size and that they exert short-ranged attractive forces on one another, he worked out a simple 'equation of state' that related pressure, temperature and volume. But instead of predicting that pressure varies smoothly with volume (or density), the equation implied that below a certain 'critical temperature' the vast crowd of particles adopts either of two stable states – one denser than the other. The denser state corresponds to the liquid, and compression or expansion can induce a 'phase transition' from one state to the other: condensation and evaporation.

See also Hydrogen and water pages 98–99, Brownian motion pages 254–255, Edge of chaos pages 490–491, A new state of matter pages 510–511

Above Heike Kamerlingh-Onnes (right) chose low-temperature work because of the research of his countryman Van der Waals (left). This led to the production of liquid helium and the discovery of superconductivity.

Right In 1910 van der Waals was awarded the Nobel Prize for Physics. The reverse side of this commemorative medal illustrates his 'equation of state'.

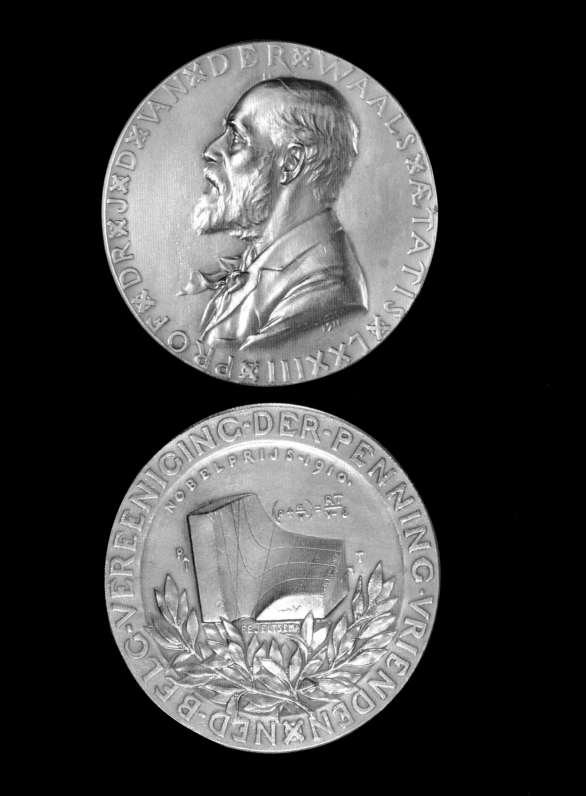

'Canals' on Mars

Giovanni Virginio Schiaparelli 1835–1910

Christiaan Huygens, in 1659, was the first to recognize any features on Mars's disc. He saw a large triangular patch (which turned out to be Syrtis Major), and measured the planet's spin period to be the same as Earth's. Six years later Giovanni Domenico Cassini discovered polar caps. Soon the dark patches were being regarded as old sea-beds filled with vegetation, and the lighter orange regions as continents.

Modern telescopic investigation started with Giovanni Virginio Schiaparelli, Director of the Brera Observatory, Milan. Between 1877 and 1890 he charted the Martian surface and recorded it as being criss-crossed with channels. Unfortunately, the Italian word *canali* also means 'canals', and this translation affected Martian studies for the next 40 years. Some 'canals' appeared double and variable. Soon people were talking about artificial waterways, intelligent populations and global irrigation.

The main advocate of Martian life was the aristocratic Bostonian Percival Lowell. From 1895, at his private observatory in Flagstaff, Arizona, he produced map after map of Mars, each showing elaborate and changing networks of canals. These, together with the seasonal variability of Martian colour, convinced him that Mars harboured life and was an agriculturally active planet. Lowell thought that the polar caps were made of water; we now know they consist mainly of frozen carbon dioxide. Lowell turned his back on dry academic research journals and published his findings in a series of bestselling books. Much of the USA's fascination for space flight and planetary exploration stems from his endeavours.

When, however, the US spacecraft Mariner 4 flew by Mars in 1965, the canals were nowhere to be seen. The human eye, looking at details on the edge of perception, had clearly joined up a series of unconnected dark patches to form linear features.

See also Heavens through a telescope pages 54–55, Alien intelligence pages 398–399, Planetary worlds pages 512–513, Galileo mission pages 514–515, Martian microfossils pages 518–519, Water on the Moon pages 522–523

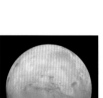

Above 'Fish-eye' view of Mars, centred on the Valles Marineris canyon system, which stretches nearly 4,800 kilometres.

Right Schiaparelli's chart of Mars, showing the markings he called the 'double channels'.

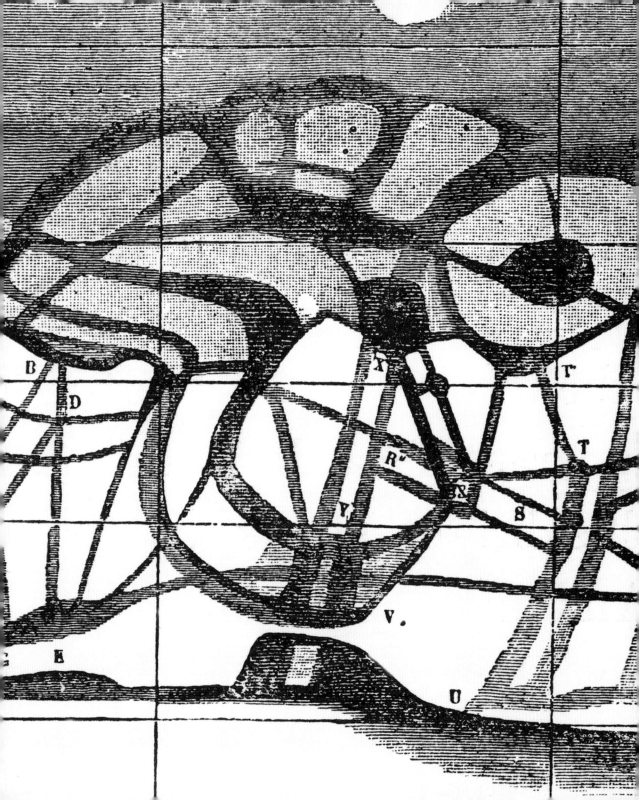

Germ theory

Louis Pasteur 1822–95

For centuries it was believed that infectious diseases were due to miasmas (poisons in the air). Although other people had suspected the role of microbes as pathogens, it wasn't until 1878 that it was fully spelt out by the French chemist Louis Pasteur. In a series of spectacular experiments he had shown that fermentation, putrefaction and infection were all due to contamination by living microbes – that microbes were the cause and not the effect of these processes. His research led to immediate practical triumphs: he saved the French silk industry by identifying the tiny parasite infecting silkworms; boosted the French wine industry by introducing heat sterilization or 'pasteurization' to prevent souring; and demonstrated the effectiveness of vaccinations against anthrax in animals and rabies in humans.

While Pasteur showed that microbes in general could cause disease, it was up to the German country doctor Robert Koch to demonstrate which particular microbe caused which specific disease. Developing laboratory techniques such as plate culture and photo-microscopy to distinguish different types of microbe, in the early 1880s Koch isolated and identified the bacilli causing tuberculosis and cholera. He also set out rules for ascribing a disease to a specific microbe: the microbe must always be associated with the disease; it must be isolated and maintained in pure culture; and microbes from the culture must reproduce the disease in experimental animals and be recovered from the infected tissues. These conditions had been previously described by Jacob Henle, but it was Koch who showed how they could be fulfilled. They are now known as 'Koch's postulates'.

Antiseptic surgery widened the application of Pasteur's germ theory. Learning of his studies of putrefaction and infection, the English surgeon Joseph Lister deduced that wound sepsis was due to bacterial contamination. In 1867 he began soaking his instruments and dressings in carbolic acid, a well-known disinfectant. Three years later he introduced a carbolic spray. Aseptic surgery, with germs excluded altogether from the site of an operation, followed swiftly.

Above Pasteur, in a tribute from the French magazine *Le Petit Journal* in 1895. The angel's ribbon carries the dedication 'from a grateful universe'.

See also Microscopic life pages 76–77, Spontaneous generation pages 90–91, Vaccination pages 102–103, Cholera and the pump pages 168–169, Antitoxins pages 214–215, Cellular immunity pages 204–205, Penicillin pages 302–303, Prion proteins pages 466–467, AIDS virus pages 472–473

Right Politicians as microbes: bizarre life-forms in a drop of water, distinct from the world of humans. From an 1883 *Punch* cartoon.

ESSENCE OF PARLIAMENT.

EXTRACTED FROM

THE DIARY OF TOBY, M.P.

HOUSE of Commons, Thursday (anticipatory).—Members all back as delighted as if they were going away. Everybody shaking hands with everybody else. PETER RYLANDS doing the honours of the place, as it were; quite in boisterous spirits.

"Another good Under-Secretaryship gone wrong," DRUMMOND-WOLFF slily whispers in his ear. "You'd better come over and join us."

"Thanks; but I'll wait a bit longer," PETER says. "CHILDERS was all very well at the War Office; it's different at the Treasury. I give him six months there, then there may be a call for a man who has finance at his finger's ends, is trusted by the country, and is a pretty fair speaker."

BRADLAUGH in high spirits. Tells me he's been round spending half an hour with GOSSET practising the steps. Sergeant-at-Arms, it seems, who has not forgotten his old skill, wants to reverse when they waltz backward from the Mace. After the practice of three Sessions, BRADLAUGH can do the forward step well enough, but finds it hard to reverse. Still means to try.

"The eyes of the country are upon us," he says, "and we must do the thing well."

Black Rod arrived shortly after two o'clock. Door shut in his face as he

Cellular immunity

Elie (Ilya Ilyich) Metchnikoff 1845–1916

In 1882, during a sojourn to Messina, Sicily, the Russian-born zoologist Elie Metchnik
witnessed the 'ingestion' of small foreign bodies by certain mobile cells in the transpar
starfish larvae under his microscope. He called these cells 'phagocytes' (Greek for 'eati
cells'). The phenomenon reminded him of something he had seen before. Almost 20 y
earlier he had observed a similar process in the cells of roundworms and compared it
the digestive mechanism of single-celled creatures such as protozoa. A committed Dar
he wanted to show the connection of simple and more complex creatures through sha
features in their embryonic development and basic life processes.

Metchnikoff reasoned that although in simple organisms phagocytes clearly have a
role in digestion, in more complex organisms they might be involved in fighting foreign
invaders such as bacteria. Marshalling evidence for this 'cellular theory of immunity',
likened the activity of the starfish cells to the white cells in animal blood (including hu
blood): microscopic studies revealed white blood cells flocking to the site of any inflam
associated with a wound or infectious disease, and then attacking and ingesting harmf
bacteria. Metchnikoff's claim that phagocytes were the basis of immunity met with su
criticism. Most bacteriologists thought that white blood cells ingested infectious agents
spread them further into the body. They preferred the alternative theory that immunit
due mainly to the cell-free part of the blood. Although poorly understood, this serum o
humoral theory was soon to be bolstered by the discovery of chemical 'antitoxins'.

Metchnikoff fought hard to satisfy his critics – even playwrights. In George Bernard
Shaw's *The Doctor's Dilemma* (1906), the pompous surgeon, Sir Colenso Ridgeon,
repeatedly urges his patients to 'Stimulate the phagoctyes!' so he might pocket fees for
unnecessary operations. But the immunological role of white blood cells was eventuall
accepted, and Metchnikoff shared the 1908 Nobel Prize for Physiology or Medicine.

tchnikoff
ngrossed in
of ageing and
ocating
e quantities
to promote
h.

See also Germ theory pages 202–203, Antitoxins pages 214–215, Transplant rejection pages 356–357, Biological self-recogr
pages 430–431, Monoclonal antibodies pages 448–449, AIDS virus pages 472–473

Right In this scanning electron micrograph of two macrophages in a human lung, the lower one is seen elongating itself to eng
round particle.

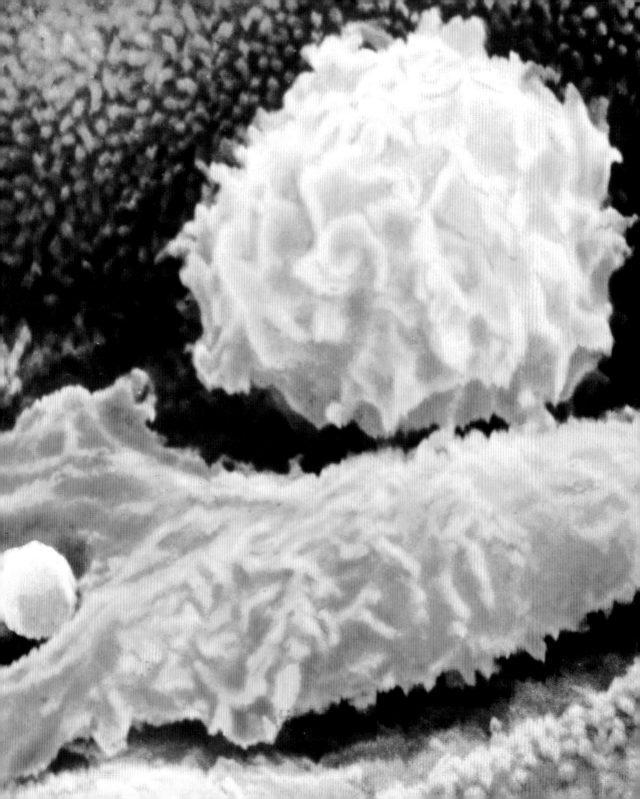

Mountain formation

Eduard Suess 1831–1914

The publication of the first volume of the Viennese geologist Eduard Suess's five-volume treatise *Das Antlitz der Erde* ('The Face of the Earth', 1885–1909) marked a new way of looking at the formation of mountains. Suess saw common features across the great Alpine ranges of Europe, and was the first to draw them together and consider a mountain belt as a geological entity. He had a global outlook and recognized similarities between the Alps and the Himalayas, for example. This led him to propose that there had once been a supercontinent in the southern hemisphere, now broken up into today's continents.

This advance was part of a general surge of interest in mountains. In the Swiss Alps, Arnold Escher documented the enormous scale of folding and the existence of great flat faults in which layers of rock are thrust over those beneath; in North America, the Rodgers brothers described tightly folded and faulted rocks in the Appalachian Mountains; in the Scottish Highlands, meticulous mapping revealed similar folds and thrusts carrying rocks many kilometres across the landscape. By 1896 one such thrust in Scandinavia had been documented that had moved a sheet of rock 130 kilometres.

All of this implied great horizontal movements of the Earth – so great that the mountains were pushed up. This was a revolutionary idea and often dismissed as ridiculous, for want of a mechanism. That did not come until almost a century later, with the advent of the theory of plate tectonics in which collisions between continents make mountain chains. Suess's triumph was to match up mountain belts across the world on the basis of the age at which they were formed. So, for example, the Appalachians and the hills of Scotland and Scandinavia were united as the Caledonian chain, as if the Atlantic were not there.

See also Earth cycles pages 100–101, Lyell's *Principles of Geology* pages 146–147, Ice ages pages 152–153, Rocks of ages pages 252–253, Continental drift pages 270–271, Plate tectonics pages 414–415

Right This sweeping view of the Himalayas takes in the plateaus of Nepal and China. The mountains were created when the area now known as India collided into the landmass of Asia over 40 million years ago.

Nitrogen fixation

Hermann Hellriegel 1831–95, Hermann Willfarth DATES UNKNOWN

Since the days of the ancient Greeks, farmers have known that crops planted in soil in which beans have recently been grown produce higher yields, and it was Theophrastus who first drew attention to this beneficial effect. A full explanation of the phenomenon – the roots of beans have nodules housing bacteria that convert atmospheric nitrogen into a form that plants can use for growth – came more than 2,000 years later, in 1886, when the German agricultural chemists Hermann Hellriegel and Hermann Willfarth identified *Rhizobium* as the bacterium that 'fixes' the nitrogen.

Plants in the legume family, which include beans, clover and alfalfa, have forged a mutually beneficial, symbiotic relationship with *Rhizobium*, which supplies them with an energy source and carbon-based molecules in exchange for nitrogenous compounds. Soluble nitrogen is also released into the soil, especially from decaying roots after the bean crop is harvested, so any following crops benefit from the residual nitrogen – a virtue exploited in organic farming systems. An alfalfa crop can contribute 350 kilograms of nitrogen to the soil per 10,000 square metres; worldwide the symbiosis of legumes and *Rhizobium* produces up to 200 million tonnes of fixed nitrogen from the atmospheric gas every year. The process is fundamental to almost all terrestrial food chains.

There is a specific relationship between various strains of *Rhizobium* and their plant symbionts, and not all combinations are equally efficient, so inoculating seeds with superior strains of *Rhizobium* can maximize crop yields. In the longer term, use of genetic engineering to transfer genes for the symbiotic association to cereals could reduce dependence on energy-expensive and pollutant inorganic nitrogen fertilizers, but this process involves many genes and will not be easy. The ecological consequences of extending the range of nitrogen-fixing plant families must also be considered.

See also Birth of botany pages 18–19, Synthesis of ammonia pages 256–257, Crop diversity pages 292–293, Citric-acid cycle pages 320–321, Photosynthesis pages 344–345, Green revolution pages 428–429, Gaia hypothesis pages 432–433

Right A nitrogen-fixing nodule on the root of the white clover *Trifolium repens*. The nodule, seen here under the scanning electron microscope, is formed by association with the bacterium *Rhizobium trifolii*.

Nervous system

Camillo Golgi 1843–1926, Santiago Ramón y Cajal 1852–1934

Cells were the latest focus of biological study. But new techniques were needed to identify the components of the cell and to show how individual cells meshed into a working system. Improvements in microscopy were essential, but the fine structure of life was not revealed until cellular staining came into prominence.

The Italian histologist Camillo Golgi led the way with his introduction of the use of silver salts. Exploration of the brain and spinal cord had previously been slow because its dense and complex structure had deterred investigators. In 1873 Golgi applied his methods to brain tissue and found that he could pick out individual nerve cells and expose their features in unprecedented detail, including the cell body and a number of delicate extensions (short dendrites as well as the longer axon). He believed that the fine terminal branches of axons fused with the branches of other axons to form a continuous communication network – the 'reticular theory'. He also established that the brain received information from the body's sensory nerves and transmitted it through the motor nerves.

The Spanish physician Santiago Ramón y Cajal improved Golgi's silver nitrate stain and used it to study the interconnections of nerve cells. By 1889 he had worked out the tightly controlled pattern of connections of cells in the grey matter of the brain and spinal cord; proposed that the dendrites of a nerve cell receive information, which is then sent through the axon; and demolished the reticular theory by showing that the terminal branches of axons do not fuse with parts of other cells. His observations supported the neuron theory, which held that nerve cells were self-contained units that formed communicating links with one another. Charles Sherrington named the connection (later found to be a gap) the synapse.

In 1906 Golgi and Cajal shared the Nobel Prize in Physiology or Medicine. Embarrassingly, Golgi used his lecture to defend the reticular theory and attack Cajal.

See also Microscopic life pages 76–77, Communities of cells pages 174–175, Conditioned reflexes pages 242–243, Neurotransmitters pages 274–275, Artificial neural networks pages 334–335, Growth of nerve cells pages 368–369, Chemical basis of vision pages 382–383, Memory molecules pages 470–471

Right Ramón y Cajal's microscope drawing of the nerve cells in a pigeon's cerebellum. Cajal was considered lazy as a child and his physician father apprenticed him as both a barber and a cobbler.

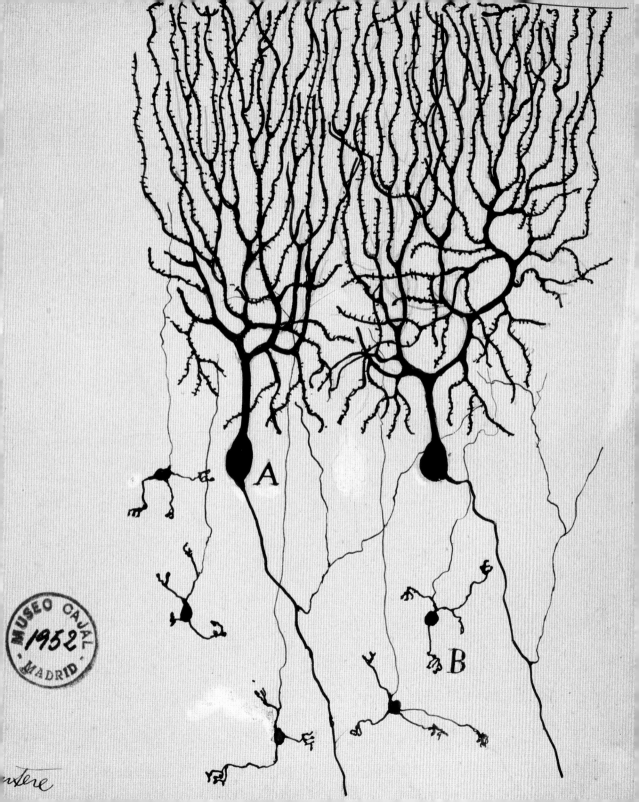

Measuring variation

Francis Galton 1822–1911

Biologists divide the differences that can be seen between members of a single species into two categories: discrete and continuous. Some attributes, such as eye colour, come in discrete types: some individuals have blue eyes, others brown. Other attributes, such as height, show more continuous variation. Probably most attributes of living species show continuous, rather than discrete, variation. We need to understand the amount of variation and how it is inherited – we need this to explain evolution, and to improve agricultural crops and livestock. It also matters in the politics of human inequality. Modern scientific research on continuous variation (called 'biometry') can be traced back in large part to the work of a Victorian polymath, and cousin of Charles Darwin, Francis Galton.

Galton found that populations often showed a 'bell curve' distribution. For example, there are many individuals of roughly average height, and decreasing numbers of individuals as we look away from the average to the tall and short extremes. Galton reported his findings in his 1889 book *Natural Inheritance* and named this distribution the 'normal distribution'. (The German mathematician Carl Friedrich Gauss had previously described the distribution mathematically.) The normal distribution arises when something is influenced by a large number of independent factors, each of small effect. Height, for instance, is influenced by many genetic factors, and by nutritional, health and disease factors. Any such factor may be a 'plus' or a 'minus'; it may add to or subtract from the individual's height. With many plus and minus factors at work in a population, few individuals end up much taller or shorter than average. Most individuals experience some mix of plus and minus factors and end up near the average height of their particular generic group.

Galton's explanation of how continuous variation is inherited turned out to be wrong, and our modern understanding of it came later, when biologists applied Mendel's theory to Galton's biometrical observations.

See also Mendel's laws of inheritance pages 192–193, Genes in inheritance pages 264–265, Neo-Darwinism pages 282–283

Right Galton's credo was: 'Whenever you can, count'. This early-20th-century photograph of 'Professor James Ricallini with Kashmir Men' shows the extremes of human variation in height.

Antitoxins

Emil von Behring 1854–1917, Shibasaburo Kitasato 1852–1931

In *Faust*, Goethe proclaims that 'blood is a very special liquid'. In 1890 Emil von Behring and Shibasaburo Kitasato's discovery of a mysterious protective property in the serum of animals immune to diphtheria and tetanus provided hard evidence that this was true.

Behring, an assistant to Robert Koch in Berlin, had been searching for an internal bodily disinfectant to cure infectious diseases. He was aware that Émile Roux and Alexandre Yersin, working in Louis Pasteur's laboratory in Paris, had isolated the poisons or 'toxins' released by diphtheria bacilli and tetanus bacilli and responsible for many of the dangerous symptoms of these diseases. Two years later, Behring, in collaboration with Kitasato, demonstrated that infected animals produce substances ('antitoxins') that can neutralize these toxins and confer immunity to diphtheria and tetanus. They also found that this immunity could be transferred via serum to other animals, so they too could withstand infection.

This opened up the potential for serum therapy. Behring's initial trials were disappointing, as his serum supplies were insufficiently potent. But Roux used horses for large-scale production of anti-diphtheria serum and successfully treated children with it in 1894. Serum therapy became popular for many other diseases such as pneumonia, bubonic plague and cholera. Although it never provided a wonder cure, it did provide critics of the 'cellular theory of immunity' with proof that germ warfare was waged less by the blood cells than by serum antitoxins – or 'antibodies' as they came to be known. This alternative 'humoral theory' was further strengthened in 1897 when Paul Ehrlich provided a chemical explanation of the interaction of cells, antibodies and antigens in terms of 'side-chain' molecules. But it wasn't until the 1930s that antibodies were identified as proteins; and in the 1950s and 1960s Gerald Edelman and Rodney Porter showed them to be large Y-shaped molecules, made of heavy and light amino-acid chains.

See also Vaccination pages 102–103, Cholera and the pump pages 168–169, Germ theory pages 202–203, Cellular immunity pages 204–205, Blood groups pages 236–237, Transplant rejection pages 356–357, Biological self-recognition pages 430–431, Monoclonal antibodies pages 448–449

Right Inoculation of a horse during the preparation of anti-diphtheria serum at the Pasteur Institute in the 1890s.

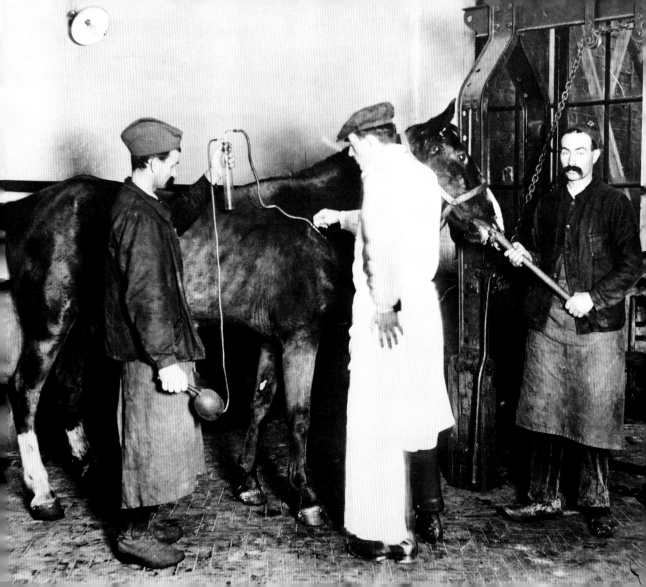

Java man

Eugène Dubois 1858–1940

In September 1891 Eugène Dubois, a Dutch army doctor, found a fossil tooth from a human-like ape buried in the banks of the Solo River near the village of Trinil on the Indonesian island of Java. Further excavations uncovered a skull cap and thigh bone. Dubois thought he had proved Ernst Haeckel's theory of the southeast Asian origin of humankind. Haeckel, a German biologist and ardent Darwinist, claimed that, of living primates, the Indonesian orang-utan is genetically closest to *Homo sapiens* and that therefore our human ancestors should be found in that region.

While lecturing at the University of Amsterdam, Dubois became inspired by Alfred Russel Wallace's writings on the islands of the Dutch East Indies. He enlisted in the army and got himself posted to Sumatra to search for Haeckel's 'missing evolutionary link' in the ape–human lineage. Plenty of fossils were found but no human-like remains. After an attack of malaria in 1890, he was transferred to neighbouring Java and placed on inactive duty, whereupon he wangled support from the colonial authorities for some large-scale excavations.

From tonnes of sediment, convict labourers recovered more than 12,000 animal fossils – including extinct elephants, hyenas and big cats – before finding what Dubois proclaimed to be the 'missing link'. He initially named it *Anthropithecus* and later *Pithecanthropus erectus* (now known as *Homo erectus*), meaning 'upright man-ape'.

On returning to Holland in 1895, Dubois tried to persuade European archaeologists of the significance of his find. Not until the late 1920s was 'Java man' eventually accepted as an extinct human species, albeit not one originating in southeast Asia. *Homo erectus* originated in Africa, as Charles Darwin had suggested, and around 2 million years ago left Africa, reaching the Black Sea by 1.7 million years ago and Java by 0.8 million years ago; here it perhaps survived until as recently as 27,000 years ago, coexisting with modern humans.

Above Haeckel's drawing of Java man's skull. Haeckel had named the 'missing link' *Pithecanthropus* long before Dubois actually found it.

See also Prehistoric humans pages 148–149, Neanderthal man pages 170–171, Taung child pages 298–299, Olduvai Gorge pages 392–393, Ancient DNA pages 476–477, Nariokotome boy pages 480–481, Out of Africa pages 492–493, Iceman pages 504–505

Right Reconstruction of Java man (*Homo erectus*) based on the bones discovered by Dubois. He carried the fossils to scientific meetings all over Europe, and once accidentally left them in a cafe.

Enzyme action

Emil Hermann Fischer 1852–1919

Enzymes are natural catalysts that increase the rate of chemical reactions. Perhaps the most popular example is found in the brewing industry in which sugars are converted into ethyl alcohol and carbon dioxide by the action of enzymes secreted by yeast cells. For many years no one knew exactly what these catalysts were made of or how they worked – as long as the beer was tasty and the alcohol content sufficiently high, no one really cared about the ins and outs of the chemistry.

Towards the end of the nineteenth century the world-famous German organic chemist Emil Fischer, who apparently had no particular interest in brewing, was pursuing the time-honoured research of all organic chemists: the elucidation of the structure of specific chemical compounds. Fischer spent ten years investigating the structures of sugars and related compounds. Hexose, for example, which is made of 6 carbon, 12 hydrogen and 6 oxygen atoms, has 16 different forms – stereoisomers – all of different shapes. During his studies Fischer made use of several enzymes to convert sugar to alcohol and discovered that each had a different specificity, recognizing one particular form of sugar but not another. Around 1894 he called this phenomenon a 'lock and key mechanism' and postulated that all enzymes were structured uniquely to recognize one particular chemical compound to the near exclusion of all others. This in turn suggested that all living cells have a multitude of enzymes, each tailor-made for a specific catalytic task. And so it has turned out.

Fischer made many contributions to chemistry beyond this fortuitous discovery, and was duly awarded a Nobel Prize in 1902. But years of working with toxic chemical substances took its toll, as did the strain of the First World War, which he had first supported but later considered as Germanic folly. In 1919, seriously depressed and in poor health, he took his own life.

See also Regulating the body pages 188–189, Benzene ring pages 190–191, Inborn metabolic errors pages 258–259, Citric-acid cycle pages 320–321

Right By the time he had received his Nobel Prize, Fischer had begun work on protein chemistry, which was to prove of equal significance to his earlier work on sugars and purines.

X-rays

Wilhelm Konrad Röntgen 1845–1923

On 8 November 1895 the German physicist Wilhelm Röntgen was experimenting with a cathode-ray tube. He had left a spare fluorescent screen on another bench, out of the way of his experiment. And yet the discarded screen lit up when he switched on the cathode tube. Röntgen realized that something was coming out of it, some invisible rays that were new to science. He found that they could penetrate all sorts of materials – wood, glass, rubber, aluminium and other metals. And when he put his hand in the beam, he saw a shadow of his bones.

The Röntgen rays were a sensation. Their property of allowing people to see through solid matter seemed like magic – black magic to some, thus the invention of lead-lined X-ray-proof underwear to repel any prurient applications.

Doctors quickly exploited their new ability to see inside people and experimented with X-rays as a cure for all sorts of illness. But then the dangers of X-rays emerged – burns and hair loss followed strong exposures, and in 1904 Clarence Dally, an assistant of the American inventor Thomas Edison, died of cancer after being severely burned. X-rays are still used to kill tumours, but the doses are now carefully controlled.

Meanwhile scientists struggled to find out what X-rays actually were. They followed straight lines, like light, but would not bounce off mirrors or bend around obstacles. Could they be waves in the ether, or bullet-like corpuscles? The question wasn't settled until 1912, when, in an experiment suggested by Max von Laue, a beam of X-rays was shown to be scattered into an intricate diffraction pattern when it passed through a crystal. This proved that they were electromagnetic waves, like light, but of very short wavelength, comparable to the distance between atoms in a crystal. X-ray diffraction became a vital probe of crystals, materials in industry, and the structure of biological molecules such as DNA.

See also Maxwell's equations pages 186–187, Radioactivity pages 224–225, The electron pages 228–229, Cosmic rays pages 268–269, The double helix pages 374–375, Structure of haemoglobin pages 390–391, Quasars pages 404–405

Right A helping hand: Röntgen's X-ray image of his wife's hand, showing the ring she was wearing.

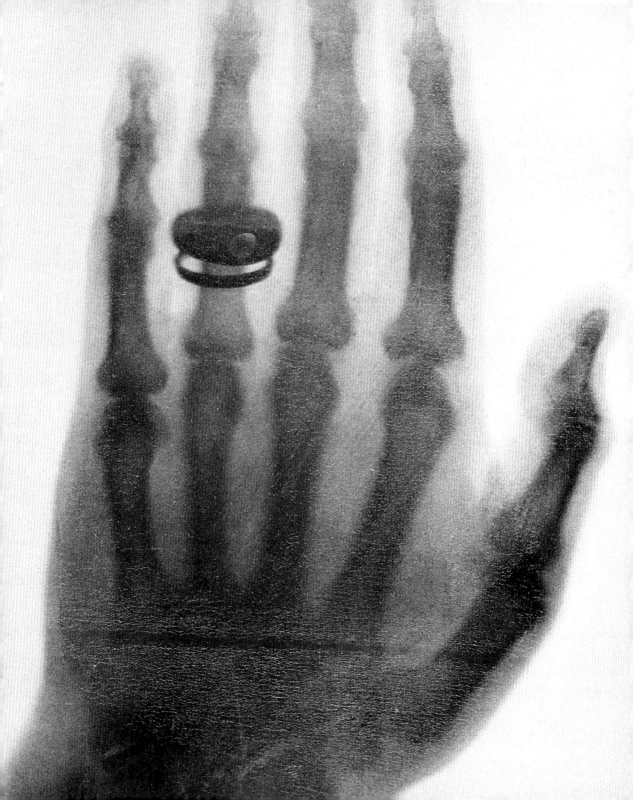

The unconscious mind

Sigmund Freud 1856–1939

'Hysteria' as a medical diagnosis has been around since the Greeks. In 1895 Sigmund Freud and Josef Breuer offered the beginnings of a new interpretation and therapy. Their *Studies in Hysteria* presented the fruits of their joint efforts at understanding and treating young middle-class Viennese women who suffered from a variety of complex physiological and psychological symptoms. They exploited the common therapeutic technique of hypnotism, but they soon split up because Freud became convinced that the cause of hysteria was always sexual trauma (real or imagined) in childhood. In the course of his original psychoanalysis – carried out by an intimate correspondence with a Berlin doctor named Wilhelm Fliess – Freud explored his own psychosexual development, wherein lay the key to understanding all neurotic illness.

That key, elaborated through a succession of publications, especially *The Interpretation of Dreams* (1900) and *Three Essays on the Theory of Sexuality* (1905), was the role of sex in both normal and abnormal development. Freud insisted that sex permeates our lives from early infancy, and that both boys and girls (he was less successful here) undergo a complicated process of differentiation of sexual identity, which leaves permanent traces in their mature identities. His concepts of the Oedipal complex, the unconscious and the tripartite nature of mental structures ('id', 'ego' and 'superego') were elaborated gradually through his contact with patients, who he believed were best helped using the technique of free association, whereby the patient simply recounted without inhibition what came into his or her consciousness. Freud later expounded the relevance of his ideas to anthropology, religion and history.

Freud always considered his methods to be valuable tools of psychological understanding, rather than simply a means of treating patients. Psychoanalysis dominated psychiatry, especially in the USA, for half a century. Its impact has now waned, but, at a popular level, we are still Freud's children.

Above Freud treated many of his patients as they lay on this couch, seen here in his last house, in London.

See also Child development pages 294–295, REM sleep pages 372–373, Images of the mind pages 478–479

Right Freud: 'My life and work has been aimed at one goal only: to infer or guess how the mental apparatus is constructed and what forces interplay and counteract in it.'

Radioactivity

Antoine Henri Becquerel 1852–1908, Marie Sklodowska Curie 1867–1934, Pierre Curie 1859–1906, Ernest Rutherford 1871–1937, Frederick Soddy 1877–1956

X-rays were only the first surprise of the 1890s. New rays were soon detected which proved to be even more profoundly alien to known science.

The French physicist Antoine Henri Becquerel thought that X-rays might be produced by fluorescence. So, one by one, he took fluorescent compounds and put them on a photographic plate wrapped in black paper. He left this outside, hoping that sunlight would make the compound fluoresce and produce X-rays, which would pass through the paper and darken the plate. In February 1896 came a seeming success: uranium potassium sulphate.

But then, one sunless day, he put a packet away in a drawer. Weeks later, when he developed the plate, it too had darkened. Becquerel had been wrong about fluorescence. Instead, uranium was spontaneously giving off penetrating rays. Other elements did this too. In 1898 Marie and Pierre Curie discovered two new radioactive elements, polonium and the ferociously active radium. Radium kicks out so much radioactive energy that if you throw a lump of it in a bucket of water, the water will boil. Where was all that energy coming from?

A worse shock was to come. Ernest Rutherford and Frederick Soddy found that radioactivity was a kind of alchemy: supposedly immutable elements were changing, transforming into other elements. Without a knowledge of relativity, quantum mechanics and nuclear physics, these were unfathomable mysteries.

Physicists identified three main forms of radioactivity. Alpha particles are bare helium nuclei; beta rays are high-energy electrons; and gamma rays are high-energy electromagnetic waves. An early fad for radioactive medicine faded when the dangers from radiation sickness and cancer became clear, but radioactivity is now used effectively in medical imaging and killing tumours, and in a thousand other ways, from dating ancient rocks and artefacts to powering spacecraft and preserving fruit.

See also Electromagnetism pages 134–135, Maxwell's equations pages 186–187, The electron pages 228–229, Rocks of ages pages 252–253, The neutron pages 312–313, Antimatter pages 314–315, Radiocarbon dating pages 346–347, Gamma-ray bursts pages 434–435, Images of the mind pages 478–479, Iceman pages 504–505

Right Pierre and Marie Curie in their laboratory. As well as confusingly suggesting that radioactivity comprised 'rays', this 1904 drawing shows that Pierre was initially considered the principal researcher.

Aspirin

Felix Hoffmann 1868–1946, Arthur Eichengrün DATES UNKNOWN

Acetylsalicylic acid is one of the wonder drugs of the twentieth century, yet its history is full of paradoxes. It was synthesized in the laboratories of the German drug company Bayer in 1897 and named and marketed two years later as 'aspirin'. We have known it thus ever since. There is, however, debate about which Bayer chemist, Felix Hoffmann or Arthur Eichengrün, was responsible for the synthesis in 1897; in any case, the compound had actually been prepared in the 1850s by a French chemist, Charles Gerhardt, and even marketed in the 1880s. The first launch failed to fly, but Bayer's research manager, Carl Duisberg, knew he was on to a good thing.

The initial animal and human testing was perfunctory, and the drug has so many side effects (gastric irritation, profound respiratory effects if taken in too large a dose) that it would never pass the requisite hurdles today. It is fortunate that it had already become a mainstay in the medicine cabinet before modern drug laws were introduced. Aspirin relieves pain, inflammation and fever, and was a godsend to sufferers from arthritis and other chronic painful diseases. Because its chemical structure was close to Joseph Lister's surgical disinfectant, carbolic acid, doctors thought it might be an 'internal disinfectant', a notion reinforced by the fact that young patients suffering from rheumatic heart disease responded positively to the drug.

Aspirin has recently fallen into disfavour with medical authorities as a casual pain-reliever, though the elucidation of its probable mode of action, as an inhibitor of natural hormones called prostaglandins, earned John Vane a Nobel Prize in Medicine or Physiology in 1982. The result of this inhibition is a blocking of platelet aggregation in blood clotting. Aspirin is now used in the treatment of heart attacks, and, as a prophylactic, in low doses, to prevent the consequences of arteriosclerosis.

See also Medicinal plants pages 28–29, Germ theory pages 202–203, A magic bullet pages 262–263, Penicillin pages 302–303

Right Since its introduction in 1899, aspirin has become the most popular drug of all time. In the USA alone, some 10,000–20,000 tonnes of aspirin are used annually.

The electron

Joseph John Thomson 1856–1940

The first piece of the modern world was found inside a cathode-ray tube in 1897. A favourite of nineteenth-century physicists, and the basis of most television screens, the cathode-ray tube is quite simple. At one end of an evacuated glass tube is a hot metal electrode. When attached to a high voltage, this electrode emits a type of radiation that is invisible until it hits a fluorescent material painted on the other end of the tube. The fluorescent material glows, often making eerie patterns in the cathode-ray tubes of the nineteenth century.

Physicists had experimented with cathode rays for decades, but no one knew what they actually were. One common view was that they were waves in the ether – a hypothetical fluid that was supposed to fill space. Joseph John Thomson supported the opposing view, that cathode rays were 'negatively charged bodies shot off from the cathode with great velocity'. Pieces of matter, in other words.

Thomson knew that their paths were bent by magnets and that they would leave a charge behind when caught in a metal container. By watching how they moved through magnetic and electric fields, he found that these particles were all identical, no matter what metal emitted them. They all had the same ratio of electric charge to mass. He wasn't the only one to make this finding, but he read far more into it. Thomson suggested that his 'corpuscles' were the universal carriers of electricity and the basic constituent of matter.

Thomson thought that atoms might be made of huge numbers of electrons, embedded in a sphere of positive charge. This model had some successes, but was being abandoned even before Ernest Rutherford discovered the atomic nucleus.

Thanks to Ernest Rutherford, we now know electrons are not the only particle, but they are still considered a fundamental part of the world, and a ubiquitous one. Electrons form all chemical bonds, holding matter together.

Right The study of the discharge of electricity through gases in cathode-ray tubes such as these led to the discovery both of X-rays and of electrons.

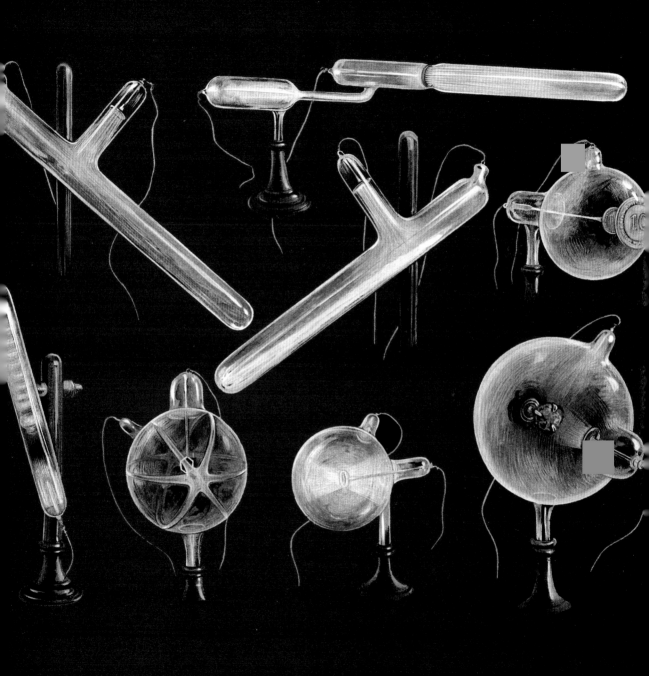

Malarial parasite

Ronald Ross 1857–1932

Above A slice through a misshapen red blood cell packed with malaria parasites.

Of all the infectious diseases, malaria (from the Italian *mal aria*, 'bad air') has affected the most people. But the true cause of this disease – which is credited with bringing down the Roman empire – was not pinpointed until the nineteenth century. The first step was Alphonse Laveran's discovery in 1880 of the malarial parasite – *Plasmodium*, a single-celled animal or 'protozoan'. In 1894 Patrick Manson suggested that infection might be transmitted by mosquitoes. And in 1897 Ronald Ross, a British medical officer working in India, finally located in the stomach wall of an *Anopheles* mosquito the eggs that are an intermediate stage of the *Plasmodium* life cycle. Devoting another year to collecting, feeding and dissecting mosquitoes, Ross traced the parasite's development through to the mature sporozoite in the mosquito's salivary glands. Here it waits to be injected into its human host when a female mosquito takes her blood meal.

Unsympathetic employers hampered Ross's work in India. Transferred to a region where human malaria was scarce, he undertook ground-breaking research on the malarial parasite of birds. In Italy, Battista Grassi stole Ross's thunder by working out the full sequence of transmission in human malaria. An ugly priority dispute followed, and, amid some controversy, Ross received the 1902 Nobel Prize in Physiology or Medicine. But his work did provide the intellectual framework for establishing tropical medicine as a distinct speciality, prompting the search for other parasite–vector pairings often responsible for diseases rife in hot climates.

Ross was one of many who developed measures to control malaria through the eradication of mosquitoes. The introduction of the insecticide DDT during the Second World War was a great boon, and in 1955 the World Health Organization judged the conquest of malaria an attainable goal. But the campaign was thwarted. Mosquitoes quickly became resistant to DDT and the insecticide proved to be harmful to the environment.

See also Microscopic life pages 76–77, DDT pages 326–327

Right The deadly disease malaria is transmitted by female mosquitoes. Ross located in the stomach wall of an *Anopheles* mosquito the oocysts that are the intermediate stage of the malaria parasite's life cycle.

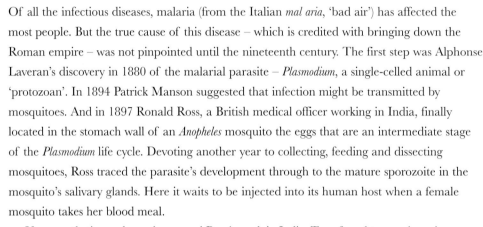

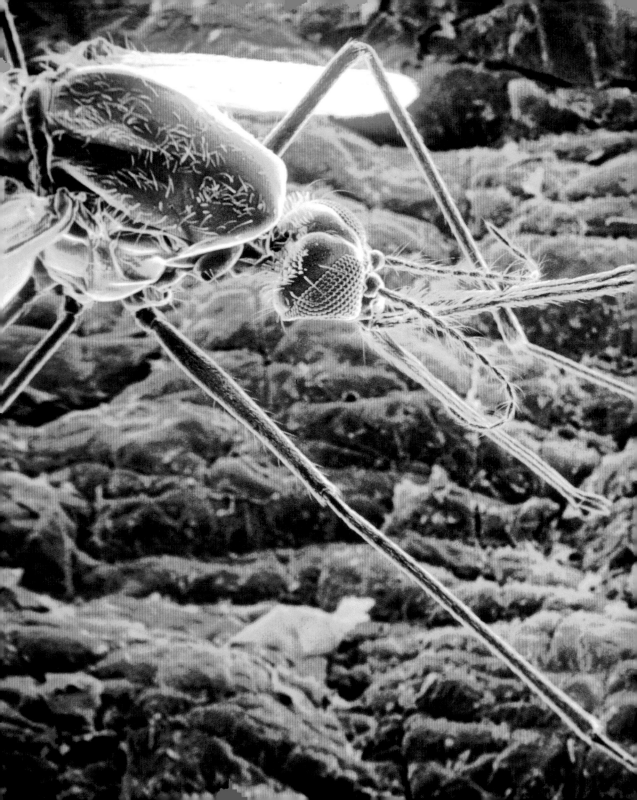

Viruses

Martinus Beijerinck 1851–1931

The founders of the germ theory began to realize that bacteria might not be the only organisms capable of causing disease. Louis Pasteur, for example, had found no causative agent for rabies and had speculated on the existence of germs too small to be seen under the microscope.

In 1895 the Dutch botanist Martinus Beijerinck turned his attention to tobacco mosaic disease, which stunted the growth of tobacco plants and mottled their leaves. He found that when he crushed up the leaves of a diseased plant and passed the sap through the finest porcelain filter, the filtrate infected healthy plants. Whatever the infective agent, it could not be grown on a culture medium or killed using chemical and heat treatments. Nor was it a toxin, as it seemed to multiply: he could infect a healthy plant, and from that infect another plant and so on. Calling the agent a 'virus' (Latin for 'poison'), he showed it could grow and reproduce only within living cells. He admitted his results were astonishing, but he stuck by them on publication in 1898. That same year Friedrich Loeffler and Paul Frosch found the virus responsible for foot-and-mouth disease in animals; in 1901 yellow fever was recognized as the first human viral disease; and in 1909 Peyton Rous identified in chickens the first tumour virus (although human cancer-causing viruses weren't isolated until the 1960s).

Specific viruses were also found to prey on bacteria. These bacteria-slayers were discovered by Frederick Twort in 1915 and Felix D'Hérelle in 1917. Despite a nasty priority dispute, the so-called 'bacteriophage' were heralded as a revolution in the treatment of infections such as typhoid and cholera. Although appearing in Sinclair Lewis's 1925 novel *Arrowsmith*, phage therapy was never proven and was quickly forgotten when penicillin became widely available. But the study of the bacteriophage did go on to provide fundamental insights into molecular biology, yielding the secrets of how genes are switched on and off and providing a vehicle for inserting foreign genes into bacteria.

Above Bacteriophage vary in size and shape. This model has an elongated icosahedral head.

See also Microscopic life pages 76–77, Cholera and the pump pages 168–169, Germ theory 202–203, Antitoxins 214–215, Genes in bacteria pages 332–333, Genetic engineering pages 436–437, Prion proteins pages 466–467, AIDS virus pages 472–473

Right Electron micrograph showing rod-shaped particles of tobacco mosaic virus. The striations along the width reflect the helical symmetry of the subunits that make up the protein coat.

The quantum

Max Planck 1858–1947, Albert Einstein 1879–1955

Quantum theory came out of a hot box. In 1900 the German physicist Max Planck was trying to explain why hot objects such as a poker glow the colours they do: from red to white hot. He wanted to work out not just the approximate colour, but the precise amount of light they emit at different wavelengths.

His idealized hot object was a black box with a small hole in it. Using ordinary classical physics, Planck could almost explain the light coming out of the box – but not quite. Experiments found that there was slightly more radiation at long wavelengths than Planck's equations predicted. To fix this, he found he had to make a peculiar assumption: that energy didn't leave the box continuously, but instead came out in chunks, or 'quanta'.

When he presented this idea on 14 December 1900, Planck wasn't sure what these energy quanta meant. But in 1905 Albert Einstein showed that light really does come in pieces – what we now know as photons.

Einstein used this idea to explain how electrons are knocked out of metal surfaces by light. Philipp Lenard had noticed in 1902 that the energy of the electrons didn't depend on how bright the light was. If light is just a smooth classical wave, brighter light should mean more energetic electrons. But Einstein realized that if each electron was being knocked out of the metal by just one photon, it would get the same kick no matter how many photons were nearby.

It was years before the idea of light quanta was accepted, but eventually quantum theory conquered the world. Physicists now believe that everything comes in irreducible quanta – not just energy, but electric charge, momentum, spin; even space and time.

Above Planck was an excellent musician who was sometimes accompanied on the violin by Einstein.

See also Wave nature of light pages 118–119, The electron pages 228–229, Superconductivity pages 266–267, Model of the atom pages 272–273, Wave–particle duality pages 300–301, Quantum electrodynamics pages 352–353, Quantum weirdness pages 464–465, A new state of matter pages 510–511

Right A scanning tunnelling micrograph reveals gold atoms on a graphite substrate. In quantum theory vibrating atoms are each characterized by a certain discontinuous set of energies.

Blood groups

Karl Landsteiner 1868–1943

Following William Harvey's discovery of the circulation of the blood in 1628, the architect Christopher Wren showed that drugs could be introduced directly into the veins, and two other early fellows of the Royal Society, John Wilkins and Richard Lower, demonstrated blood transfusion between two dogs. In France Jean-Baptiste Denys successfully transferred lamb's blood into a sick boy, but, soon after, a second patient died, and Denys was tried for murder. Although he was acquitted, the practice was banned in most European countries and no further attempt was made for 150 years. Then James Blundell of Guy's Hospital, London, showed that blood could not be safely transfused from one species to another. He gave several of his patients human blood, and transfusion became an accepted form of treatment. But many of the patients had severe reactions, sometimes fatal, so the procedure was usually one of last resort.

The Austrian physician Karl Lansteiner made transfusions safe. In 1900 he found that a sample of human serum 'clumped' red blood cells from some people but not others. He suggested in 1901 that the clumping was due to the reaction of 'antibody' molecules in the recipient's serum with 'antigen' molecules on the surface of the donor's red cells – antibodies are proteins that defend the body from foreign substances. He concluded that there were two related antigens, A and B. Some cells carried A, some carried B, some carried both and some carried neither. Hence the four blood groups A, B, AB and O.

Transfusions could be successfully made with certain combinations of blood groups, whereas with others the incoming red cells would be treated by antibodies as 'foreign', clumped together and destroyed, with dangerous results. By 1910 it was discovered that the ABO blood groups were inherited according to Mendel's laws, allowing them to be used to settle paternity disputes, map early human migrations and act as markers of suspected genetic diseases. Many other blood-group systems have since been described.

See also Circulation of the blood pages 58–59, Mendel's laws of inheritance pages 192–193, Antitoxins pages 214–215, Transplant rejection pages 356–357, Biological self-recognition pages 430–431

Right Early attempt at blood transfusion from sheep to man, c.1692. Although human–animal transfusions were initially successful (through pure luck), many deaths occurred and the practice was outlawed.

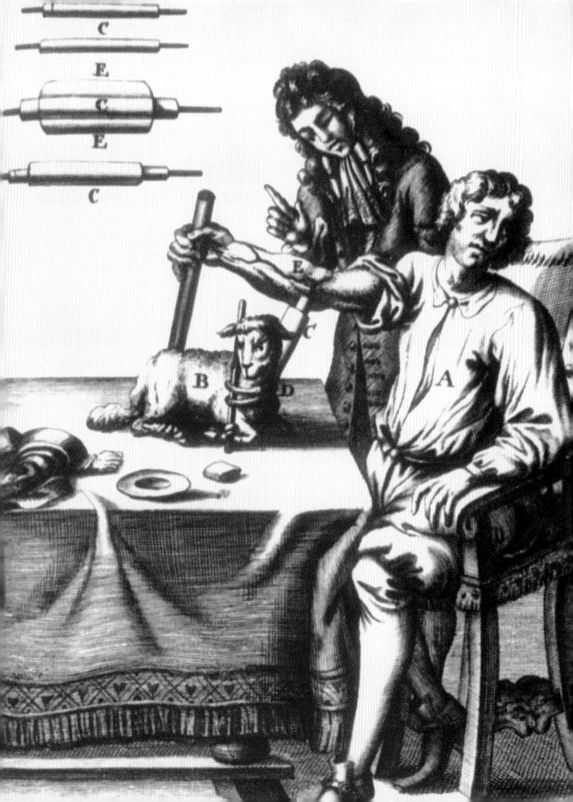

Chaos theory

Jules Henri **Poincaré** 1854–1912

An article of faith of the scientific revolution was that the world was a predictable place. Given the correct mathematical representation of a physical system, scientists believed they could map out its future and past. By the late nineteenth century this view of the clockwork universe was thrown into serious doubt. Henri Poincaré, Professor of Mathematical Physics at the Sorbonne in Paris, was studying the motion of a simplified Solar System consisting of just the Sun, the Earth and the Moon – the so-called three-body problem. He showed in 1903 that even this simple dynamical system, governed by Newton's laws of gravitation and motion, could behave in an intricate and unpredictable way. That small changes in initial conditions could lead to vast differences became the fundamental idea of chaos theory.

In 1961 Edward Lorenz accidentally found a mathematical system with chaotic behaviour in a computer model of the atmosphere. Small changes in the initial conditions produced wildly different, and so completely useless, long-range weather forecasts – a phenomenon that became known as the 'butterfly effect'. Benoit Mandelbrot, in the 1970s, extended this work into the field of fractal geometry. In classical physics the elliptical orbit of a planet is known as an 'attractor', but in a chaotic system the shape of the attractor is a fractal, so linking fractal geometry to chaotic motion.

The computer had become the laboratory and canvas of a mathematics that looked more like the real world than anything previously seen. In fact most of the real world shows chaotic behaviour. This doesn't mean that it is random; rather, that the underlying patterns are far more intricate than we had previously assumed. The theories of fractals and chaos are now a branch of the wider field of complex systems, including artificial intelligence, cellular automata and genetic algorithms. And computer simulations are providing insights into such seemingly different phenomena as air turbulence and stockmarket fluctuations. Chaotic systems remain deterministic but are also unpredictable – we will know the future only when we get there.

See also Euclid's *Elements* pages 20–21, Weather forecasting pages 284–285, Fractals pages 446–447, Edge of chaos pages 490–491

Right Computer mapping of a chaotic system known as a 'Lyapunov space'. Order is represented by the coloured forms, and chaos by the black regions.

Intelligence testing

Alfred Binet 1857–1911

Intelligence testing is taken for granted today both as a tool for applied psychologists and as a highly controversial topic giving rise to recurrent heated debate. But until the beginning of the twentieth century there was no method for measuring intelligence. There had been much effort to define intelligence and ascribe it to differences in various abilities thought to relate to psychological superiority: speed of reaction, sensory discrimination, short-term memory. But this style of work failed and was eclipsed by the simple insight of the Frenchman Alfred Binet.

Binet trained as a lawyer but became interested in psychology. He worked towards a better theoretical definition of intelligence in the 1890s, and then, in 1904 and 1905, developed practical methods for use in the French education system. He noted that as children get older they can tackle more and more difficult tasks, but that not all of them progress at the same rate. By long and painstaking observation he found straightforward tasks – repeating a short sentence, counting up to a given number – that discriminated between children at different ages and then placed them in an ordered series. By taking children through these tasks and finding the level at which they started to fail, he identified their 'mental age' – the typical age at which 50 to 75 per cent of children perform similarly. Children with a mental age greater than their chronological age were considered more intelligent; those with a mental age lower than their chronological age less intelligent.

So intelligence began to be defined in terms of observable differences in behaviour, with the attempt to explain their causes taking a back seat while the practicalities of test development and administration were worked out. Binet's method also reinforced the idea that less intelligent children were 'retarded', a term in common use until recent years. He did not invent the intelligence quotient or IQ: this was proposed the year after he died by the German psychologist Wilhelm Stern, and is the ratio of mental age to chronological age multiplied by 100.

See also Child development pages 294–295, Artificial neural networks pages 334–335, Language instinct pages 386–387

Right Slow uptake: although Binet's intelligence tests were designed for French schoolchildren, they were first adapted for use in the USA and UK. His simple insight gave rise to the psychometric testing industry, although much technical sophistication has been subsequently introduced.

Conditioned reflexes

Ivan Petrovich **Pavlov** 1849–1936

Above The conditioned salivary reflex, as worked out by Pavlov.

The Russian physiologist Ivan Pavlov was originally interested in the functions of the digestive system, in particular the regulation of the flow of saliva and digestive juices in the stomach. During experiments with dogs in which he created artificial channels (fistulae) leading from their stomachs to the outside, he found to his surprise that salivation and gastric secretion were induced by the mere sight of food or even just the sound of his footsteps. Over the following years, he explored this 'psychic excitation' more rigorously. He noticed that his dogs would begin to salivate not only by stimuli directly concerned with eating, but by any stimulus (such as the ringing of a bell or a flashing light) that they had learned to associate with food. This he called a 'conditioned reflex'.

He went on to work out the best ways of building up conditioned reflexes as well as undoing them, and framed his findings in the light of the work of Charles Sherrington on 'reflex arcs' – that is, the path by which sensory signals are transmitted to a direct connection with motor nerves in the spinal cord, as in the knee-jerk. For his research on digestion, Pavlov won the 1904 Nobel Prize in Physiology or Medicine.

During the second half of his career he extended his discoveries by proposing that innate and conditioned physical reflexes could explain all human learning and behaviour, including personality and psychic disorders. This physiological psychology was soon applied by John B. Watson, Edward Thorndike and B. F. Skinner to the theory of 'behaviourism' – a branch of psychology based on objective investigation rather than introspection. But the enduring legacy of Pavlov's work lies in his idea that common neuroses such as phobias and anxiety are the result of ill-formed conditioned reflexes – a view adopted by many psychiatrists in the second half of the twentieth century.

See also Nervous system pages 210–211, Child development pages 294–295, Behavioural reinforcement pages 322–333, Artificial neural networks pages 334–335, Memory molecules pages 470–471

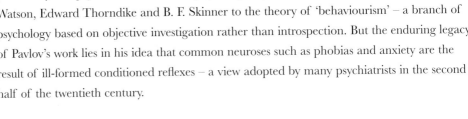

Right Pavlov's dogs during an experiment on digestion. According to the dramatist George Bernard Shaw, Pavlov would 'boil babies alive just to see what happens'.

Special relativity

Albert Einstein 1879–1955

In 1905 Albert Einstein abolished space and time. All he did was put two facts together. The first is that people moving at different velocities find the same physical laws – their experiments produce the same results. This fits our experience. We don't directly 'feel' our motion around the Sun, and we can wander around comfortably in a flying aeroplane.

But the second fact is more uncomfortable. If a hurtling spacecraft fires a laser beam, you might expect the light to go faster than if the spacecraft had been still. But it isn't so. The speed of light is the same no matter where it comes from or who measures it.

To account for these two things, Einstein had to abandon Newton's absolute space and time. Lengths and times must depend on who is measuring them. To you, the inhabitants of the hurtling spaceship seem to be squashed creatures, a phenomenon known as 'Lorentz contraction'. And they move unnaturally slowly, which is time dilation. And yet, to them, you seem just as squashed and slowed. All motion is relative in the theory of special relativity; there is no 'preferred frame'.

Time dilation and Lorentz contraction become extreme for relative speeds approaching the speed of light – the ultimate speed limit, according to the theory. Physicists see the swift particles in their accelerators moving to the prescriptions of relativity every day, and astronomers see distant galaxies receding from us rapidly, running as if they are slowed down.

Applying relativity to the idea of energy, Einstein went on to discover the most famous equation, $E = mc^2$. It means that there is energy hidden in matter – a huge amount of energy, equal to the object's mass times the speed of light squared. A kilogram of anything holds enough energy to boil a hundred billion kettles. Or destroy a city.

See also Newton's *Principia* pages 78–79, Maxwell's equations pages 186–187, General relativity pages 278–279, Stellar evolution pages 286–287, Expanding universe pages 306–307, Power from the nucleus pages 330–331

Right Einstein: 'When a man sits with a pretty girl for an hour, it seems like a minute. But let him sit on a hot stove for a minute – and it's longer than any hour. That's relativity.'

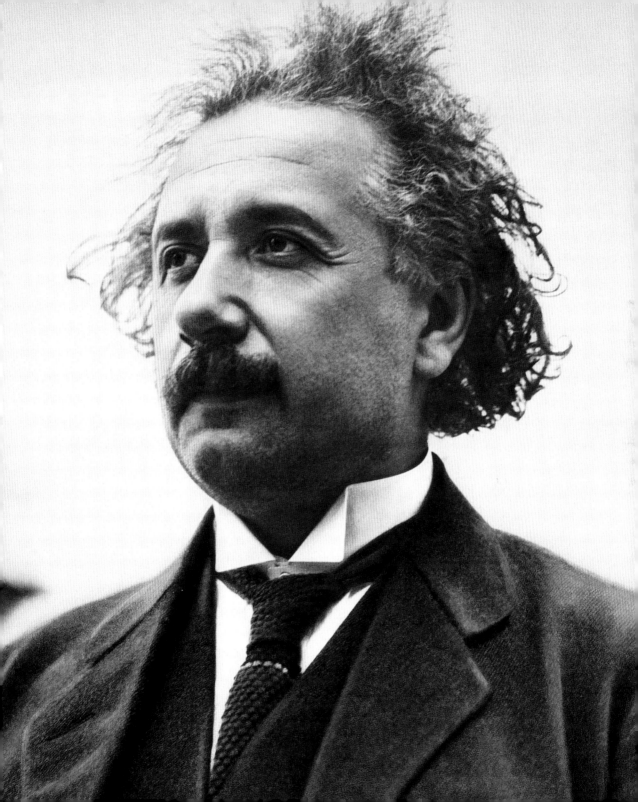

From Newton to Einstein

Martin Rees

MORE THAN TWO CENTURIES after Newton, Einstein proposed his theory of gravity known as 'general relativity'. According to this theory, planets actually follow the straightest path in a 'space-time' that is curved by the presence of the Sun. It is commonly claimed that Einstein 'overthrew' Newtonian physics, but this is misleading. Newton's law still describes motions in the Solar System with good precision (the most famous discrepancy being a slight anomaly in Mercury's orbit that was resolved by Einstein's theory) and is adequate for programming the trajectories of space probes to the Moon and planets. Einstein's theory, however, copes (unlike Newton's) with objects whose speeds are close to that of light, with the ultra-strong gravity that could induce such enormous speeds, and with the effect of gravity on light itself. More importantly, Einstein deepened our understanding of gravity. To Newton, it was a mystery why all particles fell at the same rate and followed identical orbits – why the force of gravity and the inertia were in exactly the same ratio for all substances (in contrast to electric forces, where the 'charge' and 'mass' are not proportionate) – but Einstein showed that this was a natural consequence of all bodies taking the same 'straightest' path in a space-time curved by mass and energy. The theory of general relativity was thus a conceptual breakthrough – especially remarkable because it stemmed from Einstein's deep insight rather than being stimulated by any specific experiment or observation.

Einstein didn't 'prove Newton wrong'; he transcended Newton's theory by incorporating it into something more profound, and with wider applicability. It would actually have been better (and would have obviated widespread misunderstanding of its cultural implications) if his theory had been given a different name: not 'the theory of relativity' but 'the theory of invariance'. Einstein's achievement was to discover a set of equations that can be applied by any observer and incorporate the remarkable circumstance that the speed of light, measured in any 'local' experiment, is the same however the observer is moving …

Experience shapes our intuition and common sense: we assimilate the physical laws that directly affect us. Newton's laws are in some sense 'hardwired' into monkeys that swing confidently from tree to tree. But far out in space lie environments differing hugely from our own. We should not be surprised that common-sense notions break down over vast cosmic distances, or at high speeds, or when gravity is strong.

An intelligence that could roam rapidly through the universe – constrained by the basic physical laws but not by current technology – would extend its intuitions about space and time to incorporate the distinctive and bizarre-seeming consequences of relativity. The speed of light turns out to have very special significance: it can be approached, but never exceeded. But this 'cosmic speed limit' imposes no bounds to how far you can travel in your lifetime, because clocks run slower (and on-board time is 'dilated') as a spaceship accelerates towards the speed of light. However, were you to travel to a star a hundred light-years away, and then return, more than two hundred years would have passed at home, however young you still felt. Your spacecraft cannot have made the journey faster than light (as measured by a stay-at-home observer), but the closer your speed approached that of light, the less you would have aged.

These effects are counterintuitive simply because our experience is limited to slow speeds. An airliner flies at only a millionth of the speed of light, not nearly fast enough to make the time dilation perceptible: even for the most inveterate air traveller it would be less than a millisecond over an entire lifetime. This tiny effect has, nevertheless, now been measured, and found to accord with Einstein's predictions, by experiments using atomic clocks accurate to a billionth of a second.

A related 'time dilation' is caused by gravity: near a large mass, clocks tend to run slow. This too is almost imperceptible here on Earth because, just as we are only used to 'slow' motions, we experience only 'weak' gravity. This dilation must, however, be allowed for, along with the effects of orbital motion, in programming the amazingly accurate Global Positioning Satellite system.

A measure of the strength of a body's gravity is the speed with which a projectile must be fired to escape its grasp. It takes 11.2 kilometres per second to escape from the Earth. This speed is tiny compared with that of light (300,000 kilometres per second), but it challenges rocket engineers constrained to use chemical fuel, which converts only a billionth of its so-called 'rest-mass energy' (Einstein's $E = mc^2$) into effective power. The escape velocity from the Sun's surface is 600 kilometres per second – still only one fifth of one per cent of the speed of light.

Vitamins

Frederick Gowland Hopkins 1861–1947

The ravages of life at sea were evident in the bleeding gums, skin and joints of scurvy-ridden sailors. In 1747, in a controlled trial using various foods, James Lind, a Scottish physician and former surgeon's mate in the British Navy, demonstrated that the disease could be cured by citrus fruits. In 1758 he recommended that the navy should adopt citrus fruit as a dietary staple. Where followed, his advice prevented scurvy. No other 'dietary-deficiency diseases' were recognized until the end of the nineteenth century, when the links between beri-beri and whole rice and between rickets and cod-liver oil were established. But it wasn't until the work of the English biochemist Frederick Gowland Hopkins that such diseases were subjected to scientific scrutiny.

In 1900 Hopkins showed that tryptophan, one of the amino-acid building blocks of proteins, could not be manufactured in the body and had to be present in the diet. He thus became interested in the 'synthetic diet' that supposedly contained the nutritional components essential for the support of life – purified amino acids, carbohydrates, fats and salts. His researches led him to conclude in 1906 that the synthetic diet was inadequate. His claim met with scepticism – calories mattered, not trace amounts of 'accessory food factors', as he proposed. Returning to the laboratory, he studied the growth and health of rats fed the synthetic diet, as well as those given a little milk too. His experiments were scrupulous; there were no errors through poor controls or incomplete ingestion or absorption. And they proved him right.

In 1912, the year Hopkins announced his results, Casimir Funk named the factors 'vitamines' – from 'vital amines' – believing they were chemically amines (the 'e' was dropped when it turned out that not all vitamins were amines). As each vitamin was identified and isolated, researchers gave it a new letter, although several vitamins are in fact groups or complexes of different compounds. Scurvy is due to a deficiency of vitamin C (ascorbic acid).

See also Germ theory pages 202–203, Inborn metabolic errors pages 258–259, Chemical basis of vision pages 382–383

Above Crystals of alpha-tocopherol, the most effective of the vitamin E group.

Right Advertisement for a food supplement based on cod-liver oil, c.1890. The endorsement from a satisfied customer reads: 'We cannot speak too highly of your Food, having brought up four fine children on it.' Cod-liver oil is rich in vitamins A and D.

Inside the Earth

Richard Dixon **Oldham** 1858–1936, Andrija **Mohorovičić** 1857–1936

Catastrophic earthquakes are constant reminders of pent-up dynamic forces insid
Earthquakes themselves do not kill but they generate shock waves that reverberate
the Earth, collapsing buildings and triggering lethal landslides. Frequent earthqua
death tolls in China prompted the first attempts to detect seismic activity in AD 13
destruction of Lisbon in 1755, European scientists have also tried to understand a
earthquakes – with limited success. But a by-product of seismic studies was the re
the Earth has a layered internal structure.

After the Assam earthquake of 1897, the British survey geologist Richard Dix
found he could distinguish two kinds of internal body wave in recordings made by
seismograph invented by John Milne in 1880. The existence of these primary or
compressional (P) waves and secondary or shear (S) waves had been predicted by
mathematician Siméon Denis Poisson in 1829. By 1906 Oldham had showed that
travelling through the Earth arrive opposite the earthquake epicentre later than w
expected if the deep interior of the Earth was relatively homogenous. He concluc
that a dense core (nearly 7,000 kilometres in diameter) must be responsible for slo
down some of the P waves.

Within three years Andrija Mohorovičić, a Croatian geophysicist, found subtle
changes in P and S waves that showed that the surface of the Earth is also layered
outer crust (averaging 30 kilometres thick) overlies a denser and hotter mantle (2,9
kilometres thick) and is separated from it by a seismic discontinuity called the Mol
discontinuity. It is now known that this varies in depth from 20–80 kilometres und
continents to around 7 kilometres under the oceans. In the 1960s there were plan
aborted – to drill a 'mohole' through the solid crust in order to reach the disconti
study the layer below.

See also Earth cycles pages 100–101, Catastrophist geology pages 108–109, Lyell's *Principles of Geology* pages 146–
Ice ages pages 152–153, Rocks of ages pages 252–253, Continental drift pages 270–271, Geomagnetic reversals pag
Plate tectonics pages 414–415, Eruption of Mount St Helens pages 462–463

Right Hidden crucible: 17th-century view of the Earth as a ball of solid material fissured by tubes which connect a cen

PYRO-HYLACIORUM
Subterraneorum, quorum montes
Vulcanii, veluti spiracula
quædam exstant.

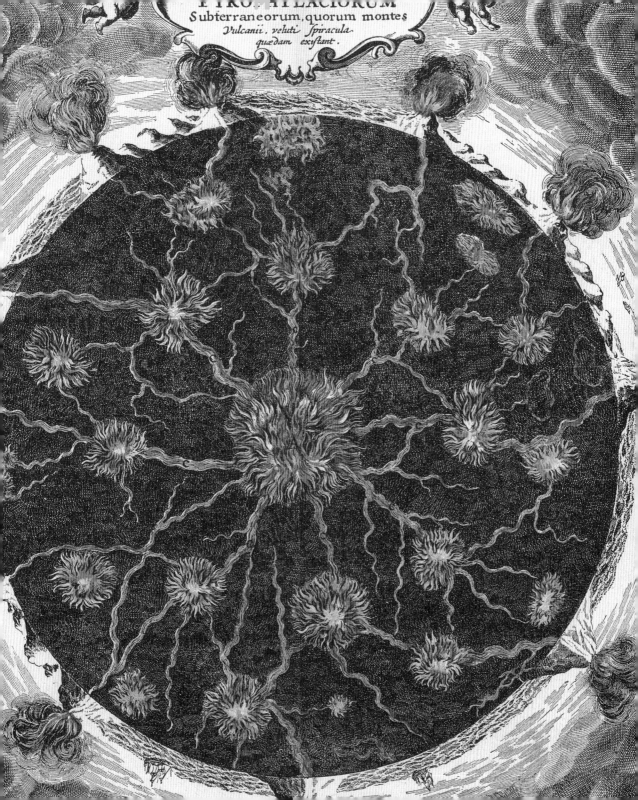

Rocks of ages

Bertram Boltwood 1870–1927

In 1907 the American chemist Bertram Boltwood measured the ratio of radioactive isotopes, or variant forms, of uranium and lead in a mineral from Glastonbury, Connecticut, and calculated that the mineral formed 410 million years ago (subsequently redated to 265 million years ago). He had developed Ernest Rutherford's work to show that uranium-rich rocks contain large amounts of lead along with helium. Boltwood postulated that the lead was a stable end-product of uranium disintegration, decaying spontaneously down through a series of radioactive isotopes. For the first time a reasonably accurate way of dating Earth's igneous rocks was established.

Two hundred years earlier, scholars such as James Ussher, Archbishop of Armagh, Ireland, had used Judeo-Christian texts to date the Creation at 4004 BC. This date was widely accepted and even printed in the Bible as an historic truth and was still being quoted in the infamous anti-evolution Scopes trial of 1925.

But by the late eighteenth century there were more scientific attempts to calculate the Earth's age. From his examination of cooling rates, George Buffon estimated 75,000 years, and James Hutton showed that geological processes were so slow that 6,000 years was quite inadequate. A century later Charles Lyell and Charles Darwin thought that the Earth was hundreds of millions of years old. The physicist William Thomson (Lord Kelvin) scorned such geological estimates. From the known melting points of rocks, he calculated that thermal diffusion required some 20 million years to cool the Earth from an initially molten state. As the role of radioactivity as an internal source of heat was not known, his was a serious underestimate.

Now we know that the Earth formed 4.57 billion years ago and that its present mass was not reached until 4.51–4.45 billion years ago. Earth's oldest known rock material is a zircon grain from Australia, dated at 4.4 billion years by the uranium–lead method in January 2001.

See also Earth cycles pages 100–101, Catastrophist geology pages 108–109, Lyell's *Principles of Geology* pages 146–147, Radioactivity pages 224–225, Stellar evolution pages 286–287, Expanding universe pages 306–307, Radiocarbon dating pages 346–347, Oldest fossils pages 410–411

Right Zircons were among the first continental minerals. Because they are highly resistant to weathering, they survive many geological cycles. Grains have been found that are 4.4 billion years old.

Brownian motion

Robert **Brown** 1773–1858, Ludwig Eduard **Boltzmann** 1844–1906,
Albert **Einstein** 1879–1955, Jean Baptiste **Perrin** 1870–1942

Most scientists were content to regard atomism as a convenient working hypothesis: it seemed a matter of faith to presume the physical reality of objects too small to observe. Some believed that energy, not atomic matter, lay at the root of all things. Yet the Austrian physicist Ludwig Boltzmann maintained that heat was nothing but molecular motion. To support his picture of a frantic microscopic world, he cited Brownian motion. A long-standing puzzle, this phenomenon was named after the botanist Robert Brown, who noticed in 1827 that pollen particles suspended in water move erratically under the microscope. The cause was debated for several decades. Vital forces were evoked, and some scientists even suggested that it was a kind of perpetual motion that violated the second law of thermodynamics.

Boltzmann believed the tiny particles were constantly struck by molecules that, although too small to see, had enough momentum to nudge the particles in a new direction. Since the molecular motions were random, the suspended particles made irregular and unpredictable changes in their course. In 1905 Albert Einstein used these ideas to provide a rigorous theoretical explanation. He realized it was futile to try to account for the velocities of the particles, since these were practically unmeasurable. Instead he deduced how the average distance travelled from a starting position varied with time. Although the particles change direction randomly, they do gradually move through the suspending medium. He saw that this was the origin of diffusion, the process that allows fluids to mix, and showed that his theory provided 'a new method of determining the actual size of atoms'.

Experimentalists immediately took up his challenge of accurately measuring the movements of suspended particles. By the end of 1908 the French chemist Jean Perrin had confirmed almost all of Einstein's predictions and calculated the size of a water molecule. Atomism was at last vindicated.

See also Atomic theory pages 124–125, Changes of state pages 198–199, Model of the atom pages 272–273

Above The atomic surface of a silicon crystal, as seen by transmission electron microscopy.

Right A laser trap confines a cloud of atoms, represented here by the red glow. Boltzmann's disagreement with the 'energists' over the reality of atoms may have been a factor in his suicide.

Synthesis of ammonia

Fritz Haber 1868–1934

Even though Earth's atmosphere is 80 per cent nitrogen, this gas is very unreactive, and only a few living species, such as bacteria and other microbes, can turn it into the useful gas ammonia (NH_3), via a process known as 'nitrogen fixation'. Other living organisms rely on this process to produce the amino acids on which all life depends. But the amount of nitrogen that microbes can fix is limited.

The chemists of the nineteenth century tried to make ammonia directly from nitrogen gas (N_2) and hydrogen gas (H_2), but the two gases would just not react together, no matter how high the temperature or the pressure. Then, in 1904, the German chemist Fritz Haber passed a mixture of the hot gases through a catalyst of iron filings and found that a tiny amount of ammonia was produced, although only 0.01 per cent of the nitrogen gas had reacted this way. It was hardly a result to interest the chemical industry, or so he said when he was consulted on the matter by BASF, one of the leading chemical companies of its day. Still, they asked him to study the process more closely, and by 1908 he had found ways to increase the yield of ammonia to 6 per cent.

Such a yield was commercially viable, said the BASF chemist Carl Bosch. The ammonia could be separated out and the unreacted nitrogen and hydrogen put through the process again and again. He encouraged Haber to scale up his research and, on 3 July 1909, an experimental plant began producing ammonia, albeit only 80 grams an hour. Soon a large-scale ammonia plant was built at Oppau, Germany, producing a tonne of ammonia an hour. Today there are thousands of Haber–Bosch ammonia plants around the world, making more than 150 million tonnes of ammonia every year and supplying the nitrogen fertilizers on which the world's food supply depends.

Above Haber, an extremely patriotic German, had laboured increasingly during the First World War on gas warfare, directing the first use of chlorine and mustard gas.

See also Synthesis of urea pages 142–143, Nitrogen fixation pages 208–209, Crop diversity pages 292–293, Green revolution pages 428–429, Gaia hypothesis pages 432–433

Right A British ammonia factory, 1928. Thanks to their earlier development of the Haber process, the Germans never ran out of nitrates for ammunition production during the First World War.

Inborn metabolic errors

Archibald Garrod 1857–1936

Around 1897 the British paediatrician Archibald Garrod diagnosed alkaptonuria in several of his patients – a disorder that turns the urine dark and often leads to arthritis. The chemical responsible for the coloration had been discovered in 1859, and it was later shown to be an acid related to tyrosine, one of the amino-acid building blocks of proteins.

Garrod's biochemical research suggested that the disorder was due to an 'inborn error of metabolism' – in other words, that it was congenital. He found that three out of four alkaptonuric children with unaffected parents were born to first-cousin marriages. In 1901 William Bateson, the geneticist and translator of Mendel, had described this pattern as characteristic of an inherited recessive trait – that is, both parents carry the abnormal gene, with a 1:4 chance that their offspring will develop the disorder. In *Inborn Errors of Metabolism* (1909) Garrod described the actual defect as the absence of an enzyme for dealing with one of the stages in the metabolism of tyrosine. The resulting metabolic block leads to the excretion of an intermediate breakdown product in the urine, which, in turn, darkens greatly as a result of atmospheric oxidation.

Garrod's work was an early application of the new Mendelian genetics, and foreshadowed the proposal of George Beadle and Edward Tatum some three decades later that each gene supervises the production of just one particular enzyme. Beadle originally studied fruit-fly eye pigments, which he considered to be generated by a series of enzyme reactions, but the system had proved too complicated for further investigation. With Tatum he turned to an even simpler organism – the bread mould *Neurospora*. As this could be grown on a medium of minimal nutrients, particular metabolic mutations could be easily detected. Together Beadle and Tatum were able to show that a different gene controlled each stage in the metabolic pathway producing the amino acid tryptophan, combining genetics and biochemistry in the way Garrod had anticipated.

See also Regulating the body pages 188–189, Mendel's laws of inheritance pages 192–193, Enzyme action pages 218–219, Genes in inheritance pages 264–265, Citric-acid cycle pages 320–321, Human genome sequence pages 524–525

Right A human male fetus in the womb, at 19 weeks.

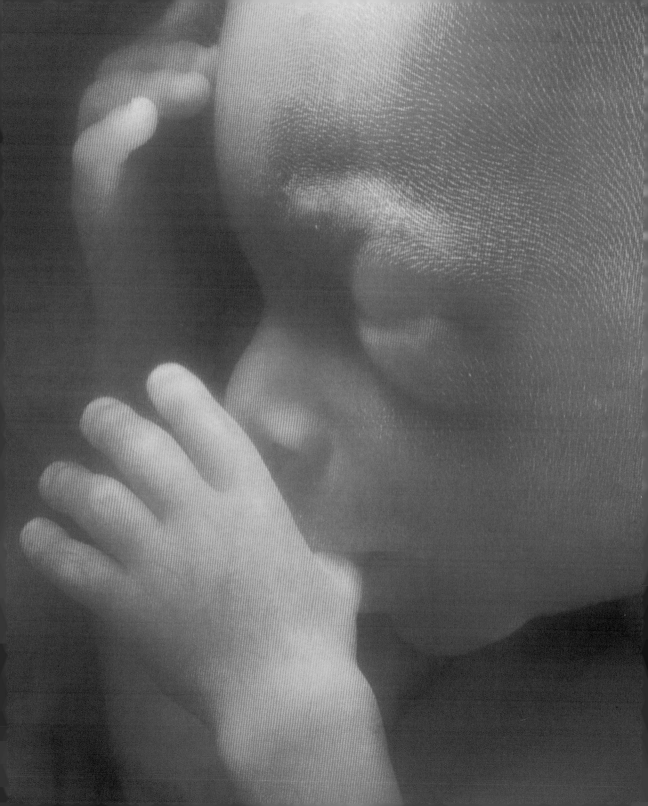

Burgess Shale

Charles Doolittle Walcott 1850–1927

The discovery by the American palaeontologist Charles Doolittle Walcott of 520-million-year-old fossiliferous rocks high in the Canadian Rockies opened one of the world's most famous 'windows' into the deep past. Known as the Burgess Shale, these ancient sea-bed sediments have yielded thousands of remarkably preserved fossils, many of which still have their soft parts intact. They provide a vivid picture of life in Cambrian seas when the early arthropods, the ancestors of invertebrates such as crabs, came to power and our most remote vertebrate ancestors were just tiny lamprey-like creatures. The locality in Yoho National Park, British Columbia, is now protected as a World Heritage Site.

Although trilobite fossils had been found in the area in 1884, Walcott discovered the Burgess Shale by chance on 31 August 1909 when crossing from Mount Field to Wapta Mountain. He spotted fossils in a block of scree and immediately recognized their importance. Over the next eight years he excavated some 70,000 specimens and sent them to the Smithsonian Institution in Washington, DC, where he worked. Administrative duties prevented him from exploiting the significance of his find. Its full wealth was revealed largely through the work of the British palaeontologist Harry Whittington, his research students and Canadian palaeontologists.

However measured, the Burgess sea-world was dominated by arthropods (more than 20 different species are known) along with sponges, echinoderms, priapulid worms, brachiopods, molluscs and a curious swimming creature called *Pikaia*, which may be one of our first vertebrate ancestors. The variety of life suggests that a division of labour similar to that of modern marine ecosystems had already evolved. This diversity and ecological sophistication so early in evolution throws doubt on the apparent explosive increase in the complexity of life near the Precambrian/Cambrian boundary. It seems more likely that a lengthy earlier development of multicellular animals stretched way back into Precambrian times.

See also Fossil objects pages 46–47, Geological strata pages 72–73, Oldest fossils pages 410–411, Five kingdoms of life pages 426–427, Extinction of the dinosaurs pages 458–459, Diversity of life pages 468–469

Right Burgess Shale community. *Hallucigenia*, shown on the sea-bed, was originally reconstructed as walking on its seven pairs of stilt-like spines. The spines are now thought to have formed a defensive array above the animal.

A magic bullet

Paul Ehrlich 1854–1915

Imagine a drug that could kill the microbes responsible for a specific disease without harming the surrounding cells of the body. The German doctor Paul Ehrlich was inspired to find such a 'magic bullet'. He had noticed that only certain cellular structures were stained by the new synthetic dyes used in the textile industry – in the 1880s he himself had stained the newly discovered tuberculosis bacillus in this way. And in the 1890s he had advanced the idea that the interaction between antibodies and antigens was chemical in nature, involving groups of 'side-chain' molecules fitting together like a key in a lock. Might not some dyes selectively destroy microbes in a similar way, he thought, just as an individual's own antibodies zero in on invading bacteria?

He began studying the effect of organic arsenic compounds on trypanosomes, the parasites that cause sleeping sickness. But after the discovery in 1905 of the microscopic spirochaete (*Treponema pallidum*) responsible for syphilis, he refocused his research team on a cure for that disease instead. It took the painstaking synthesis and testing of some 606 arsenic compounds before he was satisfied that he had found one that was lethal for *T. pallidum* but not injurious to its host. On 19 April 1910 he announced his findings at the Congress of Internal Medicine at Wiesbaden in Germany.

Initially there were severe shortages of the drug, originally dubbed 'arsphenamine' but soon renamed 'salvarsan', and several disasters arose from its careless injection. Later modified to the safer compound neosalvarsan, the drug established the field of chemotherapy and led to the search for other synthetic chemicals active against infectious or malignant diseases. In 1935 a new group of anti-microbial drugs, the sulphonamides, were introduced; and after the Second World War doctors could prescribe antibiotics and a range of increasingly powerful anti-cancer drugs – although none proved quite as magical as Ehrlich's dream.

See also Mauve dye pages 172–173, Germ theory pages 202–203, Antitoxins pages 214–215, Penicillin pages 302–303, Human cancer genes pages 456–457

Above Ehrlich smoked 25 thick cigars a day, often neglected to eat, and was venerated by younger colleagues.

Right Man suffering from tertiary syphilis, from a French medical text, c.1890. The French called it the Italian disease; the Italians returned the compliment.

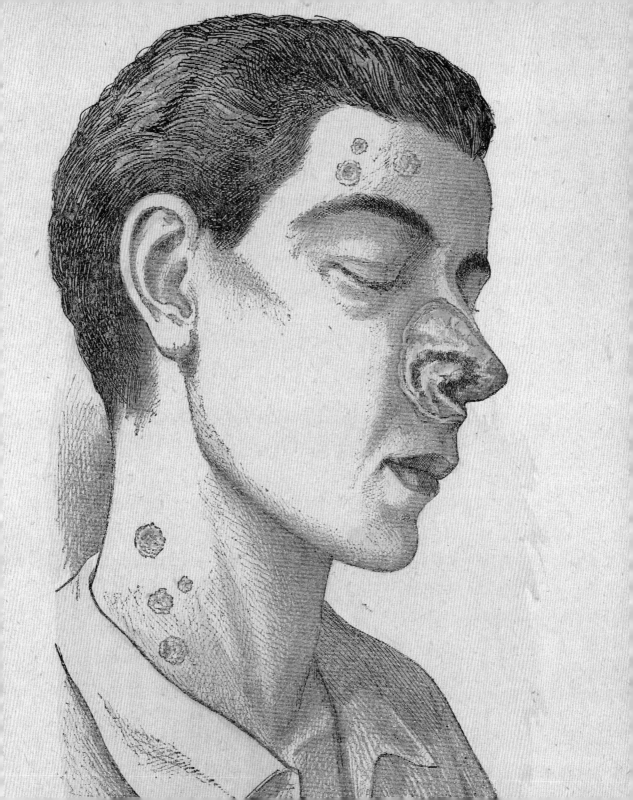

Genes in inheritance

Thomas Hunt Morgan 1866–1945, Alfred Henry Sturtevant 1891–1970,
Calvin Bridges 1889–1938, Hermann Joseph Muller 1890–1967

What is it that is passed from parents to offspring that causes the offspring to grow up like their parents? Biologists have progressively pinned down the location of the material of inheritance – to cells, then to substructures within cells, then to molecules. By the late nineteenth century it seemed likely that the rod-like structures called chromosomes, sometimes visible in the cell nucleus, were the vehicles of inheritance. Microscopic observation revealed that chromosomes divide in parental cells, and combine in offspring cells, in a way that fitted with Mendel's laws of heredity.

The next advance came in the 'fly room' at Columbia University, New York. Thomas Hunt Morgan ran the laboratory there, but his students Alfred Sturtevant, Calvin Bridges and Hermann Muller made discoveries as important as those of Morgan himself. They showed that inheritance is due to units, called genes, that are carried on the chromosomes. Morgan's first breakthrough came in 1910. He showed that a mutant form of the fruit fly, with white eyes, had an abnormal gene carried on one particular chromosome (the X chromosome). In 1911 Sturtevant, while still an undergraduate student, showed that it was possible to infer which chromosomes carried the genes that controlled several different traits in the fruit fly. He used the results of huge numbers of crosses between fruit flies with various combinations of these traits. In this way Sturtevant produced the first 'gene map', and much of genetic research since then has aimed to identify genes for particular traits and to map them on to the chromosomes.

In 1927 Muller showed that genes could be made to mutate by irradiation with X-rays. This made it possible to create more kinds of fly for genetic study and began the scientific study of mutation.

See also Eggs and embryos pages 140–141, Mendel's laws of inheritance pages 192–193, Inborn metabolic errors pages 258–259, Genes in bacteria pages 332–333, Sickle-cell anaemia pages 358–359, Jumping genes pages 362–363, The double helix pages 374–375, Human cancer genes pages 456–457, Genetics of animal design pages 460–461, Maleness gene pages 502–503, Human genome sequence pages 524–525

Right Some 100,000 scientific papers have been published on the fruit fly *Drosophila melanogaster*, and a large proportion concern genetic mutants. The mutant on the left has a disordered arrangement of eye subunits.

Superconductivity

Heike **Kamerlingh Onnes** 1853–1926

While Amundsen and Scott sought one icy extreme of the Earth, physicists were fighting to approach a far more profound cold. Absolute zero temperature is unreachable; we can only get closer and closer, never quite arriving. But this cold land holds surprises.

Heike Kamerlingh Onnes found the first of them. In 1908 the Dutch physicist had been the first to liquefy helium, by cooling it to just 4 degrees above absolute zero, or −269 degrees Celsius. But then in May 1911, the same year that Amundsen reached the South Pole, Kamerlingh Onnes saw something utterly strange. He had set two colleagues at his Leiden laboratory in the Netherlands to experiment on cold metals, measuring their resistance to the flow of electricity. When they cooled a sample of mercury to 4.2 degrees above absolute zero, its resistance suddenly dropped to nothing.

This was startling. If resistance is zero, a current can flow around a loop of wire forever. What could it mean? Kamerlingh Onnes was quick to guess that this new state of metals, which he called 'superconductivity', might have something to do with the new quantum theory, but it was not until 1957 that a full explanation emerged. John Bardeen, Leon Cooper and Robert Schreiffer worked out how electrons could join together and by a quirk of quantum mechanics ignore the metal around them.

Superconductors could save us huge amounts of energy, levitate trains or cars over rail or roadbed, and lead to far faster, smaller computers and electric motors − if we didn't have to keep them so cold. The dream is to find one that will work at room temperature or above. In 1986 Georg Bednorz and Alex Müller discovered a ceramic material that was superconducting at −238 degrees Celsius, and since then ceramics have been found that work up to temperatures approaching a cosy −100 degrees. But no one yet knows how these high-temperature superconductors work.

See also The quantum pages 234–235, Wave–particle duality pages 300–301, A new state of matter pages 510–511

Right A magnet floats freely above a superconducting ceramic cooled by liquid nitrogen. Engineers hope to use the effect to suspend trains above their tracks, so eliminating friction.

Cosmic rays

Victor Francis Hess 1883–1964

Killer rays from outer space are bombarding the Earth. In the first years of the twentieth century, this was a wild idea. Scientists had discovered ions – electrically charged atoms and molecules – appearing spontaneously in the air, and thought they were created by radioactive minerals in the Earth, whose radiation rips electrons off atoms, leaving them charged.

If this is the only source of ions, then they should become rarer as you move farther from the Earth's surface and the air gradually absorbs the radiation. Between 1911 and 1913 Victor Francis Hess made a series of balloon voyages to test the idea. He carried a device called an electroscope, which measures the build-up of charge.

Hess's electroscope showed that the number of ions did decrease as he rose through the air – but then, above about 1,500 metres, it began to increase again. He decided that they were being produced by something coming from above, some kind of radiation that could penetrate deep into the atmosphere. He had discovered cosmic rays.

Cosmic rays are fearsomely energetic charged particles, mostly protons. Almost all of them are thought to come from within our galaxy. Some stars die in huge explosions called supernovae, which create vast expanding shockwaves. Over the millennia, these shockwaves could accelerate protons and other particles up to enormous energies.

But the most energetic cosmic rays ever seen pack a billion times the punch of protons from any particle accelerator on Earth. To reach these extremes, even supernovae are too feeble. These rays probably come from outside our galaxy, perhaps from quasars, cosmic strings or exotic particles left over from the big bang. Astrophysicists are still guessing.

And they really are killers. Cosmic rays account for around 15 per cent of the average person's natural radiation dose and probably cause more than a hundred thousand fatal cancers per year.

Above Particle tracks from a cosmic-ray collision recorded in a bubble chamber.

See also Radioactivity pages 224–225, Radiocarbon dating pages 346–347, Quasars pages 404–405, Afterglow of creation pages 412–413, Supernova 1987A pages 488–489, Water on the Moon pages 522–523

Right Detection of the interactions of cosmic rays as they pass through the atmosphere involved taking equipment to the highest altitudes, often with the help of balloons.

Continental drift

Alfred Lothar Wegener 1880–1930

Ever since maps of the Atlantic borderlands were first available, the coastline of the Americas has been seen to mirror that of Africa and Europe: in the 1720s Francis Bacon noticed the jigsaw-like fit. But not until the twentieth century was there any substantial evidence to suggest that it was anything more than a coincidence. In 1911 Alfred Lothar Wegener, a German meteorologist and polar explorer, first formulated his theory of continental displacement, supporting it with a mass of evidence from several sources.

After returning in 1908 from an expedition to Greenland, Wegener became Professor of Meteorology and Navigational Astronomy at the University of Marburg. Interested in ancient climates, he was puzzled by the presence of plant fossils from the same species scattered across the southern continents of India, South America, Africa and Australia. Geologists such as Eduard Suess had called the association of southern continents 'Gondwanaland' and tried to explain the plant distribution by variously invoking the idea of land-bridges or of a contracting or expanding Earth. But the existence of tropical plant fossils in Spitsbergen coal seams within the Arctic Circle, and of glacial deposits in southern Africa near the equator, also needed explaining.

Wegener first sought a resolution in 1911, but not until 1912 did he present his theory, that all the continents were once joined as a supercontinent called Pangaea (from the Greek *pan gaia* – 'all earth') surrounded by an ocean called Panthalassa (*pan thalassa* – 'all ocean'). Pangaea moved from south to north before splitting into today's continents, but as to the mechanism he was vague: perhaps the ocean floors 'had stretched like rubber' or perhaps they had been influenced by some centrifugal force or the Moon's gravitational pull. The idea solved many geological problems, but a plausible mechanism – plate tectonics – was not proposed until the 1960s, long after Wegener's death.

Above Wegener's 1915 map shows the supercontinent Pangaea splitting up in Eocene times, some 50 million years ago.

See also Earth cycles pages 100–101, Catastrophist geology pages 108–109, Lyell's *Principles of Geology* pages 146–147, Mountain formation pages 206–207, Inside the Earth pages 250–251, Plate tectonics pages 414–415, Eruption of Mount St Helens pages 462–463

Right In 1920 one geologist dismissed continental drift as 'an explanation that explains nothing that we want to explain'. Little did he know that it had been responsible for the devastating 1906 San Francisco earthquake.

Model of the atom

Ernest **Rutherford** 1871–1937, Niels **Bohr** 1885–1962

'As if you had fired a fifteen-inch shell at a sheet of tissue paper and it came back and hit you.' This is how Ernest Rutherford described the phenomenon that led him to his model of the nuclear atom.

In 1907, one of Rutherford's students had aimed a beam of alpha particles at a piece of thin gold foil. Alpha particles are a heavyweight kind of radioactivity, and as expected most of these projectiles flew through the flimsy foil. But a few bounced straight back. This made no sense if the atoms in the foil were just arrangements of light electrons held together by a diffuse positive charge, as previously believed. Rutherford decided instead that the positive charge must be held by a nucleus in the centre of each atom. Then most of the alpha particles would miss these nuclei altogether, but the few unlucky enough to hit one would rebound. Rutherford developed the model of the atom as a tiny dense nucleus orbited by a series of much smaller electrons.

This didn't just redraw our picture of the atom. In 1913 it led the Danish physicist Niels Bohr to a still more radical theory. He took Rutherford's idea and mixed in the new quantum theory. In Bohr's model, electrons orbit the nucleus with certain fixed energies. That explained why atoms are stable – electrons can't lose all their energy and fall into the nucleus; they are only permitted to settle down into the so-called ground state.

These fixed electron orbits explained why atoms emit light in single, sharp colours called spectral lines: when an electron shifts from one orbit to another, the spare energy goes into a photon of light of an exact energy, and therefore an exact colour. So the seemingly inexplicable quantum ruled not just the volatile realm of light, but the concrete world too, the matter of which we are made.

See also Atomic theory pages 124–125, Spectral lines pages 130–131, Periodic table pages 196–197, Radioactivity pages 224–225, The electron pages 228–229, The quantum pages 234–235, Wave–particle duality pages 300–301, The neutron pages 312–313, Power from the nucleus pages 330–331, Quarks pages 408–409, Unified forces pages 416–417, Superstrings pages 474–475

Right The classic model of the atom, developed by Rutherford and Bohr: the electrons orbit the nucleus; the nucleus was later found to be made up of protons (red) and neutrons (blue).

Neurotransmitters

Henry Hallett Dale 1875–1968, George Barger 1878–1939, Otto Loewi 1873–1961

The problem of how the components of the nervous system communicated between themselves was an old one. Notions suggesting that the relationship between thought and action was instantaneous were disproved by nineteenth-century physiologists, who measured the speed of impulses along the peripheral nerves. One of the puzzles was what happened at the connections between nerve endings, which were called 'synapses' by Charles Sherrington.

An electrical model seemed likely to most scientists, but Henry Dale in England and Otto Loewi in Germany sought chemical explanations. Dale, working with chemist George Barger, was examining physiologically active chemicals in the body, including histamine, the release of which they showed to produce symptoms of allergy, and ergot, which slows uterine contractions during premature labour. In 1914 they isolated acetylcholine from preparations of the ergot fungus and demonstrated that it caused effects similar to those of the parasympathetic nerves (a branch of the autonomic system that governs non-voluntary neurological functions such as the control of blood pressure, digestion and sweating). In a series of classic experiments conducted during the late 1920s, Dale and his colleagues showed that acetylcholine is released at parasympathetic nerve endings (as well as the nerve endings of voluntary muscles).

Independently, Loewi used isolated hearts, both connected and unconnected to their normal innervations, to show that stimulation of the nerves produces substances that, transferred to the denervated heart perfusion, could slow or speed the heart rate, depending on whether the sympathetic (accelerating) or parasympathetic (slowing) nerves were operative. Between them, Dale and Loewi established the notion of chemical transmission of nerve impulses across synapses, for which they shared the Nobel Prize for Medicine or Physiology in 1936.

Acetylcholine and noradrenaline were their two basic neurotransmitters, but others have since been discovered, including serotonin and dopamine, as well as natural opium-like substances, called endorphins, which inhibit the action of pain fibres and dampen pain.

See also Regulating the body pages 188–189, Nervous system pages 210–211, Nerve impulses pages 366–367, Nitric oxide pages 496–497

Right A synapse. Nerve impulses pass across the synaptic gap through the release of chemical neurotransmitters from the end of the nerve cell. The neurotransmitters are stored in packets or 'vesicles', seen here as small red circles.

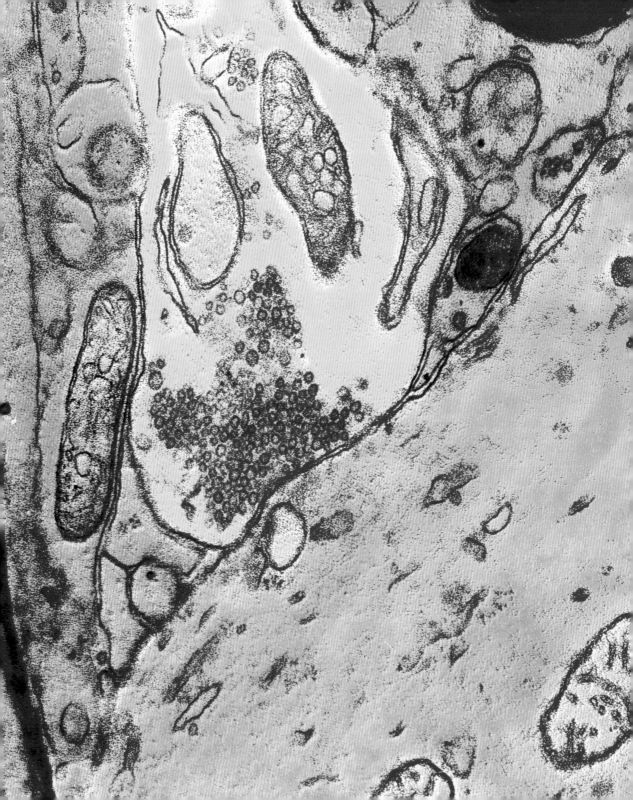

Climate cycles

Milutin **Milankovitch** 1879–1958

Working between 1914 and 1918, while being held prisoner of war in Budapest, the Yugoslav geophysicist Milutin Milankovitch became convinced that the key to past climate change was the way in which the solar radiation reaching the Earth had varied with time and latitude.

There are three main variables. First, the shape of the Earth's orbit stretches from circular to more elliptical and back again, with a periodicity of about 100,000 years. Even though the average distance from the Earth to the Sun is 150 million kilometres, for the most elliptical orbit the distance varies between limits of 140 million and 160 million kilometres every year. As the heat received by the Earth drops off rapidly relative to its distance from the Sun, this variation can produce a profound change. Second, the Earth's spin axis wobbles, or 'precesses', like a tilting spinning top, the pole position making a complete loop every 26,000 years. At the moment the northern winter is milder than it could be simply because the northern polar regions are pointing away from the Sun when the distance from the Earth to the Sun is at its least. Third, the angle between Earth's equatorial and orbital planes varies by a few degrees every 40,000 years. At the smaller value the differences between the seasons is less pronounced than at the larger value.

These three different rhythmic cycles beat together. Milankovitch realized that the resulting changes in sunshine in some geographical zones echoed the climate changes between ice ages and temperate periods. Reduce the average sunshine received by a certain area and more snow accumulates.

In the 1970s this work was continued by Anandu Vernekar of the University of Maryland. Modern knowledge of geomagnetic reversals allows us to date past glaciation periods more accurately. Milankovitch's theory still fits impressively and we are now enjoying a slight global warming as we head inexorably towards the next ice age. Similar cycles affect our neighbouring planet Mars.

See also Celestial predictions pages 26–27, Ice ages pages 152–153, Sunspot cycle pages 158–159, Greenhouse effect pages 184–185, Gaia hypothesis pages 432–433, Ozone hole pages 438–439

Right Chunks of ice fall from a glacier into the ocean. Ice ages end only when all the various climate cycles work together to make summers unusually warm, melting back the ice.

General relativity

Albert Einstein 1879–1955

Having already side-stepped absolute space and time, Einstein took the universe and twisted it into fantastic new shapes. According to Newton's law of gravity, forces work instantly over any distance. But according to special relativity, nothing travels faster than light. Einstein struggled to remove this contradiction. The moment of inspiration came in 1907, as he sat in his Bern patent office. He realized that a falling person does not feel their own weight, so acceleration and gravity must somehow be equivalent. By 1915, this idea had led him to perhaps the most revolutionary theory in the history of physics.

In the world of general relativity, space and time are curved. All objects put a dent in the otherwise flat expanse of spacetime – the Earth, the Sun, even this book. And as they move through this undulating landscape, we see their paths curve. That is why the Earth orbits the Sun. The theory predicts that light must be bent by gravity too. So when Arthur Eddington claimed to see stars shifted out of place by the Sun's gravity in 1919, it established general relativity in the minds of scientists and the imagination of the public.

But there are far more shocking predictions. When enough matter is squeezed very tightly, space is stretched to breaking point. An infinitely deep well appears in the spacetime continuum, and gravity becomes so strong that nothing can escape. This is a black hole. Astronomers now believe that the universe is littered with these monsters and that a giant black hole sits in the centre of our own galaxy. New experiments are looking for other weird effects. Huge underground detectors search for gravitational waves, ripples in spacetime created by catastrophic events such as the formation of black holes. And a spacecraft called Gravity Probe B, soon to be launched, will attempt to see spacetime dragged around by the rotation of the Earth, like syrup by a twisting spoon. General relativity can even describe the shape and evolution of the whole universe. A simple constant in its equations might explain why, according to the latest measurements, the expansion of the universe seems to be accelerating.

Above A cartoon from *The Washington Post* marking Einstein's death. The Earth in this picture carries a sign saying 'Albert Einstein lived here'.

See also Falling objects pages 60–61, Newton's *Principia* pages 78–79, Origin of the Solar System pages 104–105, Non-Euclidean geometry pages 144–145, The quantum pages 234–235, Special relativity pages 244–245, Expanding universe pages 306–307, Gamma-ray bursts pages 434–435, Black-hole evaporation pages 440–441, Superstrings pages 474–475, Great Attractor pages 498–499

Right The gravity of a massive cluster of galaxies (centre) bends light from an even more distant galaxy, dividing it into five separate images (blue). Such 'gravitational lensing' was predicted by Einstein using general relativity.

Our place in the cosmos

Henrietta Swan Leavitt 1868–1921, Harlow Shapley 1885–1972, Walter Baade 1893–1960

The universe of early astronomers, from William Herschel to Jacobus Kapteyn, had the Sun at the centre of a single flattened star system, seen from Earth as a milky band of stars girdling the celestial sphere. But the dimensions of our galaxy remained unknown, and it was an open question whether anything lay outside it apart from the two irregular star collections known as the Magellanic clouds.

Henrietta Swan Leavitt had been studying stars that periodically vary in brightness, particularly the so-called 'Cepheid variables' in the Magellanic clouds. She discovered that the brighter the Cepheid was, the longer it took to vary. As all stars in the Magellanic clouds are roughly the same distance from us, she proposed in 1912 that this relationship – between period and intrinsic brightness – could be calibrated using nearby Cepheids and used to measure distances in space.

Harlow Shapley took up the challenge. The globular clusters – densely packed aggregations of stars – were a prolific source of Cepheids. He measured the distance to the clusters by comparing the calibrated intrinsic brightness of Cepheids, as shown by the period of variation, with their apparent brightness. He concluded, in 1918, that the clusters were physically associated with the Milky Way and were distributed symmetrically in a loose sphere about its centre. As well as hinting at the true enormity of our galaxy, Shapley's findings relegated the Sun to 60,000 light-years away from the centre of our galaxy and two-thirds of the way out towards its edge.

In 1945 Walter Baade took full advantage of the enforced blackouts of the Second World War to discover that there were two populations of stars. Young Population I stars contain many more heavy elements than old Population II stars. Unfortunately, the wrong population of Cepheids was used to calibrate the relationship between period and intrinsic brightness. The correction, which Baade announced in 1952, doubled the size of the universe at a stroke.

Above The globular star cluster M13. It contains around half a million stars.

See also Heavens through a telescope pages 54–55, Spiral galaxies pages 160–161, Expanding universe pages 306–307, Great Attractor pages 498–499

Right The centre of the Milky Way. The plane of the disc of our galaxy is full of dark clouds and gas, which obscure the view, though the path of Halley's comet is clearly visible in this long-exposure photograph.

Neo-Darwinism

Ronald Aylmer Fisher 1890–1962, John Burdon Sanderson Haldane 1892–1964,
Sewall Wright 1889–1988

After Charles Darwin published his theory of evolution by natural selection in 1859,
evolution itself was soon widely accepted, whereas natural selection was almost as widely
rejected. Natural selection seemed to suffer from too many problems, not least that it made
assumptions about how inheritance happens. Biological heredity was still a mystery in 1859,
but with the revival of Mendel's theory of inheritance in 1900, the problems for natural
selection ought to have come to an end. In fact they did not: the early Mendelians were all
fervent anti-Darwinians.

One difficulty was that Mendel's theory seemed to apply only to discrete traits, such as sex,
whereas evolution consists mainly of changes in continuously variable traits, such as height. It
was not until the 1910s that mathematical biologists showed that Mendel's theory could
explain all the facts that had become known, since Francis Galton, about continuous traits.
They were then able to show that natural selection could work well with Mendelian
inheritance. This work was mainly accomplished in the late 1910s and 1920s by the British
biologists R. A. Fisher and J. B. S. Haldane and the American biologist Sewall Wright. They
were so successful that, in retrospect, Mendel's theory can be seen to have saved Darwin's
theory of natural selection. The combination of Mendel's and Darwin's theories is variously
referred to as Neo-Darwinism, the synthetic theory of evolution, and the modern synthesis.

After 1930, the modern synthesis spread into all areas of biology. In 1942, for instance,
Ernst Mayr, an émigré German biologist working in the USA, put forward a theory about the
origin of new species. He suggested that new species originate when a subpopulation of an
ancestral species becomes geographically separated. It will then evolve differently from its
ancestor. Mayr's 'geographic' theory of speciation is now supported by abundant evidence.

See also Acquired characteristics pages 128–129, Darwin's *Origin of Species* pages 176–177, Mendel's laws of inheritance pages 192–193, Measuring variation pages 212–213, Genes in inheritance pages 264–265, Random molecular evolution pages 422–423, Directed mutation pages 494–495

Right Mimicry among butterflies, described by William Bates in 1862. Neo-Darwinian genetic analysis confirmed that the growing resemblance between a poisonous species and an innocuous one arose gradually by natural selection.

Vilhelm Bjerknes was born in Christiania (now Oslo) in Norway in 1862. The son of a mathematician at the local university, Bjerknes was fascinated by the weather and became convinced that it could be accurately described by a mathematical model. He believed that by starting with a good enough picture of the current weather and putting this data into the model, it would be possible to forecast weather patterns. Because of this insight he is often regarded as the father of modern meteorology.

But Bjerknes faced a considerable challenge. Conventional hydrodynamics – the theory of fluids such as gases or liquids – could not describe the behaviour of the weather because it assumes that the density of a fluid (in this case the water-bearing atmosphere or the ocean) depends only on its pressure. In fact it depends on several factors, such as temperature and composition, that can vary from place to place.

In 1904 Bjerknes proposed a new theory in which he combined hydrodynamics with thermodynamics to take account of real variations in pressure. With the resulting mathematical model he founded the discipline that is today called numerical weather prediction. His equations, however, were too difficult to solve by hand, and it was only much later that the computing power became available to do them justice.

After the First World War, in 1920, working with his son Jacob, Bjerknes developed the theory that describes the way in which hot and cold masses of air interact to form cyclones in mid-latitudes on Earth. They used the terms 'cold front' and 'warm front' to describe the boundaries between these masses, borrowing jargon from the battlefront. They were also the first to realize that most weather activity occurs along these boundaries. The theory became known as the 'polar front theory' and forms the foundation of every modern weather forecast.

See also Trade winds pages 86–87, Foucault's pendulum pages 166–167, Greenhouse effect pages 184–185, Chaos theory pages 238–239, Climate cycles pages 276–277, Gaia hypothesis pages 432–433, Ozone hole pages 438–439

Stellar evolution

Arthur Stanley Eddington 1882–1944, Hans Bethe *b*.1906, Carl von Weizsäcker *b*.1912, Ejnar Hertzsprung 1873–1967, Henry Russell 1877–1957

The source of stellar energy was a great mystery at the end of the nineteenth century. Radioactive dating had shown that Earth, and thus the stars, were at least 2,000 million years old. So simple ideas, such as fuelling a star by allowing dust and comets to fall in, burning internal coal or converting potential energy into kinetic energy by shrinking, would not do.

In 1920 Arthur Eddington suggested a different process. He argued that under the conditions of high temperature and high pressure that exist at a star's centre, hydrogen could be slowly transmuted into helium. The Sun was just a hydrogen bomb, with a gravitational lid on; and as around 70 per cent of the Sun is hydrogen, it had sufficient 'fuel' to radiate at its present rate for 10,000 million years.

Because a helium atom is less massive than the four hydrogen atoms involved in its generation, the fusion process would convert the surplus matter to energy (in the form of radiation) in accordance with Einstein's $E = mc^2$. The exact conversion mechanisms were worked out independently by Hans Bethe and Carl von Weizsäcker in 1938, and by the mid-1950s it could be shown how stars manufacture all the massive elements.

The stellar life cycle now became a lot clearer. In the 1910s Ejnar Hertzsprung and Henry Russell had independently plotted graphs effectively showing that a star's intrinsic brightness varies with its surface temperature. These 'H–R diagrams' have become the most useful graphs in astronomy. Stars are not randomly scattered over the graph: most of them lie along a diagonal, the 'main sequence', that goes from cool dim stars to hot bright stars. Accompanying these 'dwarf' stars are a group of 'giants', which are typically 10 to 100 times bigger. Astronomers now know that main-sequence stars are stable because they shine as a result of the same hydrogen-to-helium process. When the hydrogen runs out, the star's core collapses and heats up, and then helium is converted into carbon and other high-mass nuclei. This causes the stars to expand into red giants. Eventually most turn into white dwarfs.

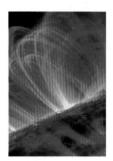

Above The Sun is an ordinary star roughly halfway through its lifetime.

See also Spectral lines 130–131, Special relativity pages 244–245, Rocks of ages pages 252–253, White dwarfs pages 310–311, Power from the nucleus 330–331, Pulsars pages 420–421, Gamma-ray bursts 434–435, Supernova 1987A pages 488–489

Right The Orion Nebula (lower centre) lies 1,500 light-years from Earth and is 2.5 light-years across. It glows owing to ionization of its hydrogen gas by hot young stars embedded in it. Eddington realized that a star's luminosity depended almost exclusively on its mass.

Insulin

Frederick Banting 1891–1941, Charles Best 1899–1978

The symptoms of diabetes were first recorded by Aretaeus in the second century AD, and the sugary taste of the diabetic's urine by Thomas Willis in the seventeenth. Around 1775 Matthew Dobson found sugar in the blood, which suggested that diabetes was not just a kidney problem, as had previously been believed, but that it affected the entire body. In the 1840s Claude Bernard studied digestion and sugar metabolism, foreshadowing the field of endocrinology by showing the influence on these processes of the body's 'internal secretions' – chemical messengers renamed 'hormones' by Ernest Starling in 1905.

Building on Bernard's work, in 1899 Joseph von Mering and Oscar Minkowski removed the pancreas from a dog and within weeks it died of diabetes, thus proving that diabetes was linked to a deficiency in this organ. But attempts at therapy with extracts of pancreatic tissue were unsuccessful. This included the 'islets of Langerhans', a cluster of pancreatic cells that degenerated in diabetics and was therefore assumed to secrete a substance that regulates sugar.

Frederick Banting, a Canadian surgeon, believed that earlier researchers had failed because pancreatic juice had deactivated this secretion. In 1921, working with the American physiologist Charles Best, he tied off a dog's pancreatic duct so that the pancreas atrophied and only the islets were left intact. They then injected an extract from these islets into a diabetic dog close to death. Within a few hours, it was restored to health.

Banting and Best, along with John Macleod, in whose laboratory they were working, turned their attention to producing a clinically safe form of 'isletin' (later called 'insulin'). The biochemist James Collip was recruited to purify insulin from fetal calf pancreases, which consist almost entirely of islet cells. After successful clinical trials, the pharmaceutical company Eli Lilly began large-scale production of insulin in 1923. Since the 1980s human insulin has been produced by genetic engineering. In a monumental oversight, only Banting and Macleod were honoured with the 1923 Nobel Prize for Physiology or Medicine.

See also Regulating the body pages 188–189, Genetic engineering pages 436–437

Right Electron micrograph of a slice through a cell of the insulin-secreting islet of Langerhans in the mammalian pancreas.

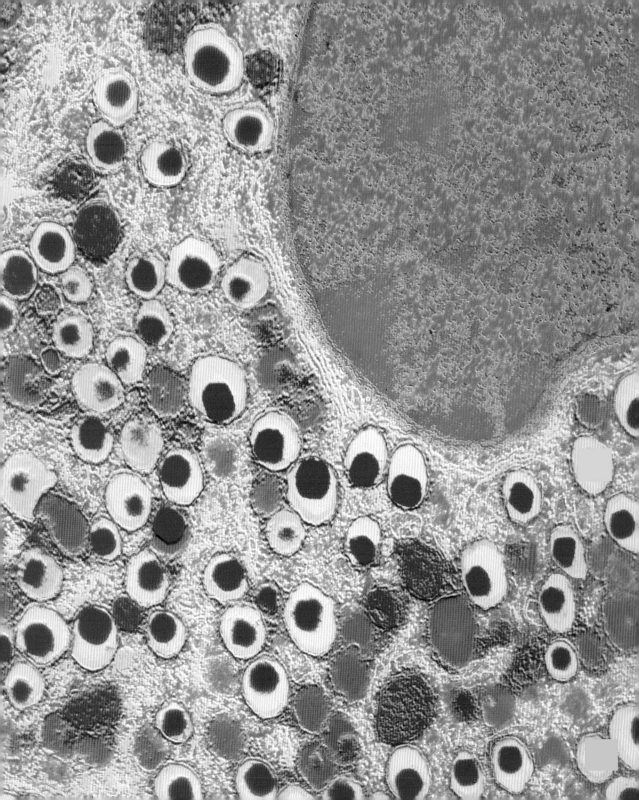

Eel migration

Johannes **Schmidt** 1877–1933

Modern research tells us that animals migrate primarily to exploit seasonal climates for breeding, feeding or hibernation, and many are known to use the Earth's magnetic field as a map. But several aspects of European eel migration still defy scientific explanation.

Aristotle believed that eels arose spontaneously from wet soil, while Pliny suggested that they grew from horsehair. It wasn't until 1893 that the Italian zoologist Giovanni Batista Grassi identified the leptocephalus – a tiny leaf-shaped transparent fish – as the oceanic larval stage of the eel. In 1922 the Danish marine biologist Johannes Schmidt found the eel's breeding ground by mapping the Atlantic distribution of decreasingly small leptocephali until he pinpointed the Sargasso Sea as the hatching site. And so a picture emerged of the life cycle of leptocephali. First they are carried in currents on a year-long journey from the Sargasso to their home rivers, where they become elvers, ascend rivers and spend a decade or more becoming fully grown eels. Then they return to estuaries, their guts wither and they swim to the Sargasso powered by stored fat, where they breed and die. Yet to this day no one has captured an eel on its journey from its home river to the Sargasso Sea, or witnessed its spawning, or identified its navigational system: stars, magnetic fields and ocean scent trails have been suggested but the question remains unanswered.

Why do eels migrate? It was believed that they left home rivers and congregated to interbreed, thus maximizing the genetic variability of the species. But the Canadian biologists Thierry Wirth and Louis Bernatchez have recently shown through molecular genetic analysis that eels from separate rivers remain in separate breeding groups in the Sargasso. How they reach the Sargasso, why they go there and how they behave when they arrive remain enigmas that have spanned the history of science.

See also Natural magnetism pages 50–51, Animal instincts pages 318–319, Honeybee communication pages 338–339

Right The eel *Anguilla anguilla*. In autumn, usually on moonless nights, the maturing eels travel down rivers – and sometimes short distances on land – to the sea.

Crop diversity

Nikolai Ivanovich Vavilov 1887–1943

Above Vavilov began the search for agricultural origins through the study of modern plants.

There is a terrible irony in the fate of Nikolai Vavilov, who devoted his life to the study of crops but died of starvation in Stalin's Saratov prison.

Vavilov undertook over a hundred expeditions to collect crop plants in 64 countries during his tenure as director of the Institute of Applied Botany and New Crops at St Petersburg. He assembled the world's largest collection of seeds, numbering some 200,000 specimens, over 40,000 of them varieties of wheat. In 1923 he set up 115 experimental stations across the Soviet Union to sow the seeds. His concept of crop species 'centres of origin', assuming that they originated in areas where they are most diverse, became important in modern efforts to collect and conserve crops, although these centres of diversity are more likely to reflect human influences than their biogeographical origins.

Loss of wild habitats and traditional cultivated crop plants was recognized as a threat to food security after the green revolution of the 1970s. Vavilov's prescience was reflected in subsequent strategies for the collection of crop seeds, pursued under the auspices of the International Plant Genetic Resources Institute in Rome. The seeds are stored in germplasm banks in or close to Vavilov's centres of diversity.

In the 1930s Vavilov fell foul of Trofim Denisovich Lysenko, a former student whose Lamarckian concepts of the inheritance of acquired characteristics were more in tune with Communist dogma than Vavilov's Mendelian inheritance, in which genes dictate an organism's fate. Lysenko denounced Vavilov to Stalin for anti-Soviet activity, leading to Vavilov's arrest in 1940, then imprisonment and death.

Vavilov, rehabilitated in the 1950s, is commemorated at the Vavilov Institute of Plant Industry in St Petersburg, which houses one of the world's greatest collections of crop plant seeds. Such centres around the globe are vital resources for breeding new crops to feed a rapidly increasing population in a changing environment.

See also Population pressure pages 110–111, Acquired characteristics pages 128–129, Mendel's laws of inheritance pages 192–193, Green revolution pages 428–429, Diversity of life pages 468–469

Right A 1946 Soviet postcard reading 'A good summer will feed us for a year!' Yet when Lysenko's crackpot method of cultivation was imposed on farmers, it led to famine.

ДЕНЬ ЛЕТНИЙ ГОД КОРМИТ!

Child development

Jean Piaget 1896–1980

Jean Piaget was a Swiss psychologist whose ideas greatly influenced our understanding of how children develop. Starting with observations of his own children, he became interested in the kinds of mistakes small children make at different ages, and began to find patterns. This was in a way the opposite of Alfred Binet, who focused on the ages at which children began to be able to do specific tasks correctly, although early in his career Piaget worked in Paris with Binet's collaborator Theodore Simon.

Piaget noticed that children sometimes seem to go backwards – perhaps having correctly said 'I went', they start saying 'I goed'. He interpreted this as the achievement of a new level of understanding: that words are not individual items, but that there are rules governing how they work. 'I went' doesn't follow the standard rule and has to be relearned as an exception. Children also have to learn that, for instance, the quantity of water in a glass does not change if it is poured into a glass of a different shape. That many adults prefer a tall thin glass of beer to a short fat glass suggests we do not always learn this adequately.

Above Charismatic and friendly, Piaget engendered intellectual disagreements.

Children develop their intelligence by developing 'schemata', structures or rule systems that enable them to deal more and more effectively with the world. They move from a view of the world entirely centred on their own sensory-motor experience, through various stages of 'accommodation' and 'assimilation', to a view that recognizes the independent existence of the rest of the world working according to objective principles.

Piaget's writing is dense and philosophical. The influence of his classic studies such as *Language and Thought of the Child* (1924) only really began to be felt in Anglophone countries in the late 1950s and 1960s. Although criticized for basing sweeping conclusions on the observation of small numbers of children, his insights had a major impact on educational thinking and practice.

See also The unconscious mind pages 222–223, Intelligence testing pages 240–241, Language instinct pages 386–389

Right Kids' stuff: a chimpanzee reputed to have the mental abilities of a five-year-old child pays a visit to a New York hospital for sick children in 1925.

Words and rules

Steven Pinker

GRAMMATICAL ERRORS LIKE *bleeded* and *singed* have long epitomized the innocence and freshness of children's minds. The errors are acts of creation, in which children lift a pattern from their brief experience and apply it with impeccable logic to new words, unaware that the adult world treats them as arbitrary exceptions. In *A Dark-Adapted Eye*, the novelist Barbara Vine introduces an unlikable child by remarking, 'He would refer to "adults" instead of "grownups" and get all his past tenses right, never saying "rided" for "rode" or "eated" for "ate".'

Children's errors with irregular verbs also have been prominent in debates on the nature of language and mind. The neurologist Eric Lenneberg pointed to the errors when he and Noam Chomsky first argued that language was innate; the psychologists David Rumelhart and James McClelland set them as a benchmark when they first argued that language could be acquired by generic neural networks. Psychology textbooks cite the errors to rhapsodize that children are lovers of cognitive tidiness and simplicity; researchers who study learning in adults cite the errors as a paradigm case of the human habit of overgeneralizing rules to exceptional cases.

Nothing is more important to the theory of words and rules than an explanation of how children acquire rules and apply them – indeed overapply them – to words. The simplicity of these errors is deceptive. It is not easy to explain why children start making them, and it's harder to explain why they stop.

Overgeneralization errors are a symptom of the open-ended productivity of language, which children indulge in as soon as they begin to put words together. At around eighteen months children start to utter two-word microsentences like 'See baby' and 'More cereal'. Some are simply telegraphic renditions of their parents' speech, but many are original productions. 'More outside!' says a tot who wants to play in the park. 'Allgone sticky!' says another after his mother has washed jam off his fingers. One of my favourites is 'Circle toast!' shouted repeatedly to parents who couldn't figure out that the child wanted a bagel. By their twos, children produce longer and more complicated sentences, and begin to supply grammatical morphemes such as *-ing*, *-ed*, *-s* and the auxiliaries. Sometime between the end of the second year and the end of the third year, children begin to overgeneralize *-ed* to irregular verbs. All children do it, though parents don't always notice it …

Jennifer Ganger and I suspected that at least some of the timing of language development, including the past-tense rule, is controlled by a maturational clock. Children may begin to acquire a rule at a certain age for the same reason they grow hair or teeth or breasts at certain ages. If the clock is partly under the control of the genes, then identical twins should develop language in tighter synchrony than fraternal twins, who share only half their genes. We have enlisted the help of hundreds of mothers of twins who send us daily lists of their children's new words and word combinations. The checklists show that vocabulary growth, the first word combinations and the rate of making past-tense errors are all in tighter lockstep in identical twins than in fraternal twins. The results tell us that at least some of the mental events that make a child say *singed* are heritable. The very first past-tense error, though, is not. When one twin makes an error like *singed* for the first time, an identical twin is no quicker to follow suit than a fraternal twin. These gaps – an average of 34 days between the first past-tense errors of two children with the same genes exposed to the same speech – are a reminder of the importance of sheer chance in children's development …

Children's speech errors, which make such engaging anecdotes in poetry, novels, television features and websites for parents, may help us untangle one of the thickest knots in science, nature and nurture. When a child says *It bleeded* and *It singed*, the fingerprints of learning are all over the sentence. Every bit of every word has been learned, including the past tense suffix *-ed*. The very existence of the error comes from a process of learning that is still incomplete: mastery of the irregular forms *bled* and *sang*.

But learning is impossible without innately organized circuitry to do the learning, and these errors give us hints of how it works. Children are born to attend to minor differences in the pronunciation of words, such as *walk* and *walked*. They seek a systematic basis for the difference in the meaning or form of the sentence, rather than dismissing it as haphazard variation in speech styles. They dichotomize time into past and nonpast, and correlate half the time line with the evanescent word ending. They must have a built-in tendency to block the rule when a competing form is found in memory, because there is no way they could learn the blocking principle in the absence of usable feedback from their parents. Their use of the rule (though perhaps not the moment when they first use it) is partly guided by their genes. They spontaneously deploy their new rule to a wide range of words coined by an experimenter or by themselves, and to verbs whose irregular forms are too faint to retrieve. Children fit the rule into its proper place in the logic of their grammatical system, keeping regular forms out of certain word structures and irregular forms out of others.

I suspect that in other parts of our psychology the interaction of nature and nurture has a similar flavour: every bit of content is learned, but the system doing the learning works by a logic innately specified. Charles Darwin captured the interaction when he called human language 'an instinctive tendency to acquire an art.' 'It certainly is not a true instinct,' he noted, 'for every language has to be learned. It differs, however, widely from all ordinary arts, for man has an instinctive tendency to speak, as we see in the babble of our young children; while no child has an instinctive tendency to brew, bake or write.'

Taung child

Raymond Arthur Dart 1893–1988

Despite Darwin's prediction that human ancestry would be traced back to Africa, it was Ernst Haeckel's alternative Asian model that had prevailed ever since Eugene Dubois found 'Java man' in the 1890s. But in 1925 Raymond Dart, an Australian professor of anatomy in South Africa, described a fossil skull destined to vindicate Darwin.

After service as a medic in the First World War, Dart became an anatomist at University College London but soon left to take up a newly established chair at Witwatersrand University in South Africa. Without specimens for teaching, he offered small cash prizes to students who could supply interesting bones. In 1924 Josephine Salomons brought in a fossil baboon skull. Intrigued, Dart requested that any other fossils from the same source, a limestone quarry at Taung in Botswana, should be sent to him.

Before long a box arrived just as Dart was dressing for a wedding. Unable to resist the temptation, he opened it and was astonished to find a natural stone cast of a brain plus facial bones, teeth and jaws. In 1925 Dart published his description of the tiny southern ape, *Australopithecus africanus*. He claimed the find – with its vertical face and small teeth – as the 'missing link' between apes and humans. Initially the reception was enthusiastic, but soon Arthur Keith, promoter of 'Piltdown man' as the missing link, was arguing that Dart's specimen was just a young ape.

Hoping to muster support in London, Dart arrived in 1930, only to be completely overshadowed by Davidson Black and his spectacular 'Peking man'. Discouraged, Dart abandoned work on the Taung child for years. Only Robert Broom, a Scottish doctor working in South Africa, believed Dart. Eventually, in 1936, Broom did find more australopithecine fossils at Sterkfontein. Even so, it was not until the 1950s and the discoveries of the Leakeys that Dart was finally vindicated.

See also Prehistoric humans pages 148–149, Neanderthal man pages 170–171, Java man pages 216–217, Olduvai Gorge pages 392–393, Ancient DNA pages 476–477, Nariokotome boy pages 480–481, Out of Africa pages 492–493, Iceman pages 504–505

Right *Australopithecus africanus* was an upright, bipedal, walking 'man-ape'. Many experts believe that australopithecines were the ancestors of modern humans.

Wave-particle duality

Werner Karl Heisenberg 1901–76, Erwin Schrödinger 1887–1961,
Louis-Victor de Broglie 1892–1987

Quantum mechanics unhinged the predictable clockwork universe, and replaced it with more subtle workings. Why? We know that light acts like a wave: it produces interference patterns, just like ripples on water. But Max Planck and Albert Einstein had established that light comes in pieces, particles we call photons. How could it be a wave sometimes and a particle at others? This and other inconsistencies of the early quantum theory led physicists to search for a fuller description of the microscopic world. Two of them succeeded.

Werner Heisenberg abandoned attempts to visualize what was going on. Instead, in 1925, he devised mathematical formulae that related observable things to each other. The next year Erwin Schrödinger took a different tack. He used the idea, suggested by Louis de Broglie, that particles of matter, such as electrons, are also waves. Schrödinger's wave equation describes how these 'matter waves' move.

The two theories turned out to be different ways of talking about the same strange microworld. According to quantum mechanics, everything is both wave and particle, and neither. A quantum entity can spread out into a blur of uncertainty, be in two places at once, even form interference patterns with itself. Only when you look to see where it is does it acquire a specific location, choosing at random from the possibilities available. So some effects have no causes; the decay of an unstable nucleus can happen at any time – we can know only the probability of its happening.

This may be an unpalatable view of the world, but it works. It explains the mundane properties of atoms, nuclei and molecules; it is behind the strange phenomenon of superconductivity, Bose–Einstein condensation, white dwarfs and neutron stars. One day soon we may use quantum computers, whose 'bits' can be both 0 and 1 at the same time, to perform calculations at unprecedented speed.

Right With the use of the scanning tunnelling microscope and the trick of quantum uncertainty, physicists can measure the location of individual atoms and recreate the three-dimensional surface topography of chemical elements and compounds.

Penicillin

Alexander Fleming 1881–1955

'Chance only favours the prepared mind.' Louis Pasteur's phrase aptly describes Alexander Fleming's accidental discovery of the antibiotic penicillin. His mind was certainly prepared, for in 1921 he had found a similar bacteria-dissolving substance in a culture of nasal mucus. The substance, which he called 'lysozyme', could not however be purified for clinical use, nor was it particularly effective against bacteria responsible for disease.

Seven years later, in 1928, Fleming noticed that staphylococcus bacilli would not grow on a culture medium accidentally contaminated with a mould, *Penicillium notatum* – colonies of bacteria around the mould growths were transparent and watery. Evidently the mould produced something poisonous to bacteria; Fleming called that something penicillin. A dilute broth of the mould prohibited the growth of a range of bacteria – staphylococci, streptococci and pneumococci – and did not appear to harm healthy tissues or interfere with the defensive role of white blood cells. But although it seemed strong and safe, penicillin was hard to produce and unstable, and in Fleming's hands remained just a laboratory aid.

The Australian pathologist Howard Florey and the German refugee biochemist Ernst Chain transformed penicillin into an antibiotic. Like Fleming, Florey was interested in lysozyme. He established that it was an enzyme that dissolved the sugar chains in bacterial cell walls, and decided to study all known substances produced by one mould or bacterium to destroy others. In 1940 Florey and Chain, assisted by the British chemist Norman Heatley, had produced enough concentrated penicillin (about 100 milligrams) for animal tests, and by 1941 they had completed the first trials on humans. The results were striking. When the USA entered the Second World War in late 1941, they also entered the race to mass-produce penicillin. Soon, new fermentation methods launched the first wonder drug onto the market. Fleming, Florey and Chain shared the 1945 Nobel Prize for Physiology or Medicine.

See also Microscopic life pages 76–77, Cholera and the pump pages 168–169, Germ theory pages 202–203, A magic bullet pages 262–263, Genes in bacteria pages 332–333, Directed mutation pages 494–495

Right A chance discovery: photograph of the original culture plate on which Fleming grew *Penicillium notatum*.

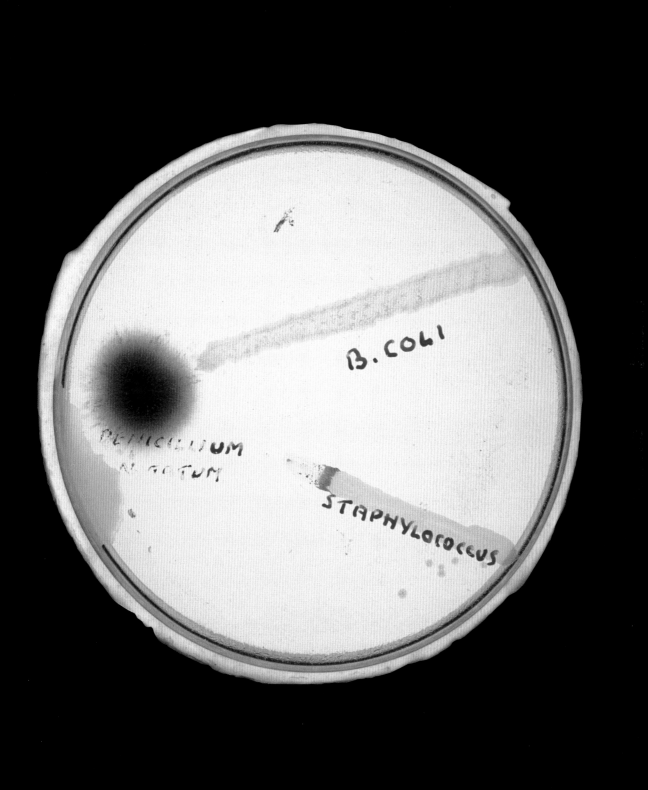

Geomagnetic reversals

Motonori **Matuyama** 1884–1958

Natural magnetism in the form of lodestone (the iron mineral magnetite) was known to the Greeks, and by 300 BC the Chinese had described lodestone as the 'south-pointer', the first compass. They found that a lodestone spoon handle, finely balanced on a polished plate, always points south, and by the beginning of the first millennium they had devised the first navigational compass.

But an understanding of the nature of the Earth's magnetic field had to wait until the seventeenth century. In his 1600 treatise *De Magnete*, William Gilbert, court physician to Queen Elizabeth I, suggested that Earth itself was a giant spherical magnet and that the compass needle points not to the heavens but to the magnetic poles of the planet. By 1635 it was known that this magnetic field is not constant across the globe, but rather it varies in strength and direction.

Early in the twentieth century it was found that many igneous rocks, when they cool and crystallize, become magnetized parallel to Earth's magnetic field. Rock magnetization was first measured by Bernard Brunhes in 1906. But not until 1929 did the Japanese geologist Motonori Matuyama show that the magnetic field has reversed polarity over the past two million years. While studying the remnant magnetization of basalt, which preserves the field especially well, he measured switches in polarity through a stratified succession of lavas. In other words, magnetic north had become magnetic south, and vice versa. This finding was later to play a crucial part in the development of the theory of plate tectonics.

In the 1960s geologists measured more than 20 reversals during the past five million years. The source of the magnetic field and its reversals is not yet fully understood, but probably has something to do with the way the electrically conductive material of the liquid outer core swirls around deep inside the Earth.

Above Many igneous rocks, when they cool and crystallize, are magnetized by Earth's magnetic field. Basalt, shown here, preserves the field especially well.

See also Natural magnetism pages 50–51, Humboldt's voyage pages 114–115, Inside the Earth pages 250–251, Rocks of ages pages 252–253, Climate cycles pages 276–277, Plate tectonics pages 414–415

Right Giant lodestone: by the mid-19th century the Earth's magnetism had been mapped over a good fraction of its surface.

MAGNÉTISME.

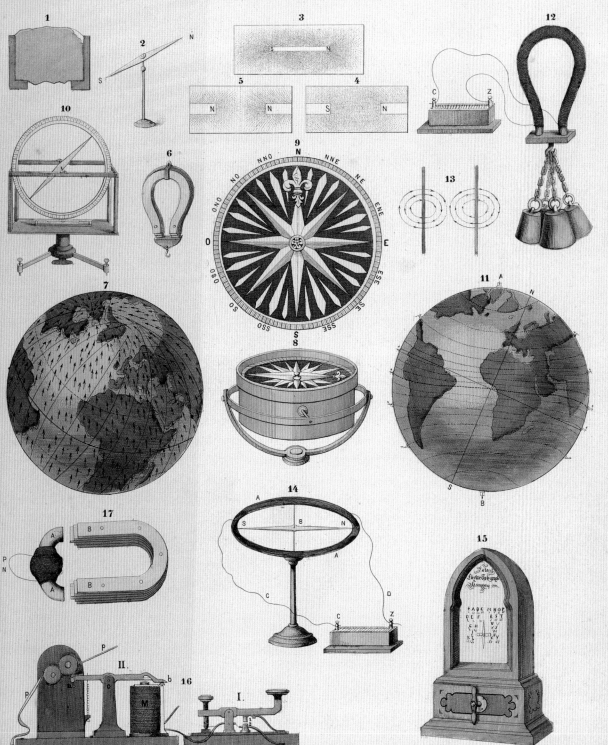

Expanding universe

Edwin Powell Hubble 1889–1953

The American astronomer Edwin Hubble benefited greatly from the completion of the 100-inch (2.5-metre) Hooker telescope at the Mount Wilson Observatory, Pasadena, just after the First World War. Using this marvellous reflector (the largest of its day), he confirmed the 1914 spectroscopic data of Vesto Melvin Slipher, which had suggested that many of the fuzzy patches of light known as nebulae were actually distant galaxies like our own Milky Way. Hubble went on to classify the galaxies as normal spirals, barred spirals, ellipticals and irregulars. Although he originally believed that his scheme represented an evolutionary sequence, he later doubted that this was the case. We now know it is not.

Slipher had spent hundreds of hours measuring spectra of light from faint distant spiral nebulae. The 'Doppler shift' of the spectral lines towards the red end of the spectrum indicated that almost all the nebulae were moving away from us, and by 1925 he had used this 'redshift' to work out 44 radial velocities. As the greatest velocities were more than 1,000 kilometres per second, Slipher knew that the nebulae were beyond the Milky Way. Hubble concentrated on estimating their distance. By 1929 he and his colleague Milton Humason had collected data for 49 spiral nebulae (now being called galaxies). Hubble was astonished to find that the more distant a galaxy was, the greater its redshift (that is, the faster it was moving away). This relationship implied that the universe was expanding and had a definite origin at a certain moment in time; these were observations that matched the predictions of Einstein's theory of general relativity, as developed by the cosmologists Georges Lemaître and Aleksandr Friedmann.

The gradient of the velocity–distance relation became known as Hubble's constant: the rate of expansion of the universe. Its inverse is a measure of the age of the universe: the time elapsed since the big bang. Unfortunately, Hubble's initial value had the age of universe at a few billion years, younger than the Earth itself. Galactic distances have since been re-estimated and the discrepancy has gone. The consensus is that the universe is 13 billion years old.

See also Spectral lines pages 130–131, Doppler effect pages 156–157, Spiral galaxies pages 160–161, Rocks of ages pages 252–253, Our place in the cosmos pages 280–281, Afterglow of creation pages 412–413, Unified forces pages 416–417, Great Attractor pages 498–499

Right Eye on the universe: Hubble looks through the Newtonian focus of the 100-inch telescope at Mount Wilson, c.1922. A Rhodes scholar at Oxford, Hubble gave up a dazzling legal career to study astronomy.

Limits of mathematics

Kurt Gödel 1906–78

Mathematics has always been seen as the most precise and logical of human endeavours. So why not attempt to place the whole of mathematics on a formal logical basis, making it fully rigorous? At the beginning of the twentieth century mathematicians tried to apply symbolic logic to the most fundamental mathematical system – arithmetic – and thus ground all other branches of mathematics, including the very concept of number. The most spectacular attempt was the monumental *Principia Mathematica* (1910–13) by Bertrand Russell and Alfred North Whitehead. And in 1900 David Hilbert had expressed the hope that someone would prove that arithmetic is also complete and self-consistent – that any mathematical proposition could be proved unambiguously true or false.

In 1931 the Austrian mathematician Kurt Gödel dealt a body-blow to both ambitions. Gödel was a professor at the University of Vienna until 1938 when, just having married, he emigrated to the USA (via Russia and Japan) and joined the Institute for Advanced Study at Princeton. His 1931 paper proved two classic results, now known as the 'incompleteness theorems'. The first theorem shows that any axiomatic system – even one as basic as arithmetic – contains propositions that the system itself cannot decide as being either true or false. Such propositions are analogous to the statement 'This sentence is false', the truth or falsehood of which cannot be established. The second theorem shows that any logical system such as arithmetic is also incomplete. It cannot prove its own internal self-consistency without assistance from outside.

Yet incompleteness did not of course mean that mathematics was invalidated. With the advent of computers – essentially arithmetical machines – mathematicians turned their attention to the more practical search for what could be computed rather than what could be philosophically decided. But as long as computers can't answer every mathematical question, mathematics will remain an essentially creative human activity.

See also Euclid's *Elements* pages 20–21, The computer pages 340–341, Four-colour map theorem pages 450–451

Above Convinced that his doctors were poisoning him, Gödel died from 'malnutrition and inanition caused by personality disturbance.'

Right Shaky foundations: an illustration of the relationship between the various branches of mathematics, taken from the 1543 edition of Luca Pacioli's *Summa Arithmetica*.

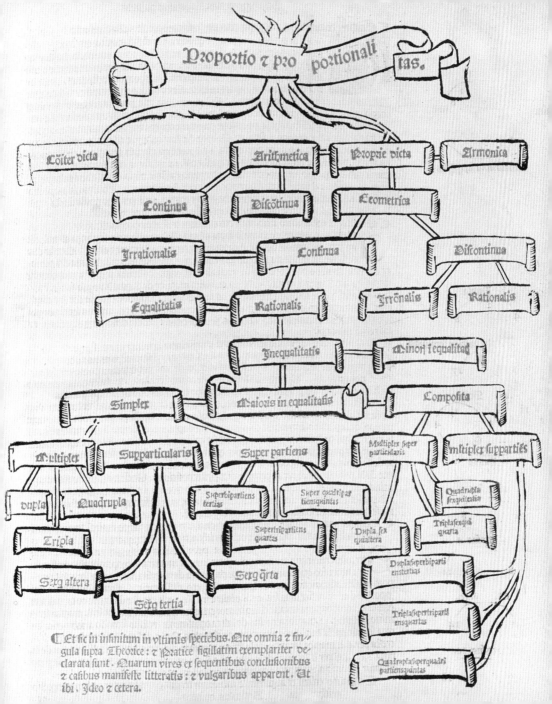

Proportio ⁊ proportionalitas.

Cõiter dicta

Arithmetica — Proprie dicta — Armonica

Continua — Discõtinua — Geometrica

Irrationalis — Continua — Discontinua

Equalitatis — Rationalis — Irrõnalis — Rationalis

Inequalitatis — Minori i equalitas

Simplex — Maioris in equalitatis — Composita

Multiplex — Supparticularis — Super partiens — Multiplex super particularis — mltiplex suppartiẽs

Dupla — Quadrupla — Superbipartiens tertias — Super quadrupartiensquintas — Quadrupla sexquintia

Tripla — Supertripartiens quartas — Dupla sex qualtera — Triplasexquid quarta

Sexg altera — Sexg qurta — Duplasuperbipartiẽstertias

Sexgo tertia — Triplasupertripartiẽsquartas

Quadruplasuperquadripartiensquintas

⫶ Et sic in infinitum in vltimis speciebus. Que omnia ⁊ singula supra Theorice ⁊ Practice sigillatim exemplariter declarata sunt. Quarum vires ex sequentibus conclusionibus ⁊ casibus manifeste litteratis ⁊ vulgaribus apparent. Ut ibi. Ideo ⁊ cetera.

White dwarfs

Subrahmanyan Chandrasekhar 1910–95

In 1844 Sirius, the brightest star in the northern sky, was found to wobble from side to side as it moved across the sky. It was being pulled by the gravitational force of a companion star too dim to be seen. Judging by its orbit, the companion ('Sirius B', Sirius being the 'A' star) had to be of about the same mass as the Sun, and it was assumed that it must be roughly the same size as the Sun too, though fainter and cooler. But in 1915 Walter Sydney Adams obtained a spectrum showing Sirius B to be as hot as Sirius A and hotter than the Sun. Coupled with its faintness, this meant it had to be a tiny star, closer to the size of the Earth than of the Sun.

Such 'white dwarfs' form from the collapse of stars such as the Sun at the end of their lives, when they are no longer supported by nuclear fusion. Arthur Stanley Eddington, in the 1920s, calculated that their density is more than 100,000 times that of water. In this strange physical state the atoms are so closely packed that all their electrons are stripped off. Eventually quantum effects prevent these 'degenerate' electrons from being squeezed any further, leading to an outward pressure that stabilizes the star.

But sometimes this pressure cannot prevent further gravitational collapse. The brilliant Indian astrophysicist Subrahmanyan Chandrasekhar was fascinated by the fact that more massive white dwarfs are smaller than less massive ones. In 1931 he determined that a star of more than 1.44 solar masses cannot become stable. Either it blows away the excess mass from its surface in a supernova explosion, or the electrons are captured by protons and produce neutrons and neutrinos. A star of between 1.44 and 3.2 solar masses can form a stable neutron star. Beyond 3.2 solar masses the star continues to collapse, forming a black hole.

ove The Helix bula shows the ults of the collision wo gases near a ng star. The gaseous pole-like objects are bbed 'cometary ots'.

See also Stellar evolution pages 286–287, Pulsars pages 420–421, Gamma-ray bursts pages 434–435, Black-hole evaporation pages 440–441, Supernova 1987A pages 488–489

Right The 'hourglass'-shaped nebula MyCn18 provides a majestic eerie view of a planetary nebula, the glowing remains of a dying

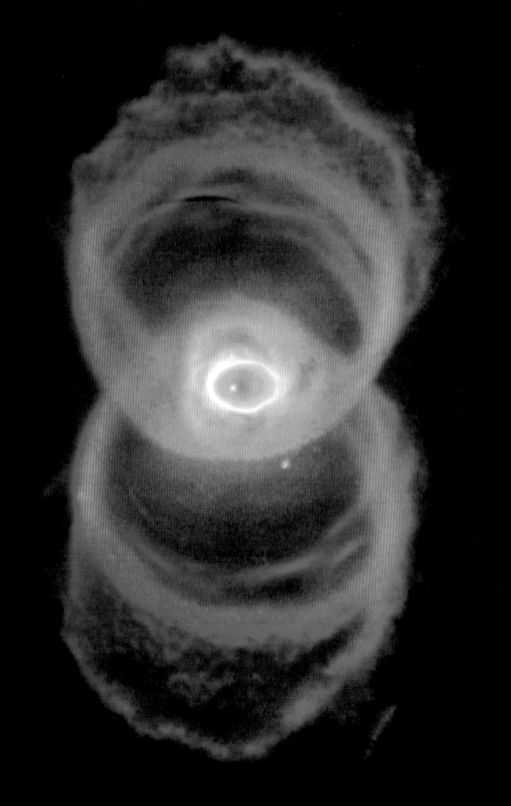

The neutron

James Chadwick 1891–1974

In the 1920s physicists thought that everything was made of just two components: electrons and protons. The prevailing theory was that, in each atom, lightweight negatively charged electrons whizzed around a tiny dense nucleus that held heavy positive protons and some more electrons.

Then, in the early 1930s, came a surprise. Physicists found that alpha-particle radiation could induce samples of the light element beryllium to give off some other form of radiation – one exceptionally good at knocking protons out of other elements. In 1932 the English physicist James Chadwick, working at Cambridge, repeated these experiments and found that he could explain the effects if the alpha particles were knocking other particles – each about as heavy as a proton, but with no electric charge – out of the beryllium nuclei. These neutral particles could in turn knock protons out of other elements.

For a while Chadwick thought that his 'neutron' was not a fundamental particle, but a tightly bound electron and proton. But by 1934 measurements showed that the neutron was slightly too heavy for that. Physicists had to live with a new basic ingredient of matter. Atomic nuclei are made not of protons and electrons, but protons and neutrons. The various isotopes (or versions) of a particular element, which have the same chemistry but different weights, all contain the same number of protons but different numbers of neutrons.

This discovery helped to drive the furiously rapid advances in nuclear physics of the 1930s. The neutron is the key to the nuclear chain reactions that drive power stations and explode atomic bombs: neutrons fly out like shrapnel from each nucleus when it splits, hitting other nuclei and causing them to break up too. They also have less violent uses now: as probes of the structure of matter, undeflected by the charges around atoms because of their electrical neutrality.

See also Atomic theory pages 124–125, Periodic table pages 196–197, Radioactivity pages 224–225, The electron pages 228–229, The quantum pages 234–235, Model of the atom pages 272–273, Power from the nucleus pages 330–331, Quarks pages 408–409, Unified forces pages 416–417, Superstrings pages 474–475

Right Fundamental force: Rutherford (with cigarette) in the Cavendish Laboratory, Cambridge. The door leads to the room in which Chadwick discovered the neutron.

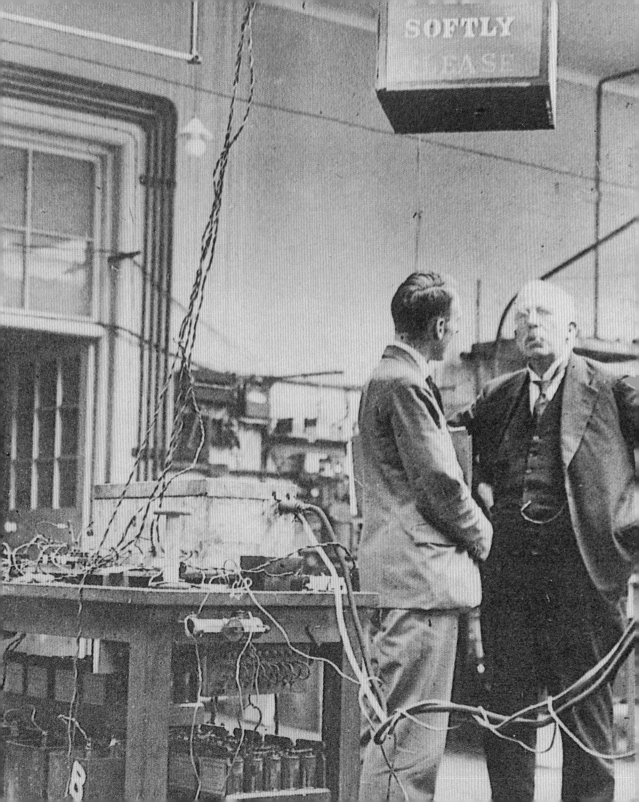

Antimatter

Paul Adrien Maurice Dirac 1902–84, Carl David Anderson 1905–91

Paul Dirac invented antimatter. In 1928 he was searching for a new version of quantum mechanics, because Schrödinger's wave equation contradicts Einstein's theory of special relativity. Dirac hit upon a more complicated equation, which fitted special relativity – but it also implied other things about the electron.

For one thing, it required these particles to have an inherent rotation. Fortunately, the electron's spin had been discovered just two years earlier. The other implication was more troubling. Dirac interpreted his equations to mean that there must be another kind of electron, a particle with positive instead of negative electric charge. But then in 1932 Carl Anderson discovered this particle appearing in cosmic-ray collisions. He called it a positron.

Dirac thought that protons should also have antimatter versions, and eventually the antiproton was discovered too. In fact, most particles turn out to have antimatter counterparts. Dirac even speculated that there might be whole stars and solar systems made of antimatter. But he was probably wrong. When a particle meets its antiparticle, they annihilate each other and create a burst of radiation. Astronomers would see this tell-tale radiation being produced at the boundary between matter and antimatter regions. So, for some reason, the universe is mostly matter.

But there is some antimatter on Earth. Positrons are emitted by some kinds of radioactive atom; the distinctive radiation produced when they annihilate is used by doctors in positron emission tomography, or PET scanning. And physicists at CERN, the European Particle Physics Laboratory in Geneva, are making a slightly more substantial form of antimatter, combining positrons and antiprotons to form a few atoms of anti-hydrogen gas. Anti-hydrogen should look just the same as hydrogen. If it turns out not to, we might need radically new physics to explain why.

See also Radioactivity pages 224–225, Special relativity pages 244–245, Wave–particle duality pages 300–301, Left-handed universe pages 380–381, Quarks pages 408–409, Gamma-ray bursts pages 434–435, Black-hole evaporation pages 440–441, Superstrings pages 474–475, Images of the mind pages 478–479

Right Invisible gamma-ray photons produce pairs of electrons (green) and positrons (red) in a bubble chamber.

Nylon

Wallace Hume Carothers 1896–1937

The chemist Wallace Carothers joined the DuPont Company in Wilmington, Delaware, USA, in 1928 and was put in charge of a group researching new polymers. These are substances consisting of tens of thousands of repeating molecular units. Organic polymers include cellulose, rubber and wool, and in 1931 Carothers came up with neoprene, a synthetic rubber, which he made by polymerizing a simple hydrocarbon – in other words, by making the simple hydrocarbon molecules link together to form long chains (up to a million atoms long).

Carothers also believed that molecules of different kinds could be joined end-to-end to form polymers. In 1934 he turned his attention to those that would produce the kind of chemical bond essential for some natural fibres such as silk, a polymer with amide bonds. The reaction of an amine and acid will produce an amide, and he showed that a good polymer could be made from molecules with chains of six carbon atoms. The amine version, with an amine group at each end of the chain, would react with an acid version, with an acid group at each end. When these chemicals were mixed they reacted immediately and the result was 6,6'-polyamide, a remarkably strong polymer that could be drawn into fine silk-like threads.

Commercial production of the product, named Nylon 66, started in 1939. The name is not derived from the names of New York and London, as is popularly supposed. The company wanted to call the new material Nulon or Nilon, but these were registered trade names, so they eventually settled on Nylon. Nylon first came to the public's notice in the form of glamorous stockings, which were exhibited at the San Francisco International Exposition and the New York World Fair, both held in 1939. Sadly, by then Carothers was dead, having committed suicide two years previously, a victim of clinical depression.

Although nylon stockings were soon on sale in the shops, their availability was short-lived; all nylon production was commandeered for making parachutes in the Second World War.

See also Mauve dye pages 172–173, Benzene ring pages 190–191

Right The coming of nylon not only boosted the stocking industry, but also ushered in a new era of synthetic fibres, with chemists learning how to make 'designer' polymers.

Animal instincts

Karl von Frisch 1886–1982, Konrad Zacharias Lorenz 1903–89,
Niko Tinbergen 1907–88

The study of animal behaviour was a late addition to biology. The anatomy of whole bodies had been studied for centuries, and micro-anatomy and molecular mechanisms were studied as appropriate technologies became available. But two problems had to be overcome before behaviour could be studied scientifically. Anthropomorphism was one: explanations of behaviour by motives that seem only strictly applicable to humans (for instance: 'Why are you going this way?' Answer: 'Because I want to go home'). Internal motives in animals are unobservable and difficult or even impossible to study scientifically.

The other problem was that behaviour is nebulous; it is difficult to define in the same way that we define legs, hands or eyes. In the 1930s three scientists, two of them Austrian (Karl von Frisch and Konrad Lorenz) and one of them Dutch (Niko Tinbergen), independently started to study animal behaviour while avoiding anthropomorphism, and by exploiting the same techniques of objective observation and experiment that are used in the rest of science.

Karl von Frisch's great discovery was the dance language of honeybees. Bees communicate to hive-mates the direction of, distance to and quality of food supplies. They do so by a special 'waggle' dance, which von Frisch decoded by careful observation and experiments in which he varied the location of a food source.

Lorenz and Tinbergen are not identified with single great discoveries; rather, each conducted a series of exemplary studies. Lorenz worked mainly with tame animals, observed close up. He studied 'imprinting', in which a young animal learns to follow a certain object – usually its parent, but it could be Lorenz if he intruded himself at the right time. Tinbergen worked primarily on non-domesticated animals in nature. He was the master of the non-intrusive experiment that revealed the mechanisms underlying behaviour. Famously, he used model gull bills to identify the stimuli that chicks use when begging food from their parents.

See also Conditioned reflexes pages 242–243, Behavioural reinforcement pages 322–323, Game theory pages 336–337, Honeybee communication pages 338–339, Chimpanzee culture pages 396–397, Evolution of cooperation pages 406–407, Memory molecules pages 470–471

Right Acting the mother: Lorenz found that newly hatched goslings tend to follow any moving object, adopting it as their mother and remaining emotionally attached to it, even if the object was himself.

Citric-acid cycle

Hans Adolf **Krebs** 1900–81

It had long been known that animals 'burn' foodstuffs to provide energy. In the second half of the eighteenth century Antoine Lavoisier had provided a convincing analysis of the animal production of carbon dioxide and water during 'respiration'. In the nineteenth century the German chemist Justus von Liebig had measured the input of fats, sugars and carbohydrates, and examined the output in terms of water, carbon dioxide and urea. But what happened between input and output had been of little concern to him.

Von Liebig's position was consolidated into the belief that plants synthesized organic molecules, while animals merely broke them down. Gradually, however, scientists came to realize that animals can synthesize complex molecules, and that what happened in the organism was much more than mere input and output. These issues of 'intermediary metabolism' exercised many biochemists in the late nineteenth century, including Otto Warburg, one of Hans Krebs's teachers. Warburg's manometer was a simple piece of laboratory equipment that could collect small amounts of waste products given off by tissue slices subjected to varying conditions and chemicals.

Using the Warburg manometer, Krebs investigated many of these metabolic pathways, first in Germany and, after the Nazi purges, in England. He was first to describe the 'ornithine' cycle, which is involved in the production of the waste product urea. He then turned his attention to carbohydrate metabolism. From the late 1930s, a series of careful experiments demonstrated the key role played by citric acid and its breakdown. Further work revealed this to be a common metabolic pathway for many complex molecules metabolized by animals. The citric-acid cycle (also called the 'Krebs cycle') could be reversible, and was thus an important key to understanding both anabolic and catabolic reactions in living organisms. Krebs shared the 1953 Nobel Prize for Physiology or Medicine with Fritz Lipmann, who had discovered one of the fundamental enzymes involved in the process.

See also Combustion pages 94–95, Hydrogen and water pages 98–99, Synthesis of urea pages 142–143, Regulating the body pages 188–189, Nitrogen fixation pages 208–209

Right Energy conversion: Roger Bannister breaks the four-minute mile in 1954. The previous year Krebs won a Nobel Prize for his work on how living creatures obtain energy from food.

Behavioural reinforcement

Burrhus Frederic Skinner 1904–90

B. F. Skinner promoted the idea that much of our behaviour is 'shaped' and 'reinforced' according to strict laws. He was a rigorous behaviourist: he did not deal with the psychological, physiological and neurological bases of behaviour, just the relations between an individual's actions and their consequences. His experimental work mainly used rats and pigeons – he famously proved that pigeons could be trained to act as bomb-aimers. And he invented the 'Skinner box', an enclosure in which a rat or pigeon had keys to peck or levers to press, lights that could be switched on or off in various combinations, and a mechanism for delivering morsels of food or an electric shock.

He showed that an animal could learn to behave in particular ways by selectively rewarding and punishing (respectively, positive and negative 'reinforcement') approximations to the desired behaviour, and worked out complex and sometimes counter-intuitive laws relating reinforcement to behaviour. For instance, behaviour established with only partial reinforcement (that is, only rewarded some of the time) tends to take longer to disappear when the reward is withdrawn than behaviour established with constant reinforcement – the 'partial reinforcement extinction effect'.

Skinner's method of 'operant conditioning', which he introduced in his 1938 book *The Behaviour of Organisms,* differs from Pavlov's so-called 'classical conditioning'. In the first, an organism acquires new behaviour by selective reward and punishment, whereas in the second, an organism responds in an already established way to a new stimulus. Pavlov's dogs learned to salivate at the sound of a bell; Skinner's pigeons learned to peck at keys in complicated patterns to get food. His ideas were massively influential and passed into everyday thinking, with his methods proving practical in a range of learning situations. Later he clashed with Noam Chomsky over whether humans have innate language skills or, as Skinner held, language is shaped and reinforced by experience.

See also Conditioned reflexes pages 242–243, Animal instincts pages 318–319, Language instinct pages 386–387, Psychology of obedience pages 402–403, Memory molecules pages 470–471

Right A rat in a Skinner box will in time learn to press the lever and be rewarded with food, thus reinforcing its behaviour.

A living fossil

Marjorie Courtenay-Latimer *b.*1907, James Leonard Brierley Smith 1898–1968

The discovery of *Latimeria chalumnae*, a surviving coelacanth, must be one of the best fishing stories ever and is certainly one of the biological finds of the twentieth century. First described as an extinct group of primitive bony fish by Louis Agassiz in 1839, coelacanths were thought to have died out 80 million years ago, in Cretaceous times.

Then just before Christmas 1938, Marjorie Courtenay-Latimer, curator of the East London Museum in South Africa, received a phone call. Captain Hendrik Goosens had some fish for her to look at. As she recounted, 'I picked away through the layers of slime to reveal the most beautiful fish I have ever seen'. Goosens agreed that the five-foot long, pale-blue fish with large scales, two pairs of limb-like fins and a strange three-lobed tail was indeed striking. He had never seen anything like it in 30 years of trawling the Indian Ocean between South Africa, Madagascar and the Comoro Islands.

Unable to preserve such a large fish, Courtenay-Latimer had it skinned by a taxidermist. Fortunately, enough remained for the fish expert J. L. B. Smith to identify it as a coelacanth, a living fossil, and to recognize its potential evolutionary importance. The hope was that its anatomy would help to establish which fish group gave rise to the first land-going, four-legged vertebrates. In the gloom of pre-war 1939, the story was front-page news around the world.

The lack of soft parts frustrated Smith's hope until the 1952 recovery of a better specimen, which showed that 'Old Fourlegs', as Smith called the coelacanth, was not our tetrapod ancestor. Coelacanths filmed in 1987 used their fins to position themselves in the water, not to walk on the seabed. A new population of coelacanths, found in 1998 in Indonesian waters, suggests that these survivors from the past will be with us for a long time to come.

See also Comparative anatomy pages 106–107, *Archaeopteryx* pages 180–181

Right Celebrated catch: the 'second coelacanth', caught off the Comoro Islands in 1952. Commonly believed to contain an elixir of life, this long-lived species is in fact merely the last survivor of a remarkably conservative lineage.

DDT

Paul Hermann Müller 1899–1965

The insecticide DDT (dichlorodiphenyltrichloroethane) was discovered in 1939 by the Swiss chemist Paul Müller while working for the J. R. Geigy company. He showed it was active against the the louse, the Colorado beetle, the mosquito and several other insect pests. What's more, DDT was (apparently) non-toxic to humans, cheap and easy to produce. Three million tonnes were manufactured during the next 30 years.

DDT's first major success came when it was used to control a serious outbreak of typhus among US troops during the Second World War. Typhus is a life-threatening infection carried by lice and has always been common during wartime. On this occasion, the usual delousing methods were not working. DDT was applied to more than a million people in January 1944, and the typhus epidemic was brought under control within three weeks – the first time this had ever happened during the winter. The application of DDT was also effective in wiping out the various species of mosquito that carry malaria. The house-fly also proved susceptible to DDT, leading to a reduced incidence of intestinal diseases such as paratyphoid and paradysentery.

Müller received the Nobel Prize for Physiology or Medicine in 1948 for the contribution DDT had made to public health. But in 1962 Rachel Carson sounded a warning note in her ground-breaking book *Silent Spring*. DDT came with an environmental price tag. Its chemical stability, initially seen as a desirable property, makes it very persistent in soil and water. All manner of wildlife was suffering toxic effects, Carson wrote. There are now suspicions that DDT accumulates in human tissue too, and may cause disease. It was banned in the USA and other developed countries from 1972, although it is still used in developing countries for malaria control. But many insects have now evolved resistance to DDT, and the search is on for safer and more effective alternatives.

See also Greenhouse effect pages 184–185, Malarial parasite pages 230–231, Green revolution pages 428–429, Ozone hole pages 438–439

Right Beach-goers on Long Island, New York, are sprayed with DDT during the testing of a new machine for distributing the insecticide (1945).

Bat echolocation

Donald **Griffin** *b.*1914, Robert **Galambos** *b.*1915

Bats are not unique in their ability to use sound rather than sight to guide their flight and to hunt in darkness, but it is in these animals that the system has reached its apex of evolution.

In the late eighteenth century, the Italian naturalist and priest Lazzaro Spallanzani used blinded bats to prove that they could fly in complete darkness without colliding with objects in their path; he also showed that they lost this ability if their heads were covered. This hinted at the involvement of another sense. Charles Jurine, a Swiss zoologist and contemporary of Spallanzani, provided the first indication that sound was the key by blocking one ear of a bat, which then lost the ability to avoid obstructions in the dark. Jurine also demonstrated that the cochlea in a bat's ear responds to sound frequencies beyond human perception.

The belief that bats use sound to map their surroundings was finally confirmed in 1940 by Donald Griffin and Robert Galambos. Their work showed that bats produce ultrasound, which we cannot hear, and that they locate objects and judge distance by using its echo. They also demonstrated that bats employ echolocation for hunting moths in darkness, but later research adds a further twist to the tale: some moths have evolved the ability to detect bat ultrasound and can take avoiding action, while others emit high-frequency sounds, confusing the bat's echolocation system – classic examples of an evolutionary 'arms race' between predator and prey.

Echolocation is not universal in bats: with a few exceptions only those of the insect-eating order Microchiroptera have a well-developed system. Echolocation has also evolved independently in unrelated birds and mammals, including oil birds and swiftlets that live in caves, some nocturnal shrews and toothed whales and porpoises, all of which operate in a sensory landscape beyond human experience.

See also Animal instincts pages 318–319, Honeybee communication pages 338–339

Right An unheard world: a greater horseshoe bat, *Rhinolophus ferrumequinum*, uses constant-frequency sonar to catch a moth in flight.

Power from the nucleus

Otto **Hahn** 1879–1968, Fritz **Strassmann** 1902–1980, Lise **Meitner** 1878–1968, Otto Robert **Frisch** 1904–79, Enrico **Fermi** 1901–54

Months before the outbreak of the Second World War, physicists found a way to release the energy of the atomic nucleus. When the German researchers Otto Hahn and Fritz Strassmann fired neutrons at a piece of uranium, some atoms were created that appeared to be barium – a far lighter element than uranium. In early 1939 Lise Meitner and Otto Frisch realized that the uranium nucleus must be breaking into two pieces. They worked out that this nuclear fission unleashes tremendous energy.

Something else escapes too. Physicists found that the fissioning uranium nucleus expels two or three neutrons – neutrons that can fly off and split other uranium nuclei, which will in turn release more neutrons. In such a chain reaction, whole chunks of uranium could be made to yield their energy.

During the war, the Allies feared that Hitler's Germany might use fission to make a devastating weapon, and they devoted huge resources to getting there first. On 2 December 1942 Enrico Fermi and his team at the University of Chicago achieved the first self-sustaining nuclear reaction. Fermi's reactor was designed to make plutonium, an artificial element that also fissions.

The first atomic bomb was based on plutonium. It exploded at Trinity, New Mexico, on 16 July 1945, with the force of 18,000 tonnes of TNT. Two more bombs, Little Boy and Fat Man, were dropped on the Japanese cities of Hiroshima and Nagasaki in August, killing hundreds of thousands of people.

Fission reactors now generate nearly a fifth of the world's electricity supply. However, most countries don't rely on them for the bulk of their power because of doubts about the safety and expense of waste disposal, and the risk of accidents like Chernobyl in 1986. The fear of climate change caused by burning fossil fuels may make nuclear power more popular.

See also Electromagnetism pages 134–135, Greenhouse effect pages 184–185, Radioactivity pages 224–225, Special relativity pages 244–245, Model of the atom pages 272–273, Stellar evolution pages 286–287, The neutron pages 312–313

Right The atomic bomb is the supreme example of the way in which the academic work of physicists became, within a few decades, applicable in ways they could never have foreseen.

Genes in bacteria

Salvador Edward Luria 1912–91, Max Delbrück 1906–81

Bacteria are ubiquitous single-celled creatures: they live on our skin, in our guts, in the sea, the earth, even in solid rock. Some are (to us) benign, others cause disease, but their main importance in science has been in the discovery of how genes work. Most of what is known about the molecular mechanisms by which genes operate was first discovered by experiments with bacteria, usually the bacterium *Escherichia coli*.

In 1940 it was not even known that bacteria had genes. Bacterial cells, unlike the cells of animals and plants, lack nuclei, and the genes of animals and plants were known to reside in the cell nucleus. Maybe bacteria were a life form based on some other hereditary mechanism than that of animals and plants. In 1943 an émigré German, Max Delbrück, and an émigré Italian, Salvador Luria, working together in the USA, performed a classic experiment. They put samples of bacteria on to a new food medium. The bacteria had to evolve new digestive skills to be able to grow in the new conditions. Luria and Delbrück could infer that the new skill arose by the same process of mutation known to be at work in plants and animals. So bacteria probably also have genes as in animals and plants. The experiment was just the start of a series of investigations that, by the late 1950s, established bacteria as the best life forms to exploit when studying the molecules of heredity.

Inheritance in bacteria does have some differences from that in plants and animals, which became clear in the 1940s, when antibiotics became available for treating bacterial diseases. It was soon seen that the bacteria rapidly evolve resistance to the antibiotics. Part of the reason is that bacteria can transfer genes between individuals without reproduction. Antibiotic resistance spreads rapidly through a bacterial population by this 'horizontal' gene transfer, of which animals and plants are incapable.

Right Three-way genetic exchange among bacteria. The male in the left of the electron micrograph is sending genes through two tubes covered by bacteriophage viruses.

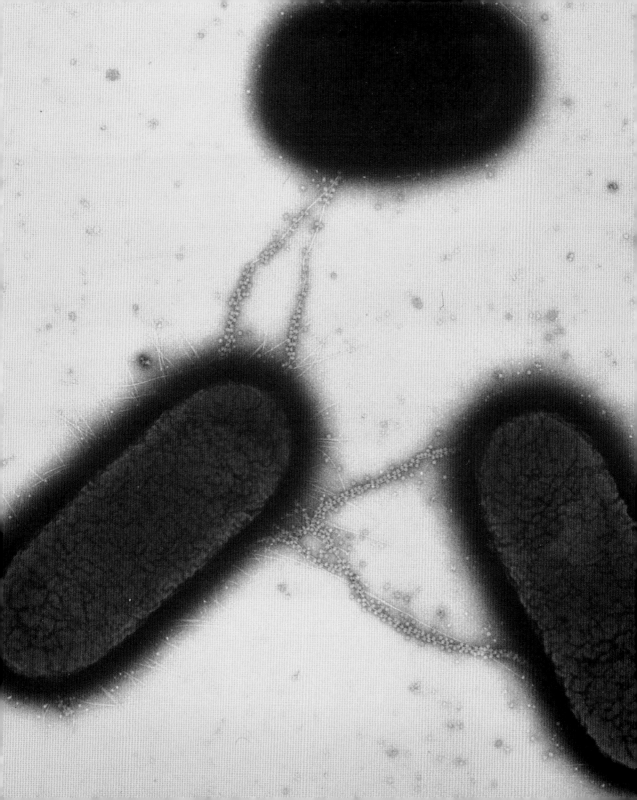

Artificial neural networks

Warren McCulloch 1898–1972, Walter Pitts b.1924

Unlike the human brain, conventional computers follow programmed instructions and do not learn by example. Artificial neural networks, on the other hand, mimic the way that the brain processes information.

Information in the brain is transferred between billions of nerve cells ('neurons') via connections called 'synapses'. As neurons can be highly branched, each forming synaptic connections with thousands of other neurons, real neural networks are enormously complicated and have great computational power.

Artificial neurons were developed in the USA in 1943 by the neurobiologist Warren McCulloch in collaboration with the logician Walter Pitts. Ironically, progress in the field was slow until the advent of cheap modern conventional computers in the 1970s. The properties of individual processing units, analogous to neurons, can be programmed and the units interconnected according to a specified network architecture. Artificial neural networks typically have an input layer, an output layer and one or more 'hidden' layers. In simple networks, signals travel in one direction only, and outputs are associated with particular inputs as required for tasks involving the recognition of patterns. Complex tasks such as voice recognition use designs that allow feedback between layers. In this case, the pattern of connectivity within the network remains in flux until equilibrium is reached – a state that represents a solution to the task.

Artificial neural networks are highly adaptable. Just as we categorize an image as a 'face' even if it is of a stranger, artificial neural networks generalize from representative examples. They are used in settings where a pattern needs to be found in complex variable data. In medicine, for example, they can be used to diagnose heart disease from electrocardiograms or cancer from digitized tissue images. They are also used in basic brain research. But whether they will ever achieve 'consciousness' remains a much-debated philosophical question.

Above Nature's neural net. The human brain contains billions of neurons, each carrying out complex computations and communicating with thousands of other neurons.

See also Nervous system pages 210–211, Behaviourial reinforcement pages 322–323, The computer pages 340–341, Information theory pages 354–355, Nerve impulses pages 366–367

Right Cog, a humanoid robot developed at the Massachusetts Institute of Technology, contains a large neural network. Is it conscious? Its engineers say: 'For the record, no'.

Game theory

John von **Neumann** 1903–57, Oskar **Morgenstern** 1902–77

In human social interactions, achieving some desired outcome usually depends on decisions made by others. In the Cold War, for example, whether or not to press the nuclear trigger depended in part on the perceived risk that the other side would launch a pre-emptive strike or retaliate with overwhelming force if attacked first.

By reducing problems to their essentials, game theorists attempt to identify key elements common to many forms of human conflict and bargaining. John von Neumann and Oskar Morgenstern laid the foundations of game theory at the Institute for Advanced Study, Princeton, USA. In a classic book of 1944, they analysed games in terms of strategies, costs and payoffs, focusing on games in which a player gains only at the expense of an opponent. In such games, strategies are found that maximize a player's minimal payoff.

But players can have mutual interests. In the 'prisoner's dilemma' game, for example, a player chooses either to cooperate with or act against an opponent. The highest mutual payoff is achieved when the two players each choose to cooperate. But a larger unilateral payoff is gained by acting against a cooperative 'sucker'. Such 'defection' is the rational strategy in a one-off game. When the game is played repeatedly and payoffs accrue over time, the best strategy now depends on the past behaviour of the opponent: if the opponent has a tendency to cooperate, it is less risky also to cooperate. With multiple players and varying degrees of knowledge about potential opponents, the iterated prisoner's dilemma has complex dynamics and often counter-intuitive outcomes.

The importance of game theory extends beyond sociology and economics. Evolutionary game theory, as pioneered by the British biologist John Maynard Smith, has illuminated many aspects of animal behaviour. This has culminated in the concept of the 'evolutionarily stable strategy', the strategy that will become predominant in the population owing to its high average payoff and consequent invulnerability to displacement by alternative strategies.

See also Animal instincts pages 318–319, Evolution of cooperation pages 406–407

Right The Americans used game theory to analyse the 1962 Cuban Missile Crisis. President Kennedy (right) persuaded the Russians that the USA would not shrink even from nuclear war.

Honeybee communication

Karl von Frisch 1886–1982

The complexities of honeybee communication and sensory perception were first revealed by Karl von Frisch, an Austrian who became one of the founders of ethology – the scientific study of animal behaviour. As early as 1919 he demonstrated that bees communicate using body movements that had been originally described in 1788 by the Reverend Ernst Spitzner, and by 1945 he had gone on to interpret a complicated series of dances that workers use to convey information to their fellow bees. In *The Dancing Bees*, first published in 1927 and translated into English in 1955, von Frisch detailed how foraging honeybees perform a 'round' or 'figure-of-eight' dance when they return to the hive. The speed and orientation of the dances in relation to the hive and the Sun, combined with waggling of the bee abdomen, inform departing workers about the distance and direction of sources of pollen and nectar.

Interpreting animal behaviour purely through human senses can give a distorted picture of the sensory world that animals inhabit. Nowhere is this more so than in the case of honeybees, the sensory capabilities of which extend far beyond human experience. Combining painstaking observation and elegant experiments, which included training bees to visit artificial food sources placed on coloured cards, von Frisch was able to show that honeybees can see ultraviolet light. We now know that this allows them to perceive patterns in petals that are invisible to humans. He also investigated their senses of smell and taste, and in 1949 proved that, in flight, they can navigate using the Sun as a reference point even when it is obscured by cloud, thanks to their use of the pattern of light polarization in the sky.

In 1973 Karl von Frisch shared a Nobel Prize for Physiology or Medicine with two other pioneers of ethology: Niko Tinbergen and Konrad Lorenz.

See also Animal instincts pages 318–319, Evolution of cooperation pages 406–407

Right On returning to the hive, a foraging honeybee performs the celebrated waggle dance, an intricate series of movements that explains to the other bees where to find pollen themselves.

The computer

Alan Mathison Turing 1912–54, John von Neumann 1903–57

In 1890 Herman Hollerith, an American engineer, devised an electromechanical machine for tabulating data for the US census. The Hollerith punch card system met with great success and became the mainstay of the office machine and data processing industries. The production in the 1940s of programmable computers using mechanical parts and electromechanical relays showed that large-scale automatic calculation was possible, but the technology was too slow for scientific and military use.

It was the development of radio tubes that led to fully electronic machines. The first general purpose electronic programmable computer was ENIAC (Electronic Numerical Integrator and Calculator), built at the Moore School of Electrical Engineering, University

of Pennsylvania, by J. W. Mauchly and J. P. Eckert. Originally intended for rapid wartime production of ballistic tables, it was not finished until 1946. Three years earlier the war effort had prompted the British to build Colossus, a digital programmable electronic computer designed specifically for cracking German codes. The theoretical background to Colossus had been worked out by the mathematician Alan Turing in 1936, when he showed that any problem could be solved mechanically if it could be expressed in the form of a finite number of manipulations that could be performed by the machine.

Turing had been a graduate student at Princeton, and it was in that year he met that other pioneer of computer theory, John von Neumann. The two would collaborate on the Allies' code-breaking activities during the Second World War. Indeed von Neuman had high hopes that ENIAC would assist in calculations necessary for building an atomic bomb. With 18,000 vacuum tubes, ENIAC could handle up to 5,000 operations a second. But programming necessitated setting switches and plugging in connections by hand. Von Neumann set to work on an improved model, and in 1945 he published a seminal paper describing a modern general-purpose, stored-program computer, which separated the control unit, arithmetic unit, memory and input and output. ENIAC was finally switched off at 11.45 pm on 2 October 1955.

See also Logarithms pages 56–57, Difference Engine pages 138–139, The transistor pages 350–351, Four-colour map theorem pages 450–451, Public-key cryptography pages 454–455, Fermat's last theorem pages 506–507

Right ENIAC, the first general-purpose electronic computer, contained no less than 18,000 radio valves and consumed up to 100 kilowatts of electricity.

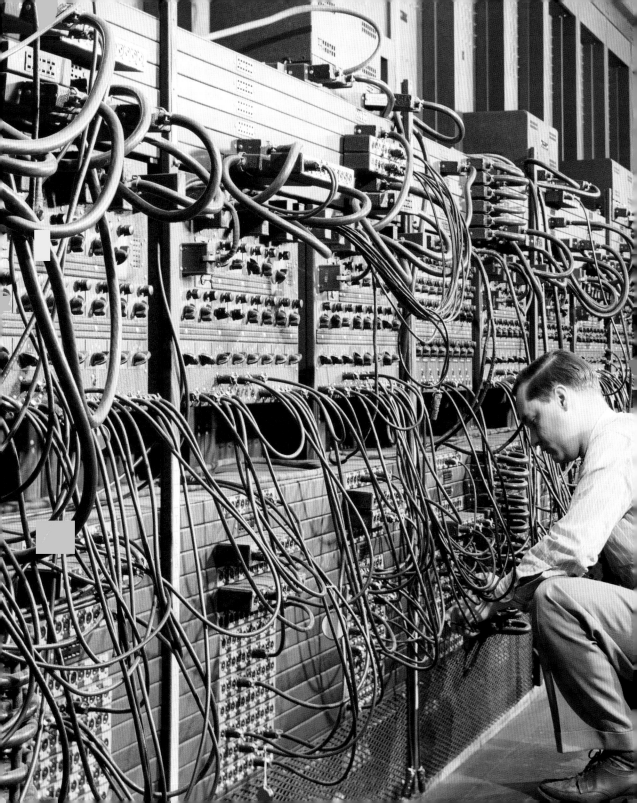

Turing machines
W. Daniel Hillis

BECAUSE COMPUTERS CAN DO SOME things that seem very much like human thinking, people often worry that they threaten our unique position as rational beings, and there are some who seek reassurance in mathematical proofs of the limits of computers. There have been analogous controversies in human history. It was once considered important that the Earth be at the centre of the universe, and our imagined position at the centre was emblematic of our worth. The discovery that we occupied no central position – that our planet was just one of a number of planets in orbit around the Sun – was deeply disturbing to many people at the time, and the philosophical implications of astronomy became a topic of heated debate. A similar controversy arose over evolutionary theory, which also appeared as a threat to humankind's uniqueness. At the root of these earlier philosophical crises was a misplaced judgement of the source of human worth. I am convinced that most of the current philosophical discussions about the limits of computers are based on a similar misjudgement.

The central idea in the theory of computation is that of a 'universal computer' – that is, a computer powerful enough to simulate any other computing device. The general-purpose computer is an example of a universal computer; in fact most computers we encounter in everyday life are universal computers. With the right software and enough time and memory, any universal computer can simulate any other type of computer, or (as far as we know) any other device at all that processes information.

One consequence of this principle of universality is that the only important difference in power between two computers is their speed and the size of their memory. Computers may differ in the kinds of input and output devices connected to them, but these so-called peripherals are not essential characteristics of a computer, any more than its size or its cost or the colour of its case. In terms of what they are able to do, all computers (and all other types of universal computing devices) are fundamentally identical.

The idea of a universal computer was recognized and described in 1937 by the British mathematician Alan Turing. Turing, like so many other computing pioneers, was interested in the problem of making a machine that could think, and he invented a scheme for a general-purpose computing machine. Turing referred to this imaginary construct as a 'universal machine', since at that time the word 'computer' still meant 'a person who performs computations'.

To picture a Turing machine, imagine a mathematician performing calculations on a scroll of paper. Imagine further that the scroll is infinitely long, so that we don't need to worry about running out of places to write things down. The mathematician will be able to solve any solvable computational problem no matter how many operations are involved, although it may take him an inordinate amount of time. Turing showed that any calculation that can be performed by a smart mathematician can also be performed by a stupid but meticulous clerk who follows a simple set of rules for reading and writing the information on the scroll. In fact, he showed that the human clerk can be replaced by a finite-state machine. The finite-state machine looks at only one symbol on the scroll at a time, so the scroll is best thought of as a narrow paper tape, with a single symbol on each line. (A finite-state machine has a fixed set of possible states, a set of allowable inputs that change the state, and a set of possible outputs. The outputs depend only on the state, which in turn depends only on the history of the sequence of events.)

Today, we call the combination of a finite-state machine with an infinitely long tape a 'Turing machine'. The tape of a Turing machine is analogous to, and serves much the same function as, the memory of a modern computer. All that the finite-state machine does is read or write a symbol on the tape and move back and forth according to a fixed and simple set of rules. Turing showed that any computable problem could be solved by writing symbols on the tape of a Turing machine – symbols that would specify not just the problem but also the method of solving it. The Turing machine computes the answer by moving back and forth across the tape, reading and writing symbols, until the solution is written on the tape.

I find Turing's particular construction difficult to think about. To me, the conventional computer, which has a memory instead of a tape, is a more comprehensible example of a universal machine. For instance, it is easier for me to see how a conventional computer can be programmed to simulate a Turing machine than vice versa. What is amazing to me is not so much Turing's imaginary construct but his hypothesis that there is only one type of universal computing machine. As far as we know, no device built in the physical universe can have any more computational power than a Turing machine. To put it more precisely, any computation that can be performed by any physical computing device can be performed by any universal computer, as long as the latter has sufficient time and memory. This is a remarkable statement, suggesting as it does that a universal computer with the proper programming should be able to simulate the function of a human brain.

Photosynthesis

Melvin Calvin 1911–97

There are some landmark experiments in science that seem in retrospect to have an elegant simplicity that belies the scale of their importance. One such is the American chemist Melvin Calvin's demonstration of the main steps in carbon fixation in photosynthesis, the process that underpins food chains for all higher forms of life on Earth and maintains our planet's climate by constantly removing carbon dioxide from the atmosphere.

It has been known since the work of Jan Ingenhousz in 1779 that sunlit green plants absorb carbon dioxide. Research over the following two centuries identified many of the key events in the process of photosynthesis, where green plants use the atmospheric carbon dioxide to make complex carbon-based molecules such as sucrose and starch which they use for growth. But it was Calvin, son of Russian émigrés, who revealed the details of the complex biochemical cycle that converts carbon dioxide into sugars.

Calvin's appointment as head of the Lawrence Radiation Laboratory at the University of California at Berkeley in 1946 coincided with the development of a readily available supply of the radioactive isotope carbon-14. Calvin immediately recognized that this could be used to follow the fate of carbon atoms in biochemical reactions in the chloroplasts of green plants. He fed radioactive carbon-14 to flasks of the green alga *Chlorella*, then stopped the reaction after various time intervals, starting at just a few seconds. As the intervals lengthened, the radioactive carbon appeared in an increasing number of compounds, showing how it had been passed along a series of reactions that Calvin was now able to describe.

The reactions were subsequently named the Calvin–Benson cycle, after Calvin and his co-worker Andrew Benson, and Calvin was awarded the Nobel Prize for Chemistry for his discoveries in 1961. In the latter years of his career he was a great advocate of the supplementation of fossil fuels with environmentally friendly plant hydrocarbons made – using solar energy and photosynthesis – by desert shrubs.

See also Greenhouse effect pages 184–185, Nitrogen fixation pages 208–209, Crop diversity pages 292–293, Citric-acid cycle pages 320–321, Symbiotic cell pages 418–419, Green revolution pages 428–429, Gaia hypothesis pages 432–433

Right Stacks of grana in a chloroplast from a leaf of maize. The sites of photosynthesis in higher plants, grana contain the light-receptive green pigment chlorophyll.

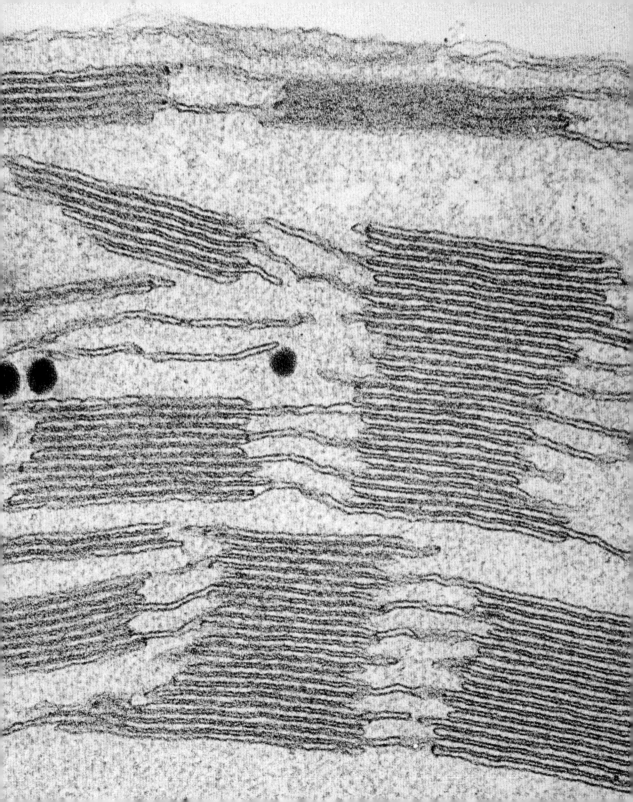

Radiocarbon dating

Willard Frank Libby 1908–80

In 1947 Willard Frank Libby developed the technique of radiocarbon dating for establishing the age of an enormous diversity of organic materials from bone to wood to fabric. More recently, accelerator mass spectrometry has extended the range of the technique, confirming it as an indispensable tool in geology, anthropology, archaeology and palaeontology.

Only small samples are now required to date invaluable items such as the famous Turin shroud. Radiocarbon analysis of the shroud in 1988 revealed that the flax it is made of was harvested around AD 1325 (plus or minus 33 years), proving the relic to be medieval and not Christ's winding sheet as was commonly believed.

Libby, an American chemist, joined the Manhattan Project during the Second World War, where he worked on the separation of uranium isotopes for the development of the atomic bomb. After the war he moved to the Institute of Nuclear Studies at the University of Chicago and soon realized that radioactive carbon, or carbon-14, could be used to date organic materials, a discovery that earned him the Nobel Prize for Chemistry in 1960.

In 1939 it was shown that carbon-14 is formed by the action of cosmic rays on atmospheric nitrogen. So carbon dioxide containing traces of this isotope is continually being incorporated into all living organisms and indeed all organic products derived from them. After death, no more carbon-14 is incorporated, and what remains begins its natural decay to the stable isotope carbon-12. As the rate of decay is known, measurement of the new ratio of the two forms of carbon establishes the time elapsed since death. But at 5,730 years, the half-life of radiocarbon is relatively short so the method is used mainly to date materials less than 40,000 years old. The discovery of fluctuations in cosmic radiation has required a slight recalibration of radiocarbon dates.

See also Radioactivity pages 224–225, Rocks of ages pages 252–253, Cosmic rays pages 268–269, The neutron pages 312–313, Photosynthesis pages 344–345, Iceman pages 504–505

Right Radiocarbon dating was used in 1988 to show that the Turin shroud is in fact an elaborate medieval forgery dating from the thirteenth century. It remains a mystery how it was made.

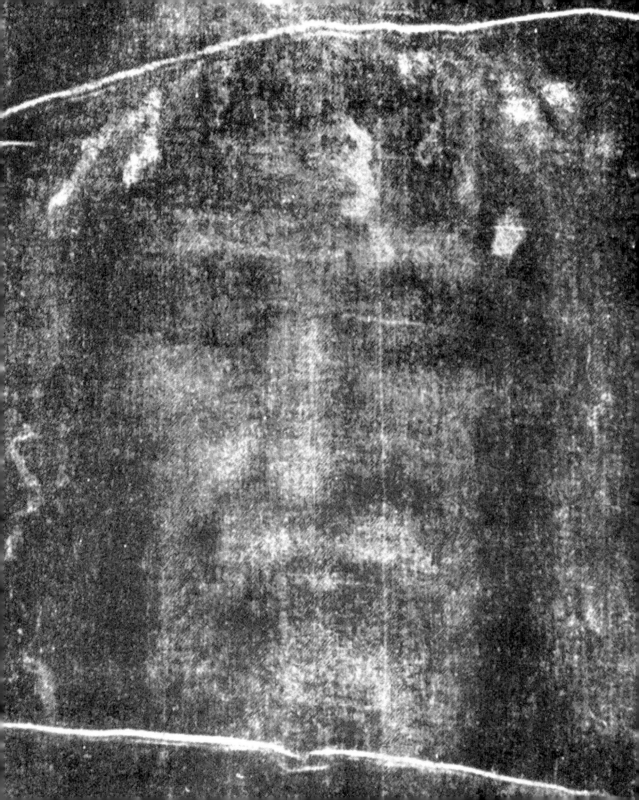

Slime-mould aggregation

John Tyler **Bonner** *b.*1920

Dictyostelium discoideum has been intriguing biologists since the early twentieth century. Although it is often called a 'cellular slime mould', it is not a mould and it is not always slimy. A better common name for it is a 'social amoeba'. What is most remarkable about it is its life cycle. In the first phase the organism consists of individual dispersed amoebas, living on decaying logs, eating bacteria and reproducing by dividing in two, like most other single-celled animals. But when starved of food, tens of thousands of these totally independent cells stream into aggregation centres to form a translucent 'slug' about a millimetre long. The slug crawls towards light, and then slowly 'differentiates' into a fruiting body consisting of a delicately tapered stalk with a terminal globe of spores at its tip, each covered with a hard case of cellulose. The spores break away, the cases crack open, and out come individual amoebas, completing the life cycle.

In this way a loose collection of individual cells transforms itself into a single structured multicellular organism – an amazing feat of self-organization. Much of the pioneering work into slime-mould behaviour was done by the American biologist John Tyler Bonner, who in 1947 established the important role of the release of a chemical messenger, cyclic AMP, in the initial aggregation. After a few hours of starvation the individual amoebas begin to secrete pulses of this chemical signal, which causes them to cluster in beautiful spiral and feathery patterns which eventually converge to form the slug.

The process is known as 'chemotaxis' – the movement of cells towards or away from higher concentrations of a chemical present in their environment – and it is perhaps the most primitive form of communication between cells. The same mechanism operates in many parts of nature and it may even be the hidden factor in morphogenesis, turning a spherical, undifferentiated fertilized egg into a complex, differentiated animal or human being.

See also Chemical oscillations pages 364–365, Growth of nerve cells pages 368–369, Genetics of animal design pages 460–461, Edge of chaos pages 490–491

Right Fruiting body of the slime mould *Dictyostelium discoideum*. The scanning electron micrograph shows the delicately tapered stalk one or more millimetres high, with a terminal globe of spores at its tip.

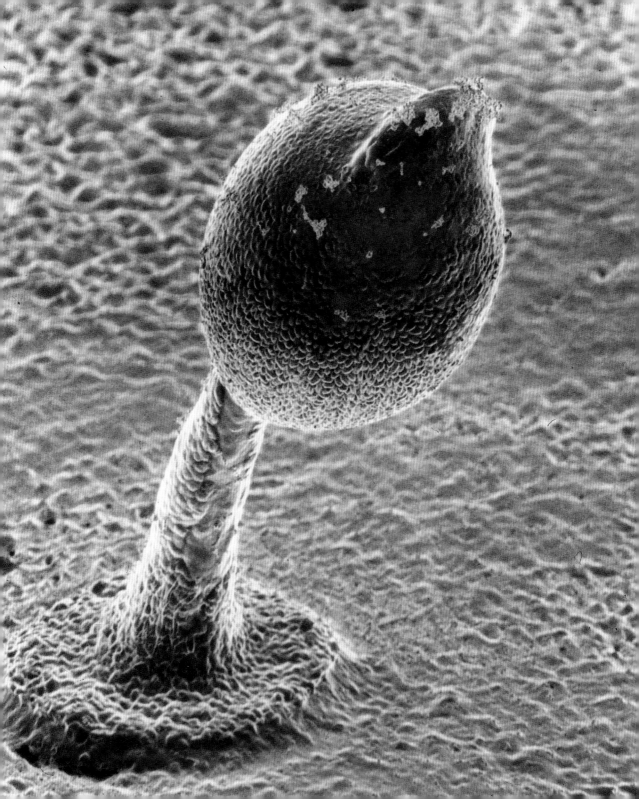

The transistor

William Bradford Shockley 1910–89, Walter Houser Brattain 1902–87, John Bardeen 1908–91

In the earliest days of radio, crystals were used as 'rectifiers', allowing alternating current to pass in only one direction. These were unreliable, however, and were soon replaced by thermionic tubes (valves) which can both rectify and amplify a current. But these too had their disadvantages: they were short-lived, power-hungry and bulky.

During the 1930s John Bardeen at the Bell Telephone Laboratories in the USA had studied the properties of semiconductors – crystalline solid materials that have an electrical conductivity between that of metals (low resistance) and insulators (high resistance) – and showed that surface effects could lead to current rectification. Seeking to ensure Bell's continued domination of the telecommunications market after the Second World War, Bardeen, together with William Shockley and Walter Brattain, embarked on a quest to find a semiconductor that could replace the radio tube. By 23 December 1947 they had discovered that a germanium crystal containing certain impurities was not only a far better rectifier than the earlier crystals or tubes, but could also behave as an amplifier. Because it worked by transforming current across a resistor they called it a 'transistor'.

The original version – a point contact transistor – was electrically 'noisy' and could control only low power inputs. It was quickly superseded by the junction transistor, which consists of a thin piece of silicon with impurities that invest different regions with different electrical properties. Typically, a 'base' region, endowed with excess positive charge carriers (holes), is sandwiched between 'emitter' and 'collector' regions rich in excess negative charge carriers (electrons). When a small voltage is applied to the base, excess holes collect at its junction with the collector, and current flows from one side of the semiconductor sandwich to the other. Unlike the old tubes, the transistor requires little power and, as it works at the molecular level, can be easily miniaturized. Today millions of tiny transistor circuits are etched on to silicon chips smaller than a fingernail, powering everything from hearing aids to supercomputers.

Above Microchips may incorporate a million transistors in one integrated circuit on a silicon sheet no larger than a fingernail.

See also Superconductivity pages 266–267, The computer pages 340–341, A new state of matter pages 510–511

Right Replica of the first working transistor built by Bardeen, Shockley and Brattain, and demonstrated on Christmas Eve, 1947.

Quantum electrodynamics

Julian Seymour **Schwinger** *b.*1918, Shin'ichiro **Tomonaga** 1906–79,
Freeman John **Dyson** *b.*1923, Richard Phillips **Feynman** 1918–88

Like charges repel. Bring two electrons close together and they will try to fly apart. Why? How do they exert this mysterious force called electricity? The theory of quantum electrodymanics, or QED, describes the process as if it were a game of catch.

The theory was hammered out in the late 1940s by Julian Schwinger, Shin'ichiro Tomonaga and Freeman Dyson, and by Richard Feynman, who invented diagrams to show how two charged particles could interact. The simplest Feynman diagram for two electrons shows one of them firing a photon at the other. The first electron recoils and the second is smacked back when it absorbs the photon. As this process is repeated the two electrons are pushed apart, like two people on roller skates playing catch with a bowling ball. But in this game you can also pull your partner towards you, so the theory can account for the attraction of opposite charges too.

It seems a weird way to describe a force – and it gets weirder. The photon carrying the force can itself be attracted or repelled by either electron, or it can briefly transform into other charged particles, which themselves can send out force-carrying photons. According to QED, every possible complication happens at once, but the more convoluted processes make a relatively small contribution to the force. Worse still, QED relies on an uncomfortable mathematical dodge called 'renormalization'.

But it works. The theory can handle all manifestations of electromagnetic forces, and with fine precision. The light emitted by hydrogen atoms, for example, has exactly the wavelength predicted by the quantum curlicues of QED. QED also says that even empty space is filled with flitting 'virtual' particles. But here the numbers don't work: the theory predicts that the vacuum should be incredibly thick with these particles, making it so dense the universe would long ago have collapsed under its own gravity. Something must be missing.

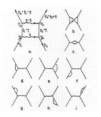

Above In Feynman's words, his famous diagrams 'represent physical processes and the mathematical expressions used to describe them'.

See also Electromagnetism pages 134–135, Maxwell's equations pages 186–187, The electron pages 228–229, Unified forces pages 416–417, Superstrings pages 474–475

Right Feynman said that the theory of quantum electrodynamics matched experiment as closely as if one predicted the distance from New York to Los Angeles and was off by the thickness of a human hair.

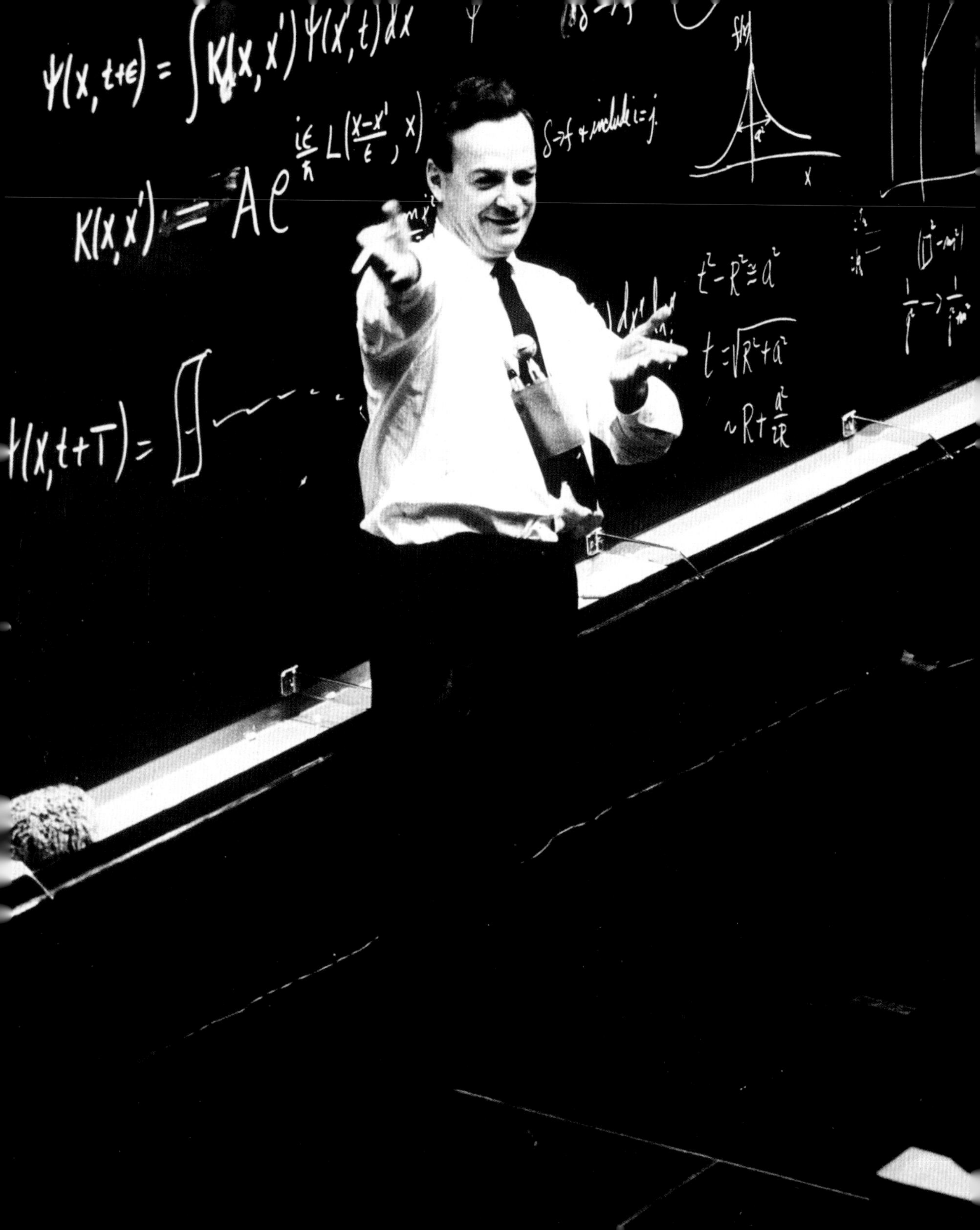
ψ(x, t+ε) = ∫K(x, x')ψ(x', t)dx
K(x, x') = Ae
t² − R² ≅ a²
t = √(R² + a²)
≈ R + a²/2R
ψ(x, t+T) =

Information theory

Claude Elwood Shannon 1916–2001

We cruise along the information highway, immerse ourselves in the information explosion and embrace the latest information technology. But few of us realize that the man who laid the foundation of the information age was the American mathematician Claude Shannon, who in 1948 published his mathematical theory of communication, which we now call information theory.

Shannon gave a precise mathematical meaning to information. He began his seminal paper by observing that 'the fundamental problem of communication is that of reproducing at one point either exactly or approximately a message selected at another point'. In his view the information content of a message consisted of a combination of the binary digits 0 and 1 ('bits'). These can be thought of as representing a sequence of 'yes–no' situations. All communication channels today are measured in bits per second, reflecting what Shannon called 'channel capacity'. He showed that the likelihood of information being garbled could be measured through loss of bits, distortion of bits, addition of extraneous bits and so on. Upper limits could now be put on communication rates; and concepts such as redundancy, noise and even entropy (as a measure of the amount of information) could be defined with mathematical precision. This has allowed engineers to improve the reliability and speed of message transmission in everything from deep-space communication and the Internet to compact-disc players and wireless phones.

The importance of Shannon's work was recognized immediately, and information theory was soon invoked in biology, linguistics, psychology, economics, physics and even art and literature. According to a 1953 issue of *Fortune*, 'It may be no exaggeration to say that man's progress in peace, and security in war, depend more on fruitful applications of information theory than on physical demonstrations, either in bombs or in power plants, that Einstein's famous equation works'.

Above Shannon was renowned for his eclectic interests. He designed and built chess-playing, maze-solving, juggling and mind-reading machines.

See also Wave–particle duality pages 300–301, The computer pages 340–341, The double helix pages 374–375, Quantum weirdness pages 464–465

Right Well connected: with its myriad electrical cables, this 1883 French vision of the 20th century accurately predicted the rise of the information age, if not the modes of communication.

Transplant rejection

Peter Medawar 1915–1987, Frank Macfarlane Burnet 1899–1985

The transfer of a tissue or an organ from one person to another had been a centuries-old dream. Thanks to the suturing techniques pioneered by Alexis Carrel and Charles Guthrie at the start of the twentieth century, transplants of organs such as kidneys, hearts and spleens were a technical possibility. But the organs of different individuals often proved incompatible; rejection of the transplant was frequent and invariably led to death.

An understanding of the process of rejection and its suppression started to crystallize as the Second World War was coming to an end. The British zoologist Peter Medawar had treated those burned in the Blitz, and after the war he carried out carefully controlled skin-graft experiments on animals. His experience led him to conclude that graft rejection was a direct result of the recipient's immune system attacking the transplant: the recipient's immune system recognizes antigens on the graft as foreign and produces antibodies that target the graft, which is then destroyed by white blood cells. The effects of rejection became graphically clear after the first human kidney transplant, which took place in the USA in June 1950. Eight months later, surgeons discovered why the kidney was failing. They found a tiny shrunken organ rendered useless by the patient's immune system.

Medawar's studies of the mechanism of rejection were influenced by the Australian biologist Frank Macfarlane Burnet, who in 1949 suggested that an individual's ability to recognize cells or tissues as 'self' was acquired during embryological development ('acquired immunity'). In the early 1950s Medawar showed that mice could tolerate skin grafts if, as embryos, they were exposed to cells from the donor of the graft ('acquired immunological tolerance'). Successful transplantation therefore relies on the induction of tolerance in adults or the suppression of immunity. In 1962 Roy Calne used immunosuppressive drugs to prolong the life of a kidney-transplant patient, although the real breakthrough came in the late 1970s with the introduction of the powerful immunosuppressor cyclosporine.

See also Cellular immunity pages 204–205, Antitoxins pages 214–215, Blood groups pages 236–237, Biological self-recognition pages 430–431, Monoclonal antibodies pages 448–449

Right A nurse attends to the first heart transplant patient in 1967. Sadly, he died of pneumonia after 18 days. The heart had not been rejected – it maintained good circulation to the end.

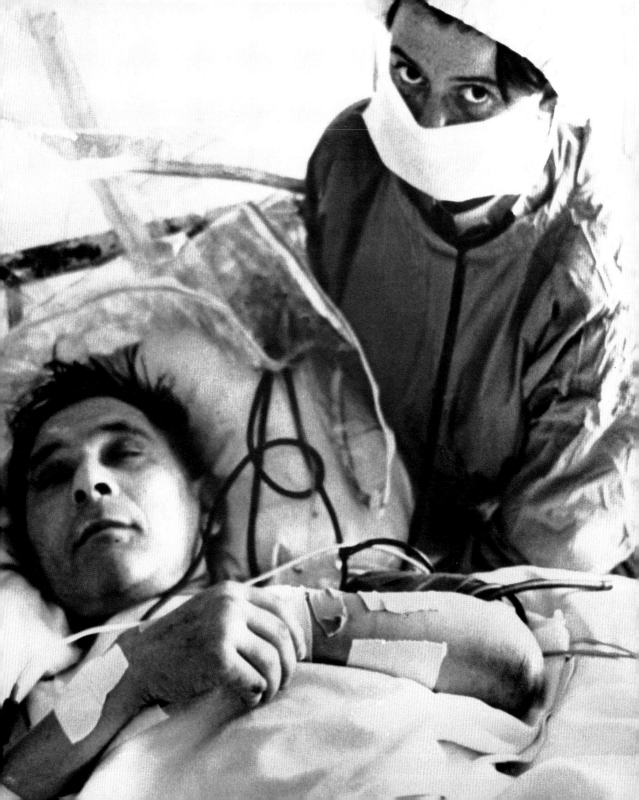

Sickle-cell anaemia

Linus Carl **Pauling** 1901–94

A healthy red blood cell is disc-shaped and concave on both sides. In sufferers of the human genetic disease sickle-cell anaemia, the cells dramatically distort into a crescent (or sickle) shape. These deformed cells block blood vessels, leading to circulatory problems and kidney and heart failure. They are less able to carry and release oxygen than normal red blood cells and have a shorter lifetime – all of which leads to severe anaemia. First described in 1910 by the American physician James Bryan Herrick, the disease is most common among African populations, although it is also prevalent in India and in some southern European countries.

The discovery of the cause of sickle-cell anaemia was a landmark in the history of molecular medicine. In 1949 the American chemist Linus Pauling showed that the haemoglobin molecules – the oxygen-carrying proteins that give blood its red colour – in people with sickle-cell anaemia were fundamentally different from normal ones. The source of this variation was traced in 1954 by another chemist, Vernon Ingram, who used a technique called chromatographic fingerprinting to separate proteins according to their chemical properties. Ingram revealed that sickle-cell haemoglobin differed from normal haemoglobin in just one of its amino-acid building blocks. And this change, in turn, results from a so-called 'point mutation' in the haemoglobin gene on chromosome 11. Changing just one 'letter' in the DNA code can, therefore, mean a lifetime of human suffering.

Sickle-cell anaemia is intriguing for another reason. People born with two mutated haemoglobin genes get the disease, but those born with one mutated gene and one normal gene – who are said to have sickle-cell trait – can lead reasonably normal lives and also be relatively resistant to malaria. This probably explains why the mutation has persisted in populations where malaria is particularly prevalent: the benefits of a single dose may outweigh the down side of a double dose.

See also Circulation of the blood pages 58–59, Mendel's laws of inheritance pages 192–193, Inborn metabolic errors pages 258–259, Structure of haemoglobin pages 390–391

Right The sickle-cell mutation causes red blood cells to collapse in the absence of oxygen. The disease is the high price paid for malaria resistance in the past.

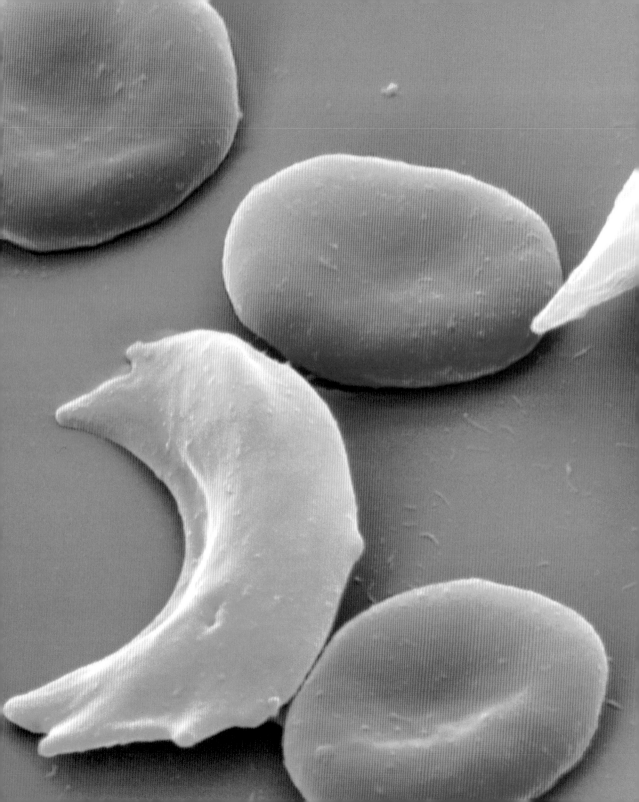

A cometary reservoir

Jan Hendrik Oort 1900–92

Astronomers divide comets into two groups depending on the time it takes them to orbit the Sun. Long-period comets have periods greater than 200 years; short-period comets, less than 200 years. Planetary perturbations should stir up comets into a uniform distribution, scattered randomly to both larger and smaller orbits. But the Danish astronomer Jan Oort found a dramatic excess of comets with periods exceeding a million years. In 1950 he proposed that the source is a vast spherical cloud enveloping the Solar System and extending 10,000–100,000 astronomical units from the Sun (the nearest star is about 270,000 astronomical units away, where 1 astronomical unit is the mean Earth–Sun distance). He calculated that the cloud should contain about 10^{12} comets.

Nearby passing stars disrupt the Oort cloud through gravitational influence or even collision, knocking some comets out of the Solar System and pushing others in towards the Sun on very long orbits. Some of the latter pass close to the major planets and become captured in the inner Solar System. As short-period comets now trapped in new orbits, they decay quite quickly, forming meteoroid dust streams. If Earth passes through one of these streams we see shooting stars in the upper atmosphere. There may be an intermediate stage between long- and short-period comet populations.

In 1951 Gerard Kuiper suggested that the Solar System did not finish abruptly at Pluto's orbit. Beyond Pluto there should be some 'dirty snowball' planetesimals, pieces of icy matter similar to Halley's comet, on nearly circular orbits. These small bodies are so thinly spaced that they cannot conglomerate to form a planet. Physically and chemically, they are nearly indistinguishable from cometary nuclei. The first member of this Kuiper Belt was discovered in August 1992. The belt probably contains more than a million objects of up to 1,000 kilometres in diameter that feed the short-period comet population.

See also A new star pages 48–49, Halley's comet pages 84–85, Origin of the Solar System pages 104–105, Comet Shoemaker–Levy 9 pages 508–509, Water on the Moon pages 522–523

Right This 18th-century illustration shows the observed paths of the periodic comets and discusses the comet theories of such astronomers as Kepler, Cassini and Halley.

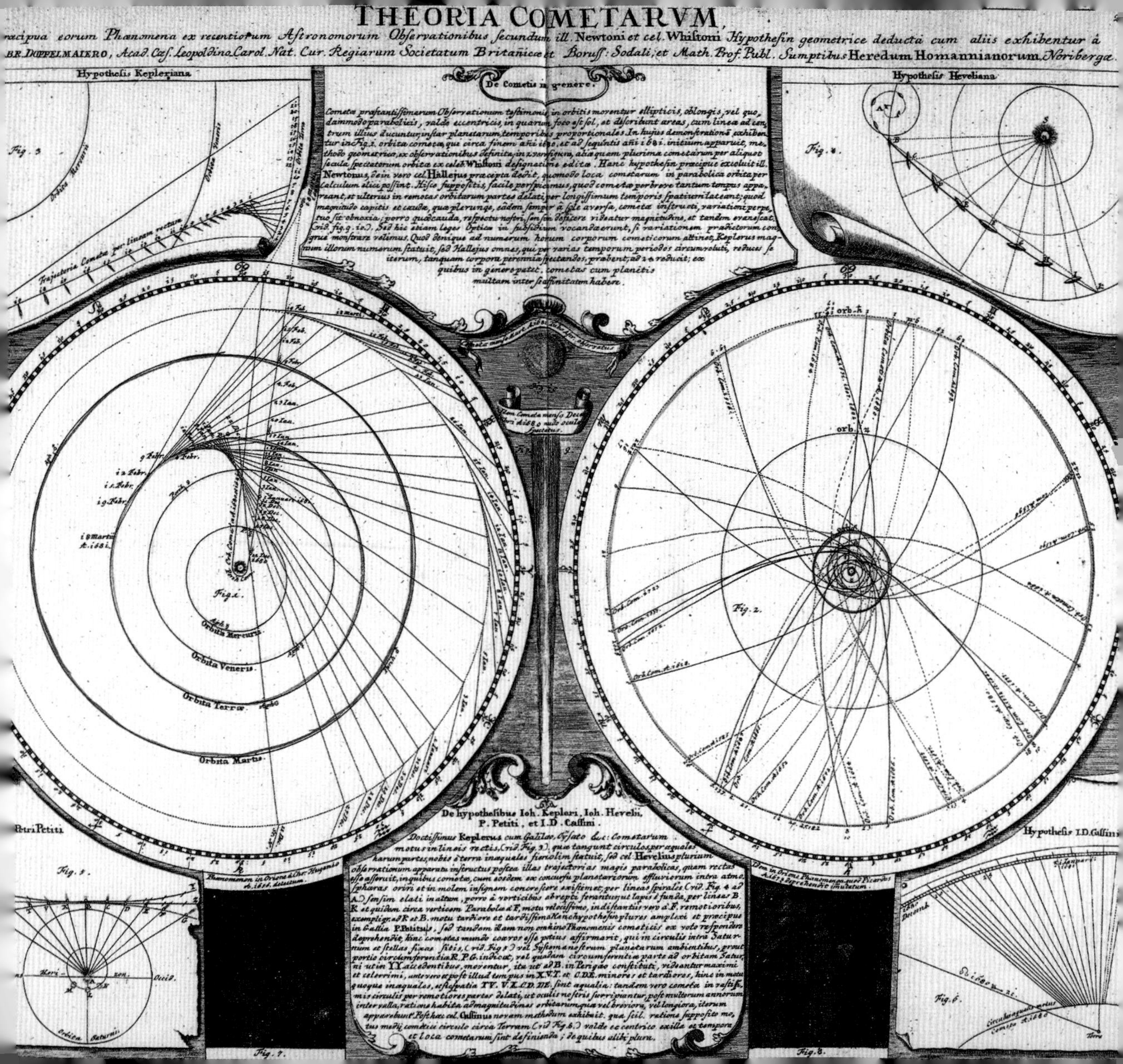

THEORIA COMETARUM,

pracipua eorum Phænomena ex recentiorum Astronomorum Observationibus secundum ill. Newtoni et cel. Whistoni Hypothesin geometrice deducta cum aliis exhibentur à
BR. DOPPELMAIERO, Acad. Cæs. Leopoldino Carol. Nat. Cur. Regiarum Societatum Britanicæ et Borusf. Sodali, et Math. Prof. Publ. Sumptibus Heredum Homannianorum, Norimbergæ.

Hypothesis Kepleriana

Fig. 3

Hypothesis Heveliana

Fig. 4

De Cometis in genere.

Cometas præstantissimarum Observationum testimonio in orbitis moventur ellipticis, oblongis, vel quo, summopto parabolicis, valde excentricis in quorum foco est sol, ut disformat areas, cum linea ad centrum illius ducuntur, ita ut planetarum temporibus proportionales. In hujus demonstrationi exhibetur in Fig. 1. orbita cometæ, quæ circa finem anni 1680, et ad sequentis anni 1681. initium apparuit, methodo geometrico, ex observationibus definita in xensisum, alia quam plurima cometarum per aliquot facile spectatorum orbita ex cel. Whistoni designatione edita. Hanc hypothesin præcipue excoluit ill. Newtonus, dein vero cel. Hallejus præcepta dedit, quomodo loca cometarum in parabolica orbita per calculum illius possint. Hisce suppositis, facile perspicuinus, quod cometæ perbreve tantum tempus appareant, et ulterius in remotas orbitarum partes delati per longissimum temporis spatium latitent; quod magnitudo copitis et caudæ, quæ plerumque, idem, longior et sit aversa, cometis inservit, variationes perspicuæ.

Sed hic sciam leges Optica in subsistibus revocandæ erunt, et variationem practicorum congruæ monstrare volemus. Quod denique ad numerum horum corporum cometicorum attinet, Keplerus magnum illorum numerum statuit, sed Hallejus omnes, qui per varias temporum periodos circumvolvit, reduces et iterum, tanquam corpora perennia spectandos, grabent; et à reducit; ex quibus in genere patet, cometas cum planetis
multam inter se affinitatem habere.

Fig. 1

Orbita Mercurii
Orbita Veneris
Orbita Terræ
Orbita Martis

Fig. 2

Petri Petiti

Fig. 5

De hypothesibus Ioh. Kepleri, Ioh. Hevelii, P. Petiti, et I. D. Cassini.

Doctissimus Keplerus cum Galilæo, &c. Cometarum
motus in lineis rectis, (vid. Fig. 3.) quæ tangunt circulos per æquales
harum partes, nobis è terra inæquales suriolim statuit, sed cel. Hevelius plurimum
observationum apparatu instructus postea illas trajectorias magis parabolicas, quam rectas
esse asseruit, in quibus cometæ, cum eodem in conceptu planetarum xensium essurissem circa solem
fastuerus orvin et in molem insignem concresere consistere, per lineas spirales (vid. Fig. 4. et
A) sensim elati in altum, porro à vorticibus abrepti ferantur; ut lapis è funda, per lineas B
R et quibus circa vorticem Parabola à F, motu velocissimo, in distantius vero à F remotioribus,
exempli gratis R et B, motu tardiorus et tardissima Kanchypothesin plures amplexi et præcipue
in Gallia P. Petitus, sed tandem Kam non omnibus Phænomenis cometicis eos voto respondens
deprehensisse sine cometa, mundo convex asserunt, qui in circulis into Saturnum
et stellas fixos fixis, (vid. Fig. 5.) vel systema nostrum planetarum ambientibus, prout
portio circumferentiæ R. P. G. indicat, vel quadam circumferunt à parte ad orbitam Satur-
ni ut in ŲŲ ascendentibus, merentur, ita ut à P. D. in Perigeæ consistit, videantur maximi
et celerrimi, ut vero à P et sse illud tempus in XVI et CDE minores et tardiores, hinc in orem
quoque inæquales, effugiatis ŲŲ V L E D, sint sequalia: tandem vero cometa in vastis-
simis circulis pro remotioribus præter solari, ut oculis nostris fere quasi per post multum annorum
intervalla, ratione habita ad magnitudines orbitarumque vel breviora, vel longiora, iterum
apparebunt. Post hæc cel. Cassinus novam methodum exhibet. quæ scil. ratione supposito me-
tu modi cometici circulo circa Terram, (vid. Fig. 6.) valde excentrico et concentrico exilia et tempore
et loca cometarum sint definienda; de quibus alii plura.

Hypothesis I.D. Cassini

Fig. 6

Fig. 7

Fig. 8

Jumping genes

Barbara McClintock 1902–92

After gaining her PhD in botany at Cornell University, USA, in 1927, Barbara McClintock began to study the genetics of maize. At the time, most geneticists used the fruit fly as their 'model' organism, but maize was preferred at Cornell. The colour of the kernels on a cob of maize are a clear expression of its genetic inheritance, while the plant's large chromosomes, which carry the genes, are easier to study under the microscope. And the slow maturation of maize allows the researcher more time to reflect on a genetic experiment.

By 1931 McClintock had shown that the exchange of genes during the production of germ cells – known as 'meiosis' – is accompanied by an exchange of chromosomal material. The experiments are regarded as a milestone in the history of genetics, for they establish the link between chromosomes and genetic inheritance.

But McClintock is probably better known for her work on 'jumping genes'. In 1941 she moved to the Cold Spring Harbor Laboratory in New York state, which was to become a famous gathering place for pioneers in molecular biology. Noting the occasional appearance of odd-coloured spots and splashes on the leaves and kernels of her maize plants, she began to wonder about mechanisms that controlled the genes for colour. She developed the idea that there were mobile genetic elements that could jump around the chromosome. When they jumped into a gene, they disrupted its switching on and off. The genome – the total complement of genetic material in a cell – was far more fluid than anyone had ever imagined.

When McClintock presented this work to the genetics community in 1951, she was met with blank stares and indifference – even whispers that she was a little mad. By the 1970s, however, McClintock's mobile genetic elements, named 'transposons', had been discovered in a number of organisms. She was rewarded for her pioneering work with the Nobel Prize for Physiology or Medicine in 1983.

Right Barbara McClintock, aged 81, on winning the 1983 Nobel Prize for Physiology or Medicine. She holds corn of the type she used in her studies on mobile genetic elements.

Chemical oscillations

Boris Pavlovitch **Belousov** 1893–1970,   Anatoly **Zhabotinsky** *b.*1938

In 1951 the Soviet biochemist Boris Belousov observed an oscillating chemical reaction. He was attempting a 'test-tube' version of a metabolic process. At first yellow, his solution of inorganic ingredients turned clear. But within moments it turned yellow again, then clear. It seemed unable to settle into a steady state. His claim received short shrift, and for good reason. The lack of a preferred direction – an arrow of time – implied violation of the second law of thermodynamics: that entropy must always increase.

Even so, the American chemist William Bray had also observed an oscillatory reaction in 1921. What's more, he had tried to explain it using the work of the mathematician Alfred Lotka which showed theoretically how a chemical reaction might develop temporary, damped oscillations. Yet it was not until the 1960s that the careful experiments of Anatoly Zhabotinsky, a biochemist in Moscow, showed that Belousov's oscillations were real. He modified the reaction mixture so that it changed colour more dramatically, from red to blue.

This so-called Belousov–Zhabotinsky (BZ) reaction came to the attention of Western scientists in 1968, and became a topic of international research. Lotka's scheme included the crucial element of feedback – one of the products of the reaction catalysed its own formation (autocatalysis). This 'nonlinear' behaviour (meaning that effects do not change in direct proportion to causes) can have striking consequences. Researchers showed that a hypothetical autocatalytic reaction can account for the BZ reaction. The second law survives because the chemical oscillations happen out of equilibrium: they will eventually disappear unless the mixture is fed with fresh reagents and end-products are removed.

Under certain conditions the colour change can spread through the medium in concentric ripples. These can be mutated into spirals as well as stationary patterns. A similar chemical process might induce animal spots and stripes during embryo growth. The BZ reaction is also analogous to the way electrical signals coordinate tissue contraction in the beating heart.

See also Laws of thermodynamics pages 164–165, Slime-mould aggregation pages 348–349, Genetics of animal design pages 460–461

Right Chemical waves in the BZ reaction. Dismissed as artefacts of poor experimental technique, Belousov's findings were almost buried from sight in obscure conference proceedings.

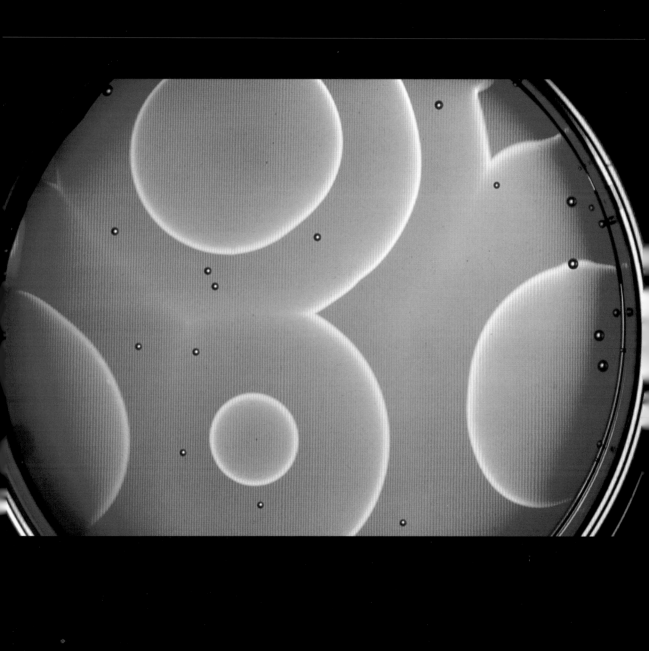

Nerve impulses

Alan Lloyd Hodgkin 1914–98, Andrew Fielding Huxley *b*.1917

Information is carried along nerve fibres, or 'axons', by electric impulses called 'action potentials'. In 1952 the English physiologists Alan Hodgkin and Andrew Huxley (grandson of Charles Darwin's 'bulldog', Thomas Henry Huxley) provided a detailed explanation of how action potentials arise and pass along axons. Along with John Eccles, Hodgkin and Huxley won the 1963 Nobel Prize for Physiology or Medicine.

Hodgkin and Huxley worked on the giant axon of the squid as its size and accessibility make it a convenient model for studying the electrical and physiological changes during an action potential. When in its resting state, the membrane of the axon is polarized so that the inside is slightly negatively charged relative to the outside. Depolarization of the membrane causes positively charged atoms, or ions, of sodium to flood in through specific channels in the membrane. A transitory positive feedback mechanism causes more of these sodium-ion channels to open, leading to an even greater flow into the cell. As depolarization increases, the sodium-ion channels start to close, and other channels open that pump potassium ions out of the cell, eventually returning the membrane to its resting potential. The channels mediating these changes open and close according to the voltage across the membrane.

Hodgkin and Huxley went on to derive equations that predict almost perfectly the speed and strength of an action potential as it travels along the axon. Their studies also helped to explain why action potentials travel long distances without decay, a property crucial to the efficient transmission of information in the nervous system. Their model is very general, applying not only to the squid's giant axon, but also to many other types of excitable cell, as well as to other active nerve structures such as dendrites and synapses.

See also Nervous system pages 210–211, Neurotransmitters pages 274–275, Growth of nerve cells pages 368–369, Nitric oxide pages 496–497

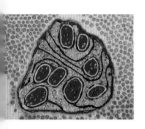

Above Electron micrograph of a section through nerve fibres (red).

Right 'Animal electricity': although his experiments didn't quite prove it, the 18th-century Italian anatomist Luigi Galvani rightly thought that frog tissue could produce electricity, and that such electricity could excite muscles.

Growth of nerve cells

Rita Levi-Montalcini *b*.1909, Stanley Cohen *b*.1922

A home laboratory in Italy during the Second World War seems an unpromising environment for a brilliant female scientist living in fear of anti-Semitic persecution – and hardly an ideal one for pursuing research that would ultimately lead to a Nobel Prize. So it was for Rita Levi-Montalcini, a medical graduate of the University of Turin, now famed for her work on the growth of nerve cells.

The growing cells of the peripheral nervous system send out projections, called axons, to specific targets, such as muscles, guided by chemical cues and growth factors, the nature of which were totally mysterious until the late 1940s. Working in secret, and using equipment sometimes cobbled together from kitchen utensils, Levi-Montalcini showed how the growth of peripheral nerve cells in chicken embryos was affected by limb amputation, research she continued after the war in the laboratory of Viktor Hamburger at Washington University, St Louis. An important discovery was that nerve-cell death, as well as growth and specialization, is a normal aspect of embryonic development, and that the number of surviving nerve cells is related to the size of the target tissue. Removal of an embryonic limb, for example, greatly increases nerve-cell death in the dorsal root ganglion, a region that normally supplies nerve fibres to the developing limb.

By 1952 Levi-Montalcini had discovered that mouse tumours implanted into chick embryos produce a diffusible substance that promotes nerve growth near the tumour, a phenomenon that could be mimicked in cultures of nerve cells. The substance – dubbed 'nerve growth factor' – was isolated and chemically analysed by Stanley Cohen in 1954. Cohen and Levi-Montalcini shared the 1986 Nobel Prize for Physiology or Medicine.

Nerve growth factor belongs to a family of so-called 'neurotrophic factors', each acting through specific receptors on the cell surface. Therapies based on these factors show great promise for promoting recovery in damaged adult nervous systems.

See also Nervous system pages 210–211, Neurotransmitters pages 274–275, Slime-mould aggregation pages 348–349, Chemical oscillations pages 364–365, Genetics of animal design pages 460–461

Right A mature nerve cell grown in culture. Under the influence of nerve growth factor, an immature nerve cell develops a thick mass of extensions.

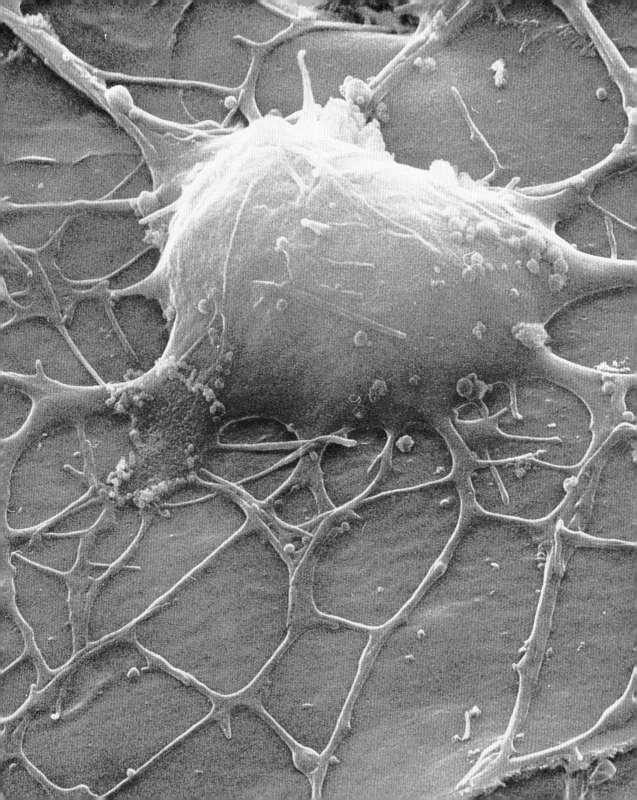

Origin of life

Stanley Lloyd Miller *b.*1930, Harold Clayton Urey 1893–1981

Living creatures are now reproduced by other living creatures, but how did life first originate? The Russian biologist Aleksandr Ivanovich Oparin had suggested in 1924 how the molecular building blocks of life could have formed from simpler chemicals. These ideas were taken further by the American atmospheric chemist Harold Urey. Urey suggested that the early, prebiotic atmosphere of Earth would have lacked oxygen, but contained ammonia, methane, water vapour and hydrogen. Electric discharges from lightning, or ultraviolet radiation, could produce simple biological molecules – molecules such as amino acids and sugars that are now used by living organisms – from these prebiotic precursor molecules. But with Urey, these ideas remained theoretical. They were not experimentally tested until 1953 when a student, Stanley Miller, asked Urey if he could test his ideas as a thesis project. Urey's interests had moved on, but he allowed Miller to do the work in his laboratory.

Within only a few months, Miller enjoyed spectacular success. He produced a wide range of biological molecules in a test-tube 'atmosphere' of simpler chemicals. His results caused an uproar because he had undermined one of the remaining religious challenges to the scientific worldview. People who accepted evolution might, before Miller, have still believed that God had acted to create biological molecules from chemicals. Miller had demonstrated that natural processes could do the job and helped to break down the distinction between the living and the non-living.

Since then much research on the origin of life has followed Miller's basic experimental set-up, and all the building blocks have been synthesized in this way. But the investigation is unfinished. We also need to know how the molecular building blocks of life (which Miller's experiments produce) could assemble into a self-reproducing system. No one has yet managed to crack that problem.

See also Spontaneous generation pages 90–91, Synthesis of urea pages 142–143, Alien intelligence pages 398–399, Oldest fossils pages 410–411, Symbiotic cell pages 418–419, Five kingdoms of life pages 426–427, Martian microfossils pages 518–519, Water on the Moon pages 522–523

Right Vital spark: Miller recreates his famous experiment on the origins of life. According to Urey, Miller's research advisor, 'If God did not do it this way, then he missed a good bet'.

REM sleep

Nathaniel **Kleitman** 1895–1999, Eugene **Aserinski** 1921–98,
William Charles **Dement** *b.*1928

In 1868 Wilhelm Griesenger not only observed twitching eyelids in sleeping humans and animals, but also thought they were somehow related to dreaming. George T. Ladd claimed that during deep dreamless sleep we turn our eyeballs upwards and inwards – the position 'most favourable to the disappearance from consciousness of all disturbing visual images' – but that during vivid visual dreams the eyeballs move 'gently in their sockets', taking various positions 'induced by retinal phantasms'.

E. Jacobson, the author of a popular book on the 'ABC of Restful Sleep' that appeared in 1938, was the first to suggest that these eye movements could be measured electrically and recorded photographically. But it wasn't until 1953 that Nathaniel Kleitman of Chicago, with Eugene Aserinski and William Dement, opened an era of laboratory techniques for studying dreams. In a series of controlled experiments they found a link between distinctive patterns of brain electrical activity, characteristic rapid eye movements (REM) and dreaming. When people were awakened during periods of REM sleep and low-voltage, high-frequency brain waves, they were much more likely to say they had been dreaming than at other times. During these periods, breathing, heart beat and blood pressure increased to waking levels.

Sleepers have several periods of REM sleep throughout the night, each lasting about 10–20 minutes. If continually disturbed during these periods, then they begin to suffer psychological distress, and the periods of REM sleep mount up during successive nights to make up for the lost dreaming. Even so-called 'non-dreamers' have REM sleep just like those of dreamers – Aristotle was in fact the first to suggest that sleepers always dreamed, whereas non-dreamers simply had poor dream recall. The fact that all animals and birds do it – even fetuses in the womb – argues that dreaming sleep must have an important evolutionary role, although exactly what this is remains a matter of debate.

See also Aristotle's legacy pages 16–17, Regulating the body pages 188–189, The unconscious mind pages 222–223

Right 'The Dream of Reason Produces Monsters' by Francisco de Goya, 1799. Francis Crick and Graeme Mitchison have proposed that dreams erase unwanted associations and memory traces.

The double helix

Francis Harry Compton Crick *b.*1916, James Dewey Watson *b.*1928

Deoxyribonucleic acid (DNA) is the most important molecule of our time. That this is so can be traced back to one key discovery: the structure, or shape, of DNA. A knowledge of a molecule's structure does not always lead to an understanding of how the molecule works, but it did in the case of DNA. James Watson, a young American, arrived in Cambridge, England, in 1951 and began working on the structure of DNA together with Francis Crick, a British PhD student. It was a hot research topic because DNA had only recently been shown to be the molecule of biological inheritance.

Watson and Crick deduced the structure of DNA using one chemical clue and the method of X-ray diffraction. DNA is too small for its structure to be observed directly, and X-ray diffraction is an indirect method of working out the structure of very small entities. The chemical clue came from a rule that had been noticed by Erwin Chargaff. DNA contains four kinds of subunit, symbolized by the letters A, C, G and T. Chargaff found that the amount of C equals the amount of G, and that the amount of A equals the amount of T. This suggested to Watson and Crick that DNA has two strands, with G in one strand bound to C in the other, and A bound to T. X-ray diffraction told them that the strands had a helical shape: DNA is a double helix.

The structure, which Watson and Crick announced in *Nature* in 1953, immediately suggested how the molecule could be reproduced (unwind the strands and one strand can act as a master-copy to produce a new strand) and how it could contain biological information (the sequence represented by the letters A, C, G and T makes up a code). In the next decade or so biologists cracked this 'code' and set the stage for modern molecular genetics.

Above DNA consists of an outer sugar-phosphate backbone (light blue), with four subunits or 'bases' (spheres) making up the genetic code.

See also X-rays pages 220–221, Genes in inheritance pages 264–265, Genes in bacteria pages 332–333, Information theory pages 354–355, Sickle-cell anaemia pages 358–359, Structure of haemoglobin pages 390–391, Random molecular evolution pages 422–423, Genetic engineering pages 436–437, Human cancer genes pages 456–457, Genetics of animal design pages 460–461, Directed mutation pages 494–495, Maleness gene pages 502–503, Human genome sequence pages 524–525

Right James Watson (left) and Francis Crick with their model of DNA. In the words of the biologist Max Perutz, '1953 became the *annus mirabilis*. The Queen was crowned, Everest was climbed, DNA was solved'.

The digital river
Richard Dawkins

OUR GENETIC SYSTEM, WHICH is the universal system of all life on the planet, is digital to the core. With word-for-word accuracy, you could encode the whole of the New Testament in those parts of the human genome that are at present filled with 'junk' DNA – that is, DNA not used, at least in the ordinary way, by the body. Every cell in your body contains the equivalent of forty-six immense data tapes, reeling off digital characters via numerous reading heads working simultaneously. In every cell, these tapes – the chromosomes – contain the same information, but the reading heads in different kinds of cells seek out different parts of the database for their own specialist purposes. That is why muscle cells are different from liver cells. There is no spirit-driven life force, no throbbing, heaving, pullulating, protoplasmic, mystic jelly. Life is just bytes and bytes and bytes of digital information.

Genes are pure information – information that can be encoded, recoded and decoded, without any degradation or change of meaning. Pure information can be copied and, since it is digital information, the fidelity of the copying can be immense. DNA characters are copied with an accuracy that rivals anything modern engineers can do. They are copied down the generations, with just enough occasional errors to introduce variety. Among this variety, those coded combinations that become more numerous in the world will obviously and automatically be the ones that, when decoded and obeyed inside bodies, make those bodies take active steps to preserve and propagate those same DNA messages. We – and that means all living things – are survival machines programmed to propagate the digital database that did the programming. Darwinism is now seen to be the survival of the survivors at the level of pure, digital code.

With hindsight, it could not have been otherwise. An analog genetic system could be imagined. But we have already seen what happens to analogue information when it is recopied over successive generations. It is Chinese Whispers. Boosted telephone systems, recopied tapes, photocopies of photocopies – analog signals are so vulnerable to cumulative degradation that copying cannot be sustained beyond a limited number of generations. Genes, on the other hand, can self-copy for ten million generations and scarcely degrade at all. Darwinism works only because – apart from discrete mutations, which natural selection either weeds out or preserves – the copying process is perfect. Only a digital genetic system is capable of sustaining Darwinism over eons of geological time. Nineteen-fifty-three, the year of the double helix, will come to be seen not only as the end of mystical and obscurantist views of life; Darwinians will see it as the year their subject went finally digital.

The river of pure digital information, majestically flowing through geological time and splitting into three billion branches, is a powerful image. But where does it leave the familiar features of life? Where does it leave bodies, hands and feet, eyes and brains and whiskers, leaves and trunks and roots? Where does it leave us and our parts? We – we animals, plants, protozoa, fungi and bacteria – are we just the banks through which rivulets of digital data flow? In one sense, yes. But there is, as I have implied, more to it than that. Genes don't only make copies of themselves, which flow on down the generations. They actually spend their time in bodies, and they influence the shape and behaviour of the successive bodies in which they find themselves. Bodies are important too.

The body of, say, a polar bear is not just a pair of riverbanks for a digital streamlet. It is also a machine of bear-sized complexity. All the genes of the whole population of polar bears are a collective – good companions, jostling with one another through time. But they do not spend all the time in the company of all the other members of the collective: they change partners within the set that is the collective. The collective is defined as the set of genes that can potentially meet any other genes in the collective (but no member of any of the thirty million other collectives in the world). The actual meetings always take place inside a cell in a polar bear's body. And that body is not a passive receptacle for DNA.

For a start, the sheer number of cells, in every one of which is a complete set of genes, staggers the imagination: about nine hundred million million for a large male bear. If you line up all the cells of a single polar bear in a row, the array would comfortably make the round trip from here to the Moon and back. These cells are of a couple of hundred distinct types, essentially the same couple of hundred for all animals: muscle cells, nerve cells, bone cells, skin cells and so on. Cells of any one of these distinct types are massed together to form tissues: muscle tissue, bone tissue and so on. All the different types of cells contain the genetic instructions needed to make any of the types. Only the genes appropriate to the tissue concerned are switched on. This is why cells of the different tissues are of different shapes and sizes. Most interestingly, the genes switched on in the cells of a particular type cause those cells to grow their tissues into particular shapes. Bones are not shapeless masses of hard, rigid tissue. Bones have particular shapes, with hollow shafts, balls and sockets, spines and spurs. Cells are programmed, by the genes switched on inside them, to behave as if they know where they are in relation to their neighbouring cells, which is how they build their tissues up into the shape of ear lobes and heart valves, eye lenses and sphincter muscles.

Contraceptive pill

Gregory Pincus 1903–67, Min-Chueh Chang 1908–91, John Rock 1890–1984

In the first half of the twentieth century biologists unravelled the complicated hormonal control of mammalian reproduction. Meanwhile, the exploitation of steroids as therapeutic agents gained ground with the mass production of cortisone. Equipped with a new understanding of the female menstrual cycle, scientists perceived a solution to period problems, pre-menstrual tension and infertility in the oral administration of sex hormones. But before research could continue they needed cheap, reliable, synthetic sex hormones, such as progesterone, as the purification of hormones from animal ovaries was expensive and inefficient.

In 1943 Russell Marker extracted the steroid diosgenin from a wild yam growing in Mexico and transformed it in the laboratory into progesterone (or 'progestogen', as the synthetic variety became known). The drawback was that huge amounts were required for an effective oral dose. In 1951 Luis Miramontes, working under the direction of Carl Djerassi, modified Marker's progestogen to form norethisterone ('norethindrone' in the USA), which was far more active than human progesterone when taken by mouth. A year later Frank Colton developed a similar compound, norethynodrel. Both products were released onto the market for the treatment of gynaecological disorders.

In 1951 the American biologist Gregory Pincus and colleagues established that the newly synthesized progestogens could inhibit ovulation. The implications of this work were soon recognized by the veteran campaigner on women's issues, Margaret Sanger. With the help of the wealthy Katherine McCormick, Sanger arranged for Pincus, Min-Chueh Chang and John Rock to receive a large grant for research into effective hormone-based birth control. A large clinical trial was mounted among the poor of Rio Pedras, Puerto Rico, in the mid-1950s. In May 1960 the US Food and Drug Administration licensed norethynodrel as an oral contraceptive under the name of Enovid. By 1965 the sexual revolution was well underway, and more than 6.5 million American women were 'on the pill'.

See also Regulating the body pages 188–189, Insulin pages 288–289, Nitric oxide pages 496–497, Maleness gene pages 502–503

Right Post-pill paradise: women in the swinging sixties could take contraception into their own hands.

Why can we make generalizations about nature? Because some things don't matter. You can often change certain things such as location or direction, and objects still behave in the same way – so if your car can reach 170 kilometres an hour driving north, you expect it to achieve the same speed driving east. Similarly, it seems common sense that the world should be mirror symmetric: if you can spin a coin clockwise it should be possible to spin it anticlockwise.

And yet there is a point at which the mirror breaks. Tsung Dao Lee and Chen Ning Yang realized in 1956 that some reactions between subatomic particles made it look as though one of the forces of nature, the weak nuclear force (responsible for the decay of neutrons), violates mirror symmetry. Experiments soon proved them right. Neutrons can spontaneously decay into sets of three particles: a proton, an electron and a neutrino. When this happens, the neutrino is always left-handed: that is, it spins around its direction of travel like the action of a corkscrew turning anticlockwise. Startled by this, physicists started to question other symmetries of nature. They soon found that the weak nuclear force breaks another old rule – it affects matter and antimatter in a slightly different way. Unexpected as this was, it could have a profound significance.

Without some natural bias between matter and antimatter, we ought not to be here. Long ago, all the matter in the universe would have met all the antimatter and exploded in a shower of radiation. But if, in the cauldron of the big bang, matter was slightly more likely to emerge from the constant collision of subatomic particles, there would have been some of it left over after all the antimatter had been mopped up. Those dregs are us.

See also The electron pages 228–229, The neutron pages 312–313, Antimatter pages 314–315, A subatomic ghost pages 384–385, Afterglow of creation pages 412–413, Unified forces pages 416–417, Superstrings pages 474–475

Chemical basis of vision

George Wald 1906–97

When packets of light – called photons – enter the eye and strike the retina, the energy they contain has to be converted, via a series of complex steps, into an electrical signal that passes from the retina to the optic nerve and then to the brain. The first part of this process involves visual pigments in the rod and cone cells, the retina's light receptors.

In 1956 the underlying chemical mechanisms were worked out by the biochemist George Wald, an émigré from Poland to the USA. During the First World War it was noted that dietary deficiency of vitamin A caused blindness, which pointed to a crucial role of vitamin A in vision. In 1933, working at Harvard University, Wald succeeded in isolating vitamin A from the retina. Vitamin A is used in the retina to form rhodopsin and other related visual pigments. These always consist of two parts: a colourless protein called an opsin that spans the membranes of disc-like structures in rod and cone cells; and, deep within the opsin and joined to it, a derivative of vitamin A called retinal.

When struck by a photon, the retinal absorbs the light energy and changes shape from 'kinked' to 'straight'. This converts the opsin part of the molecule into an active enzyme, sparking a cascade of reactions that results in the release of a transmitter at the junction between the light-receptor cell and an optic-nerve cell.

Exposure to light quickly releases the retinal from the opsin. Because some of this retinal is destroyed and is not recycled, it has to be replenished, using up stocks of vitamin A. Humans cannot make their own vitamin A, but plants produce it as a part of the molecule carotene. That explains why eating carrots is good for your eyesight, and also why a shortage of fresh vegetables can lead to vitamin A deficiency and blindness.

See also Nervous system pages 210–211, Vitamins pages 248–249, Neurotransmitters pages 274–275, Nerve impulses pages 366–367, Right brain, left brain pages 400–401, Nitric oxide pages 496–497

Right Rod cells in the human retina. Numbering around 130 million, rods are sensitive to very dim light and are therefore considered to be the receptors for night vision.

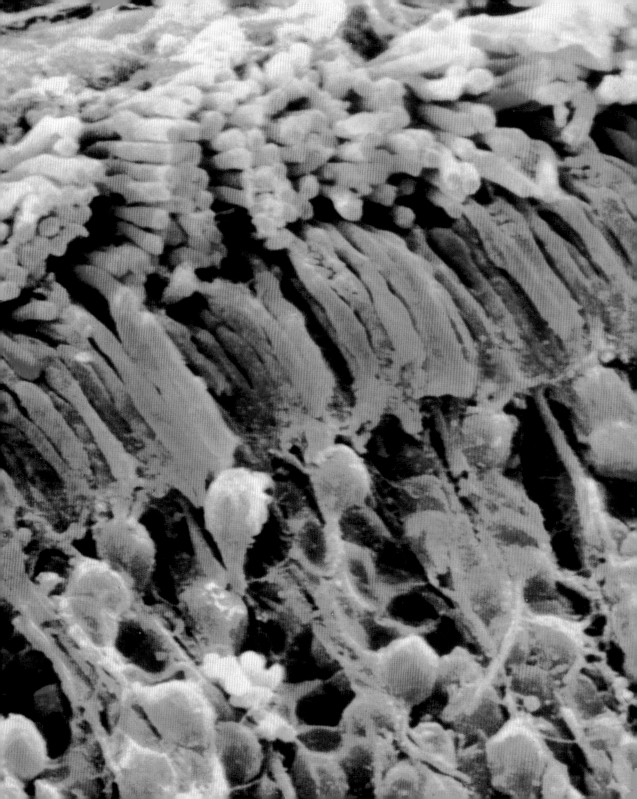

A subatomic ghost

Wolfgang Pauli 1900–58, Frederick Reines 1918–98, Clyde Cowan 1919–74

When a nucleus decays by emitting an electron, energy disappears. Or so it appears. This disturbed physicists so much that, in 1930, Wolfgang Pauli invented a new subatomic particle just to account for it. This particle could spirit away the energy from this process of beta decay. It sounds like a fix, but it was a fix that worked very neatly.

Enrico Fermi named this particle the neutrino, which is Italian for 'little neutral one'. He developed a detailed theory of beta decay that involved an entirely new fundamental force, the so-called weak force. The neutrino is unusual because it feels only gravity and the weak force; it is impervious to electromagnetic and strong nuclear forces. That makes it slippery enough to fly straight through the Earth – so it can easily steal away undetected with the spare energy from beta decay.

Nevertheless, the weak force does make neutrinos collide with other particles occasionally. When, in 1956, Fred Reines and Clyde Cowan set up their detector next to a nuclear reactor, they saw distinctive patterns of gamma rays caused by neutrinos hitting protons.

Neutrino telescopes – large underground tanks of fluid monitored for the by-products of these weak interactions – now routinely detect neutrinos produced by fusion reactions in the Sun's core. In fact, they see fewer than astrophysicists predict they should, a shortfall that is still unexplained. In 1987 the same telescopes detected a vast stellar explosion, a supernova, in a galaxy near ours. That blast of neutrinos confirmed the theory that a tiny super-dense neutron star was created in the explosion.

According to the standard model of particle physics, neutrinos have no mass. But in 1998 the Super Kamiokande neutrino detector in Japan found evidence that they actually have some small mass. This could be a sign of a new fundamental physics, and it could mean that the gravity of neutrinos has affected the formation of galaxies.

See also Radioactivity pages 224–225, Left-handed universe pages 380–381, Quarks pages 408–409, Afterglow of creation pages 412–413, Unified forces pages 416–417, Gamma-ray bursts pages 434–435, Supernova 1987A pages 488–489, Great Attractor pages 498–499

Right Buried a kilometre under a mountain near Tokyo, Japan, the Super Kamiokande neutrino detector contains 50,000 tonnes of purified water and 13,000 sensors that pick up the distinctive light bursts from neutrino collisions.

Language instinct

Noam Chomsky *b.*1928

Notable also for his political activism, the American Noam Chomsky introduced ideas about language that have important implications for psychology and linguistics. Contrary to what behaviourists such as B. F. Skinner had proposed – that we learn language by accumulating examples from experience – Chomsky set out to show that language is a skill that human beings are innately predisposed to acquire.

With the publication of his 1957 book *Syntactic Structures*, he started the movement known as 'generative linguistics', the proponents of which claim that we understand and produce language according to rules, and that at their most basic these rules are common to all human languages. Despite the apparently vast differences between, say, English and Chinese, these languages share a 'deep structure', which is why any human infant can acquire any human language. It is at the level of 'surface structure' that languages vary so much.

Chomsky's 'generative grammars' were designed to explain how it is that we can recognize, understand or produce countless sentences we have never heard before, make partial sense of sentences that contain words we don't know, and recognize that some words are probably made up, such as 'fteggrup' or 'nganga'. We can recognize a well-formed sentence in English and understand the relationship between its parts even if it is meaningless: 'Blotherasts argle contornaceously bethwart mungled chardwicks and fintipled mesterlinks'.

Chomsky proposed rule systems that account rather simply for the way in which sounds combine and change, and how words change their form, even so-called 'irregular' verbs. Part of the controversy surrounding his work is related to the extent to which he claims an innate rather than a learned basis for language – a claim that has been hotly contested, not least by behaviourists such as Skinner. Although the human brain has areas of specialized function, a 'language organ' that imposes structure on incoming speech has yet to be identified.

See also Mapping speech pages 182–183, Animal instincts pages 318–319, Behavioural reinforcement pages 322–323, Artificial neural networks pages 334–335, Right brain, left brain pages 400–401, Images of the mind pages 478–479

Right 'Tower of Babel' by Pieter Brueghel, 1563. Chomsky aimed to establish a 'universal grammar' that would account for the range of linguistic variation in what is humanly possible.

In 1958 Eugene Parker proposed that the Sun's two-million-degree-Celsius corona is expanding in all directions, producing a stream of electrically charged particles (plasma), mainly electrons and protons, which blows outwards through the Solar System. Initially treated with considerable scepticism, his 'solar wind' theory was soon established as astronomical fact. But the subject had a long history. At the start of the twentieth century George FitzGerald and Oliver Lodge hinted that magnetic storms experienced on Earth were related to increased solar-flare and sunspot activity a few days prior to the storms. It seemed that something was emitted by the flares that eventually reached Earth; and from the time lapse, FitzGerald estimated that it travelled at roughly 500 kilometres per second.

Later findings offered support for the solar-wind theory. Fred Hoyle showed that the magnetic connections between an expanding solar plasma and the surrounding galactic magnetic fields would slow down the solar rotation, and thus remove a long-standing problem with the nebular hypothesis of planetary origins; Sydney Chapman proposed that the impingement of a stream of solar material on the open field lines of Earth's magnetosphere (which are in effect belts of high concentrations of charged particles that extend out into space) would cause the aurorae (the northern and southern lights); Cuno Hoffmeister calculated the velocity of a putative solar wind from the 'wind-sock' angle of comet tails; and Ludwig Biermann postulated that it would account for the breaking up of comet tails by producing accelerating knots of plasma (mainly electrically charged carbon monoxide).

Parker's solar wind coupled with the solar rotation draws out the solar magnetic field lines into an Archimedes' spiral. Starting with the Soviet space probes Lunik 3 and Venera 1 of 1959, a host of spacecraft has investigated the interaction of the solar wind with the magnetospheres of the planets. The steady expansion of the solar corona is sometimes supplemented by energetic mass ejections during solar flares.

Structure of haemoglobin

Max Ferdinand **Perutz** 1914–2002

X-ray crystallography is a powerful technique for working out the three-dimensional structure of a molecule. It involves directing a beam of X-rays at a crystalline specimen. As the X-rays pass through, they are deflected by the regularly spaced layers of atoms inside the crystal, just as light is diffracted by a fine grating. The pattern of the diffracted X-rays can be recorded on a photographic plate and used to calculate the position of each atom.

The method was developed between 1912 and 1915 by Max von Laue and by the father-and-son team William Henry and William Lawrence Bragg. It was initially used on simple substances such as salt. The Austrian-born biochemist Max Perutz was one of the first to apply it to protein molecules, which contain thousands of atoms and have structures far more complex than those of minerals. In 1937, at Cambridge, England, he began work on haemoglobin, the protein in red blood cells that carries oxygen from the lungs to the tissues. He grew large crystals of horse haemoglobin while his colleague John Kendrew turned to the related but simpler protein myoglobin.

Perutz and Kendrew gradually refined their techniques. Crucially, they found that the addition of an atom of a heavy element such as gold or mercury to the molecule would improve the results of X-ray diffraction. By 1957 Kendrew had cracked the structure of myoglobin and Perutz followed with haemoglobin two years later. Perutz went on to show how the four subunits of haemoglobin alter their structure when they take up oxygen. He also discovered that haemoglobin molecules collapse into a sickle shape in the blood disorder sickle-cell anaemia. Perutz's and Kendrew's work opened the door to the study of the structure and function of proteins – the master molecules of the cell – and their methods are still very much in use today. The two scientists shared a Nobel Prize for Chemistry in 1962.

See also Antitoxins pages 214–215, X-rays pages 220–221, Sickle-cell anaemia pages 358–359, The double helix pages 374–375

Right Computer graphic of the haemoglobin molecule. Perutz recalled that 'the first protein structures revealed wonderful new faces of nature'.

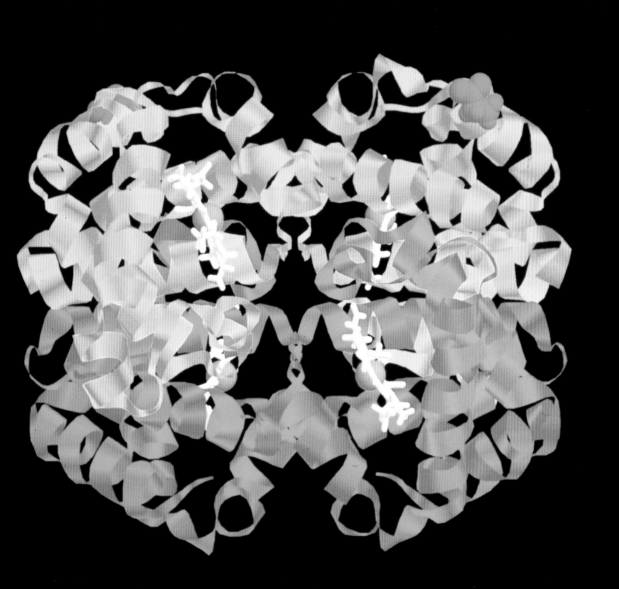

Olduvai Gorge

Louis Leakey 1903–72, Mary Leakey 1913–96

In 1959 Louis Leakey and Mary, his second wife, found a 1.75-million-year-old skull of *Zinjanthropus* (now *Paranthropus*) *boisei* at Olduvai Gorge, Tanganyika (now Tanzania). Louis read French and Kikuyu at the University of Cambridge but also studied anthropology and joined a British Museum expedition to Tanganyika. Persuaded by Darwin's prediction that evidence for human ancestry was to be found in Africa, Louis went on to spend more than 30 years looking for human-related fossils in East Africa before finding the robust australopithecine that he nicknamed 'Nutcracker Man'. Mary was also interested in archaeology and attended lectures at University College London before joining Louis' 1935 expedition to Olduvai.

Their 1959 find reinforced Raymond Dart's 1924 claim for the importance of the Taung child skull (*Australopithecus africanus*) and Robert Broom's later australopithecine finds (*A. africanus* and *Paranthropus robustus*) in South African caves. By the 1950s, Broom claimed that the 2.5–3-million-year-old *A. africanus* stood just over 1.2 metres (4 feet) tall and could walk fully upright more or less like a human.

Importantly, the Leakeys' East African finds were within stratified sediments whose relative age could be established from accompanying animal fossils, and later from the radioactive dating of inter-bedded volcanic lavas and ashes. Broom's Olduvai find *P. robustus* was the first hominid to be reliably dated by potassium–argon isotopes. And in 1960 the Olduvai excavations also produced the remains of another hominid along with primitive stone tools. Named *Homo habilis* in 1964 by Louis Leakey, Phillip Tobias and John Napier, it was then the earliest known tool-making and relatively large-brained hominid.

Mary led the 1976 expedition to Laetoli, Tanzania, which uncovered the oldest known hominid footprints (made around 3.75 million years ago) and verified Broom's claim that australopithecines could walk fully upright.

Above Hominid footprints, fossilized in volcanic ash. This 70-metre trail was found by Mary Leakey's expedition at Laetoli, Tanzania. It dates from 3.75 million years ago.

See also Prehistoric humans pages 148–149, Neanderthal man pages 170–171, Java man pages 216–217, Taung child pages 298–299, Radiocarbon dating pages 346–347, Ancient DNA pages 476–477, Nariokotome boy pages 480–481, Out of Africa pages 492–493, Iceman pages 504–505

Right Remains of the day: Mary and Louis Leakey study fossilized skull fragments in Tanganyika, 1959.

Hayflick limit

Leonard Hayflick *b.*1920

For many years biologists believed that cells were immortal – it was being part of an organism that made them die. This idea arose from the work of the French surgeon Alexis Carrel, who began to culture cells from a chick's heart to see how long they could survive outside the body. The cells outlived Carrel, who died in 1944, and were finally discarded two years later.

But in 1961 the American biologist Leonard Hayflick suggested that Carrel was wrong and that cells do have a limited lifetime after all. He grew several types of human cell in culture and showed that they always died after dividing about 50 times. The older the cells were at the start of the experiment, the fewer cycles they went through before they expired. Later studies revealed that the number of divisions a cell goes through is related to an organism's life span. Mice, which live for up to three-and-a-half years, manage 14 to 28 cell divisions. But Galapagos tortoises, with a life span of 175 years, boast 90 to 120 rounds. The 'Hayflick limit' gives humans a potential life span of about 120 years. But most of us don't survive that long, because our cells accumulate damage as we age. The ends of chromosomes – known as telomeres – get shorter as cells divide. Researchers are looking into ways of blocking telomere shortening as a potential way of extending the lives of cells and, by implication, the human life span. (Intriguingly, Dolly, the sheep cloned from an adult cell in 1996, had shorter telomeres than expected.)

Death, then, is inevitable. The only cells that are immortal are cancer cells, which multiply without check. In a healthy state, genes are programmed to kill off damaged cells, a process known as apoptosis. Again, researchers are working on various therapies based on apoptosis with the aim of limiting tissue damage caused by conditions such as stroke.

Above In 'apoptosis', the cell's nucleus fragments and its cytoplasm shrinks to form grape-like clusters.

See also Communities of cells pages 174–175, Human cancer genes pages 456–457, Dolly the cloned sheep pages 516–517

Right HeLa cells. In 1951 Henrietta Lacks, a 31-year-old cervical cancer patient, died in Baltimore. Her tumour cells have been continuously cultured ever since as a tool for cancer research.

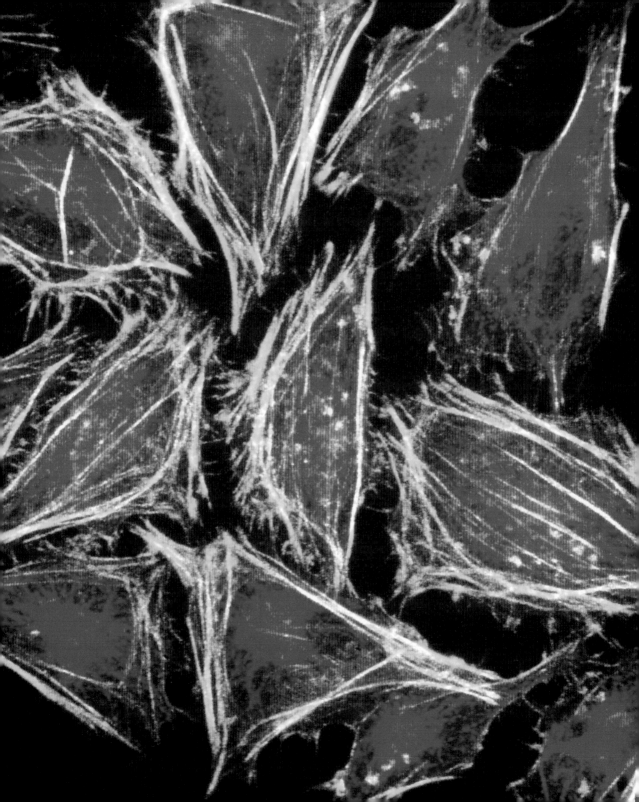

Chimpanzee culture

Jane Goodall *b.*1934

Humans and chimpanzees are descended from a common ancestor that lived about five million years ago. Yet despite millennia of coexistence, by the middle of the twentieth century humans still knew next to nothing about the everyday lives of their closest sister species. Then, in 1961, Jane Goodall set up camp at the Gombe Stream Reserve in Tanzania. Slowly and patiently, she accustomed the chimpanzees of Gombe to her presence and logged details of their behaviour. Her work not only taught us about chimpanzees but also challenged our perception of our own uniqueness. Perhaps the most startling discoveries were that chimpanzees not only use but actually fashion a range of tools with which to fish and probe for food; that they hunt in groups and share the spoils; and that the males make lethal war-like expeditions against their neighbours.

As others followed Goodall's pioneering work and developed long-term studies in different parts of Africa, the full range of the chimpanzee's behavioural capacities was recognized. The latest work has brought together the results of a total of 151 years of observation at seven long-term study sites to reveal that, like ourselves, chimpanzees show great cultural variation. Different communities, for example, vary in the tools they use. In West Africa chimpanzees crack nuts using stone hammers, a practice absent in East Africa despite the availability of the same materials. Among over 40 other local customs are a variety of courtship and grooming patterns.

Studies with captive chimpanzees show that they recognize themselves in mirrors, an ability shared with other great apes (gorillas and orang-utans). The mental sophistication implied by findings in the wild and in captivity, and in particular the similarities to our own psychology, have led some countries to give apes legal protection from use in biomedical experimentation. There is even a movement to award apes the equivalent of 'human rights'. This could scarcely have been conceived of half a century ago.

See also Genetic cousins pages 442–443

Right Jane Goodall encounters a curious chimpanzee at the Gombe Stream Research Centre in 1972.

Alien intelligence

Frank Drake *b*.1930

In 1961 Frank Drake, a radio astronomer at the National Radio Astronomy Observatory, Green Bank, West Virginia, was organizing a small meeting of scientists to discuss how radio astronomers, with a new 26-metre (85-foot) dish telescope, might search for signals from extraterrestrial civilizations. The project was called 'Ozma' after the queen of the fictional city of Oz. Drake built the meeting's agenda around a simple equation.

The equation gave the number of detectable civilizations in the Milky Way galaxy. This number is equal to a series of factors, each of equal importance, simply multiplied together. Start with the rate at which stars are formed in the galaxy and multiply this by the fraction of stars that have planetary systems. Then multiply by the fraction of those planets hospitable to life and by the fraction of those planets on which life has actually developed. Multiply by the fraction of 'living' planets that evolve intelligent life forms and by the fraction of intelligent life forms that want to communicate and are capable of doing so. Finally multiply by the length of time that such a communicating civilization remains detectable. (The recent development of the hydrogen bomb worried many participants that a civilization that could communicate would also possess the means to destroy itself.)

The answer then, as now, was a minimum of one thousand and a maximum of one hundred million. The pioneer of the search for extraterrestrial intelligence decided that an extraterrestrial life form might send a message using one of the quieter regions of the radio spectrum, between the 1,420-megahertz hydrogen (H) line and the 1,665-megahertz hydroxyl (OH) line. This became known as the 'water hole'. The form of the communication might be a series of Morse-code-like dots and dashes. Telescopes have been pointed in the direction of likely Sun-type stars and hours of radio hiss have been scanned. No message has been picked up... yet.

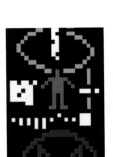

Above Part of a pictographic radio message transmitted into space in 1974. Earliest response, if any, isn't expected until 50,000 years' time.

See also 'Canals' on Mars pages 200–201, Origin of life pages 370–371, Planetary worlds pages 512–513, Galileo mission pages 514–515, Martian microfossils pages 518–519, Water on the Moon pages 522–523

Right First impressions: would we recognize extraterrestrials if we met them?

Right brain, left brain

Roger Wolcott Sperry 1913–94

The human brain looks symmetrical, and for many years it was thought that the two hemispheres performed roughly the same functions. But, beginning in 1962, the American neuroscientist Roger Sperry pioneered elegant studies in humans which showed that certain functions are located predominantly in one or other side of the brain, work for which he shared the Nobel Prize for Medicine in 1981.

The hemispheres are interconnected by two bands of nerve fibres, one of which is known as the 'corpus callosum'. To study the role of these interconnecting fibres in animals, Ronald Myers and Roger Sperry cut the corpus callosum and limited the visual input to the hemispheres by disconnecting fibres running from one eye to the opposite brain hemisphere. Animals trained to respond to a feature presented to one eye failed to respond when tested with the other eye. Clearly, the corpus callosum was needed to transfer visual information from one hemisphere to the other.

Extending the work to humans, Sperry and colleagues investigated epileptic patients whose interconnecting hemispheric fibres had been sectioned to stop epileptic activity spreading across the brain. Superficially, these patients appeared to behave normally, and previous researchers assumed that the hemispheres worked essentially independently. But Sperry showed not only that the corpus callosum had a similar function in humans and animals, but also that certain mental activities were performed better by one hemisphere than the other.

In fact, although the two sides of the brain do indeed have many functions in common, the left is specialized for language, whereas the right is specialized to deal with spatio-temporal problems, such as the mental rotation of geometric forms. It is as if we have two minds, one verbal and one non-verbal, residing in the opposite sides of our brains. This has profound philosophical implications for the nature of consciousness and our concept of 'self'.

See also Mapping speech pages 182–183, Nervous system pages 210–211, Artificial neural networks pages 334–335, Chemical basis of vision pages 382–383, Images of the mind pages 478–479

Right A drawing of the human brain and spinal column, c.1714. In everyday life Sperry's patients showed a unified psyche, but in the laboratory their separated hemispheres functioned as two minds.

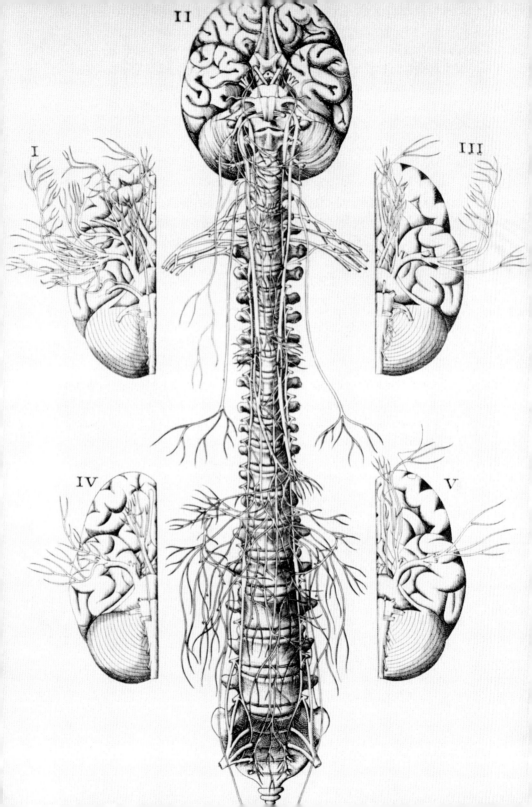

Psychology of obedience

Stanley Milgram 1933–84

Obedience to authority is instilled in us from birth – obedience to our parents, to our teachers, to our bosses, to the law. In fact it is a prerequisite for the functioning of any human society. But can this tendency to obey and conform explain why so many otherwise decent, law-abiding German citizens committed atrocities in the Second World War? Does their behaviour reveal a potential in us all?

In 1961–2 the American psychologist Stanley Milgram ingeniously explored the response of ordinary people to immoral orders. A volunteer would go to a laboratory ostensibly to take part in an experiment investigating the effect of punishment on learning. As the teacher, he was taken to a room where another person – the pupil – was strapped into a chair and electrodes were placed on his wrists. The teacher was then instructed to read a list of word pairs, ask the pupil to recall them and give an electric shock of increasing intensity every time the pupil made a mistake. Of course no shocks were actually delivered, and the pupil was merely a stooge acting as if in pain.

Appallingly, in the first experiment 25 out of 40 teachers continued delivering shocks until they reached the highest level – 450 volts – marked 'Danger: Severe Shock'. Worried teachers often turned to the experimenter, saying, 'Am I responsible?' And as soon as he said they were not, they seemed to proceed more easily, although many became extremely nervous and stressed during the experiment and begged the experimenter to stop. Follow-up experiments showed that teachers were less obedient when in the same room as the pupil or separated from the experimenter. And women were just as obedient as men.

It seems that many people do not have the resources to resist authority, even when directed to act callously and inhumanely against an innocent victim – an example of obedience unrestrained by conscience. This raises the age-old question: what is the correct balance between individual initiative and social authority?

See also The unconscious mind pages 222–223, Behavioural reinforcement pages 322–333

Right Milgram's subjects obeyed the experimenter no matter how vehement the pleading of the person being shocked or how painful the shocks seemed to be.

l'Étonnement

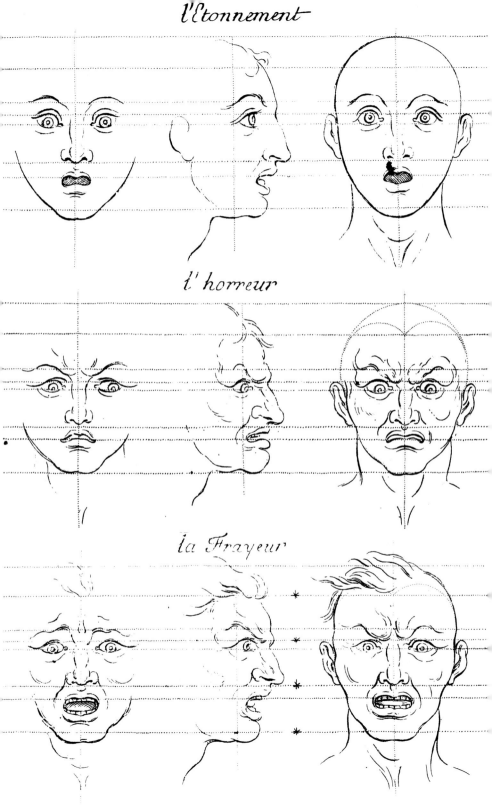

l' horreur

la Frayeur

Quasars

Maarten Schmidt *b.*1929

By the mid 1950s radio astronomers had an ever-growing list of radio sources that could not be identified. Photographic plates obtained using the Earth's largest reflecting telescopes were scanned diligently for matching optical stars or galaxies. In 1960 Thomas Matthews and Allan Sandage pinpointed a strange faint blue variable 'star' in the position of source 3C 48. And in 1963 another dim 'star' with a jet of material coming out of it was identified as 3C 273. These unusual objects were termed 'quasi-stellar radio sources', or 'quasars' for short.

The spectra of quasars were baffling: the broad emission lines were unfamiliar. Later in 1963 the Dutch-born American astronomer Maarten Schmidt realized that the lines were produced by nothing more than hydrogen, but that they were shifted way towards the red end of the spectrum. This enormous redshift meant that 3C 273 and 3C 48 had respective velocities of 15 per cent and 30 per cent that of light, and so were exceedingly far away (and old). To be visible at all, the optical sources must be extremely luminous. We now know that they must radiate about 100,000 times the energy of the entire Milky Way galaxy. Quasars are parts of extremely active galaxies. In fact they are the highly energetic cores, or central nuclei, of these galaxies and are known as 'active galactic nuclei' (AGN).

Two clues as to their structure come from nearby objects. The strange galaxy Centaurus A, with its thick dust lane and prominent regions of radio emission, is the closest example of an AGN. Martin Ryle noted that the quasar radio sources tended to be doublets, with the optical region lying between the two radio regions. The optical region is the site of energetic explosions, and these eject huge (200,000-solar-mass) clouds of material in diametrically opposite directions, which produce radio waves when they interact with the surrounding medium. The radio regions, like the coincident intense X-ray-emitting regions, are about a light-day across, compared with a typical galaxy which is tens of thousands of light-years across.

See also Spectral lines pages 130–131, Doppler effect pages 156–157, Stellar evolution pages 286–287, Black-hole evaporation pages 440–441, Great Attractor pages 498–499

Right Quasars are the cores of very active galaxies, and they are probably powered by massive central black holes. Pictured here is the nearest example of an active galaxy, Centaurus A.

Evolution of cooperation

William Donald **Hamilton** 1936–2000

How in the Darwinian view of life could one animal evolve to help another animal? Natural selection favours those individuals who leave the most offspring. But if one animal helps another, the one who helps will leave fewer offspring and the one who is helped will leave more offspring: natural selection, it appears, works against all kinds of altruistic and cooperative behaviour.

But animals do help each other. Ants, bees and wasps are dramatic examples. A honeybee immediately dies after it uses its sting, as the back end of its body is ripped out to pump poison into its victim after the bee's death. The hive-mates of the dead bee benefit from its suicidal act. 'Altruistic' behaviour such as this was for a long time a puzzle for Darwinians.

The puzzle was solved in 1964 by Bill Hamilton, an English biologist then just beginning his PhD. Hamilton realized that it can be advantageous for an individual to sacrifice itself for others, provided that it is sacrificing itself for the benefit of genetic relatives. Any gene in an individual is also likely to be in its brothers and sisters. Natural selection actively favours altruism if the altruistic act provides enough benefit to the recipients, relative to the cost to the altruist.

Hamilton's theory has been strikingly successful in predicting when animals will behave altruistically. The theory works wonderfully well for ants, bees and wasps because these insects have a peculiar system of genetic inheritance. An ant may share more genes with its sister than with its own offspring, which explains the high development of social behaviour (among sisters) in these insects. Hamilton's theory provides the foundations of all modern research on social behaviour. The theory has also been applied to human behaviour (though this was not Hamilton's main interest), particularly after Edward O. Wilson promoted the controversial theory of human sociobiology in the 1970s and 1980s.

See also Darwin's *Origin of Species* pages 176–177, Animal instincts pages 318–319, Game theory pages 336–337, Honeybee communication pages 338–339

Right In it together: a leafcutter ant carries a cut piece of leaf back to its nest, while several smaller worker ants protect it from parasitic flies.

Quarks

Murray **Gell-Mann** *b.*1929

Physicists had learned, by the 1930s, to build all matter out of just three kinds of particle: electrons, neutrons and protons. But a procession of unwanted extras had begun to appear – neutrinos, the positron and antiproton, pions and muons, and kaons, lambdas and sigmas – so that by the middle of the 1960s a hundred supposedly fundamental particles had been detected. It was a mess.

By inventing a new, deeper level of existence, the American theoretician Murray Gell-Mann tidied up. In 1961 he noticed patterns in the properties of many of these particle types that hinted at an underlying structure, a handful of more fundamental particles that in 1964 he called 'quarks'. There are now thought to be six quarks, called 'up', 'down', 'strange', 'charm', 'bottom' and 'top'. Protons are made of two ups and a down, neutrons of two downs and an up, and they are all held tightly together by other particles called 'gluons'. Combinations of the other quarks make up all those exotic short-lived compound particles. All of them are bound together by a tremendously powerful but short-ranged force called the 'colour force', which is carried by the gluons.

Less than a second after the big bang, the universe was so hot and dense that it was an amorphous plasma of quarks and gluons – a state that physicists are trying today to reproduce in their particle accelerators.

Not everything is made of quarks. Electrons and neutrinos, plus the two electron-like particles muons and taus, are in a different class of fundamental particles called leptons. Then there are the particles that carry forces: photons, gluons, W, Z and Higgs particles. Altogether these three main particle types, quarks, leptons and the force-carrying particles, make up everything that exists – as far as we know. But this 'standard model' of particle physics probably isn't the last word. There are yet more levels of reality to be explored.

See also The neutron pages 312–313, Antimatter pages 314–315, Afterglow of creation pages 412–413, Unified forces pages 416–417, Superstrings pages 474–475

Right The bubble chamber came to dominate images of subatomic particles in the 1960s. By interpreting the beautiful swirling pictures, physicists build theories that describe the behaviour of the particles and the forces that control them.

Oldest fossils

Elso **Barghoorn** 1915–84

Earth's 4.6-billion-year history is divided into two major eons: the Phanerozoic (meaning 'evident life') and the Precambrian or Cryptozoic (meaning 'hidden life'). During the 545 million years of Phanerozoic time, abundant life is shown by fossil discoveries. The four billion and more years of Precambrian time seemed to be devoid of any traces of life. But in the 1960s Elso Barghoorn discovered two-billion-year-old Precambrian microfossils in a layer of exposed rock in the Gunflint Ironstone, Western Ontario, Canada. Charles Darwin had been all too aware of the missing Precambrian record of life. Puzzled by the way fossils representing 'several of the main divisions of the animal kingdom suddenly appear in the lowest known [Cambrian] fossiliferous rocks', he could 'give no satisfactory explanation'.

Organic-like structures found in ancient Laurentian limestones of the Canadian Shield were named 'Eozoon' ('dawn-animal') in 1865 by the Canadian geologist John W. Dawson. Regarded as evidence of Precambrian life, they were later shown to be inorganic. However, Canadian rocks did eventually provide the first genuine Precambrian fossils to be widely publicized, although little-known Precambrian spore-like fossils were found earlier in Russia.

The advent of the transmission electron microscope revealed well-preserved organic-walled microfossils in the silica-rich chert (fine-grained quartz) deposits of the Gunflint Ironstone. Described by Barghoorn and Stanley Tyler in 1965, the microbes were associated with laminated mound structures called stromatolites dating from around 2,000 million years ago (mya).

We now know that microbial life began some 3.8 billion years ago with photosynthesizing prokaryotes tolerant of an oxygen-free atmosphere. Some built stromatolites, and by 2,100 mya, increasing oxygen levels allowed more complex eukaryotes to evolve. Sexually reproducing multicelled animals (metazoans) appear 1,200 mya, and by 610 mya macroscopic life included a variety of marine soft-bodied organisms. Known as Ediacarans, they include ancestors of the major Cambrian fossil groups.

See also Rocks of ages pages 252–253, Burgess Shale pages 260–261, Genes in inheritance pages 264–265, Origin of life pages 370–371, Symbiotic cell pages 418–419, Five kingdoms of life pages 426–427, Life at the extreme pages 452–453, Martian microfossils pages 518–519

Right Stromatolites – the structures generated by bacterial mats, the most ancient community. This photograph shows recent stromatolites forming in Shark Bay, Western Australia.

Afterglow of creation

Arno Allan Penzias *b*.1933,   Robert Woodrow Wilson *b*.1936

In 1917 Albert Einstein introduced a constant into his theory of general relativity to 'force' the universe to be static, but the observations of Edwin Hubble in the 1920s showed that he was wrong. The universe started very small and was getting bigger. In 1931 George Lemaître suggested that the universe's contents were initially packed into a spherical 'primeval atom', about 30 times larger than the Sun.

George Gamow, Ralph Alpher and Robert Herman investigated the temperature and energy of this initial super-dense state. In 1949 they suggested that the universe originated in a huge hot explosion. They showed how nuclear reactions in this primordial fireball could have converted hydrogen nuclei (protons) and neutrons into helium, explained how this accounted for the proportions of these elements in very old stars, and predicted that the radiation produced would have thinned out and cooled as the universe expanded, making the sky 'glow' with a faint background of microwaves at a temperature of about five kelvin (K).

In 1950 the British astronomer Fred Hoyle dismissed this theory as a 'big bang'. He supported a 'steady state' model of the universe, in which the continuous formation of new galaxies feeds local regions of expansion, and for nearly 15 years the two cosmological theories battled it out. The conflict ended in 1965 when Arno Penzias and Robert Wilson accidentally found the microwave background radiation while trying to eliminate hiss from their radio equipment. Its temperature turned out to be 2.7 K.

In 1992 the Cosmic Background Explorer satellite found the background radiation to be a near-perfect fit to a temperature of 2.726 K, but its differential microwave radiometer detected slight ripples in the radiation – temperature deviations of one part in 100,000. If the radiation had been uniform, it would have been difficult to explain how matter in the early universe clumped together into individual stars and galaxies some 300,000 years after the big bang.

See also General relativity pages 278–279, Expanding universe pages 306–307, A subatomic ghost pages 384–385, Quarks pages 408–409, Unified forces pages 416–417, Great Attractor pages 498–499

Right The famous 'ripples in space' were detected in 1992 by the COBE satellite. In this view of the whole sky, the cool (blue) patches reveal where gas is coming together under gravity to form the seeds of the first galaxies.

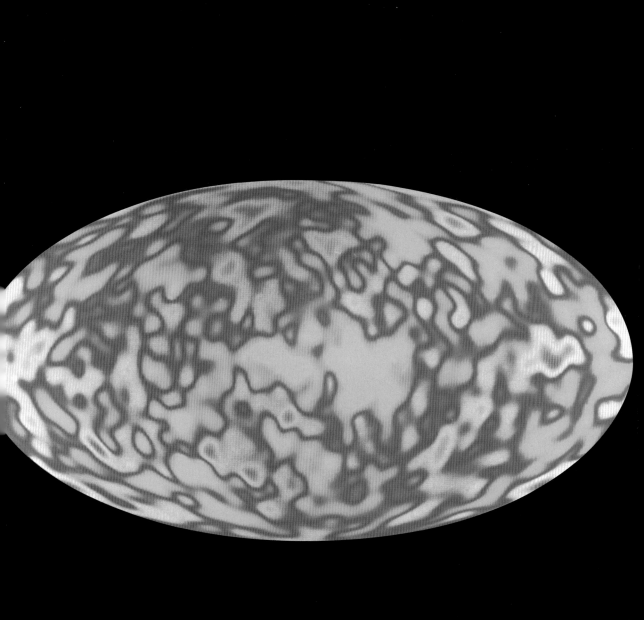

Plate tectonics

Drummond Hoyle Matthews 1931–99, Frederick John Vine 1939–88,
Dan Peter McKenzie *b*.1942

From 1925 to 1927 the German oceanographic vessel SS *Meteor* crisscrossed the Atlantic taking sonar echo-soundings of the sea-bed. These revealed the Mid-Atlantic Ridge, a submarine mountain chain running the length of the ocean, but its significance was not realized until after the Second World War.

The American geologist Maurice Ewing pioneered marine seismic techniques that showed the ocean crust to be much thinner (7 kilometres) than the continental crust (20–80 kilometres). In 1953 Ewing and Bruce Heezen, an American oceanographer, discovered the global extent of mid-ocean-ridge systems and later found a central depression in the Mid-Atlantic Ridge. This rift coincided with unusually heavy submarine earthquake and volcanic activity and indicated that the ridge is being pulled apart.

A full explanation awaited the work of their colleague Harry Hess, an American geologist whose study of the Pacific Ocean floor led to his theory of sea-floor spreading. In 1962 he suggested that mid-ocean ridges form over rising convection currents in the hot mantle which force molten rock upwards. This is extruded from the rift valleys, spreading in all directions to form new ocean-floor material. The farther away from the ridge, the older the ocean floor.

In 1963 two British geologists, Drummond Matthews and Frederick Vine, found repeated symmetrical patterns of polarity reversals in magnetized lavas on either side of the mid-ocean ridges. The patterns recorded the history of reversals in the Earth's magnetic field. This symmetry confirmed Hess's theory of sea-floor spreading. Finally, in 1967, the British geophysicist Dan McKenzie synthesized continental drift and sea-floor spreading into the theory of plate tectonics, in which the Earth's crust is broken down into large mobile plates: most volcanoes and earthquakes are at plate boundaries, and where the plates collide, mountain ranges form.

See also Earth cycles pages 100–101, Catastrophist geology pages 108–109, Lyell's *Principles of Geology* pages 146–147, Mountain formation pages 206–207, Inside the Earth pages 250–251, Rocks of ages pages 252–253, Continental drift pages 270–271, Geomagnetic reversals pages 304–305, Eruption of Mount St Helens pages 462–463

Right The San Andreas Fault is where the edge of the North American continent slides past the plate that makes up the Pacific Ocean floor. The plates move relative to each other at a rate of one centimetre a year.

Unified forces

Sheldon Lee Glashow *b.*1932, Abdus Salam 1926–96, Steven Weinberg *b.*1933

Nature is fond of secret identities. Electromagnetic forces, for example, are strong and long ranged, whereas the feeble 'weak nuclear force' can't reach outside a tiny atomic nucleus. And yet they can be considered the same.

The theory called quantum electrodynamics, QED, says that electromagnetism is exerted by photons. Because of the theory's great successes, physicists searched for a similar description of the other forces of nature, and in 1967 Sheldon Glashow, Abdus Salam and Steven Weinberg finally managed to work one out for the weak nuclear force. They got a bonus. Their theory required four carrier particles. First there were three new particles called W+, W and Z bosons to take care of the weak force. The Ws and Z were duly discovered in the giant particle accelerators at CERN, the European Laboratory for Particle Physics in Geneva, in 1983, and the theory was proved. But the three physicists found that their theory included the old familiar photon, the carrier of electromagnetism. The two forces were handled by exactly the same equations. So why are they so different?

The W and Z bosons are extremely heavy. To pop into existence and carry the weak force, they have to 'borrow' a lot of energy. This is allowed by quantum mechanics, but only rarely, which makes the weak force weak; and only for a short time, which makes it short ranged. At very high temperatures these constraints disappear: the electromagnetic and weak forces remove their disguises and merge into a single 'superforce'. They may have been unified in the hot young universe, less than a trillionth of a second after the beginning of time. The strains produced when the forces split apart may even have put the bang in the big bang by setting off the strange process called 'inflation': a brief exponential expansion of the universe.

See also Expanding universe pages 306–307, Quantum electrodynamics pages 352–353, Afterglow of creation pages 412–413, Superstrings pages 474–475

Right LEP (Large Electron-Positron Collider) is the world's largest scientific instrument. Located 50 metres below the surface at CERN, in a tunnel 27 kilometres long, it accelerates particles to 99.9999 per cent of the speed of light.

Symbiotic cell

Lynn **Margulis** *b.*1938

Every cell in a human body is descended from an ancestral merger between two simpler cells that took place 2,000 million years ago. We can see that historic event in the structure of our cells. The genes in our cells are arranged in two locations. Most of our genes are found in the cell nucleus; but a small number of genes are found outside the nucleus, in another structure called the mitochondrion. Why do our cells have these two gene sets?

In a paper published in 1967 the American biologist Lynn Margulis argued that our mitochondria are ultimately descended from free-living bacterial cells. In the past, a larger cell engulfed a smaller cell, perhaps in order to digest it. The smaller cell managed to survive inside the larger cell, and the two-cell combination made a successful team. They may have had complementary skills. The small cell may have been able to burn food-fuel in oxygen to create energy (this is what mitochondria do in modern cells) and the larger cell may have produced the fuels to be burned. Over time, the two partners evolved into the type of cell that our bodies are built of today. Mitochondria still look like bacteria, and the few genes that remain in mitochondria are similar to bacterial genes. An analogous symbiotic event led to the evolution of chloroplasts, the structures that perform photosynthesis in green plants.

Our cells, with both mitochondrial and nuclear gene sets, are called 'eukaryotic' cells. Bacterial cells lack both distinct nuclei and mitochondria and are called 'prokaryotic' cells. Prokaryotic cells dominated life on Earth from soon after life's origin 4,000 million years ago until the evolution of the eukaryotic cell 2,000 million years ago. All modern plant and animal life is built of eukaryotic cells. The symbiotic merger identified by Margulis may have been the great breakthrough in the evolutionary rise of complex life on Earth.

See also Communities of cells pages 174–175, Viruses pages 232–233, Burgess Shale pages 260–261, Citric-acid cycle pages 320–321, Genes in bacteria pages 332–333, Photosynthesis pages 344–345, Origin of life pages 370–371, Oldest fossils pages 410–411, Five kingdoms of life pages 426–427, Out of Africa pages 492–493

Right Mitochondria look like bacteria, and even grow and divide in two at their own pace within the larger cell. This electron micrograph of a mitochondrion (green) shows the internal membranes where respiration takes place.

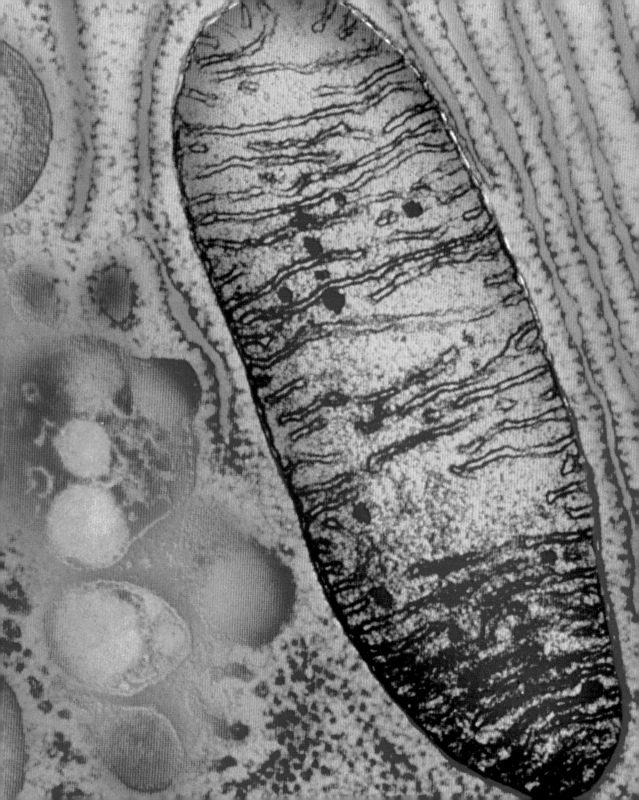

Pulsars

Susan Jocelyn **Bell Burnell** *b.*1943, Antony **Hewish** *b.*1924

Jocelyn Bell, a research student at the University of Cambridge, and her supervisor Antony Hewish were using a huge 3.7-metre-wavelength radio telescope with 2,048 separate receivers (covering some 1.6 hectares) to study the flickering of quasar radio waves as they pass through the solar wind. In July 1967 Bell discovered that every sidereal day (23 hours 54 minutes) the receivers were picking up a strange signal. A faster pen recorder showed that it consisted of a series of extremely regular radio pulses every 1.33730113 seconds. They thought they might have detected the first signals of an extraterrestrial civilization. But soon another source in a different direction convinced Bell that 'it was highly unlikely that two lots of Little Green Men could choose the same unusual frequency and unlikely technique to signal to the same inconspicuous planet Earth!'.

Other 'pulsars' were soon discovered, and now about 600 are known. It is generally accepted, as proposed by Thomas Gold in 1968, that pulsars are rapidly spinning magnetic neutron stars: the condensed cores of old supernovae. These 10–20-kilometre-diameter stars act like celestial lighthouses with beams of 'synchrotron radiation' issuing from the magnetic poles – electromagnetic radiation produced by charged particles such as electrons moving at high speed through a magnetic field. The sweeping of this beam across Earth produces a pulse lasting tens of milliseconds. The arrival of the pulse varies slightly with wavelength, and this enables the distance of the source to be calculated, assuming we know the density of the electrons along the line of sight.

In November 1968 a 33-millisecond pulsar was found at the centre of the Crab Nebula, the expanding cloud of debris left behind by a supernova observed from Earth in 1054. Comparisons with typical older pulsars indicate that pulsars slow down with age. In September 1969 the period of the Crab pulsar suddenly decreased by 3×10^{-10} seconds as a result of the flattened neutron star readjusting its shape as its rate of spin decreases.

See also Stellar evolution pages 286–287, White dwarfs pages 310–311, Solar wind pages 388–389, Gamma-ray bursts 434–435, Supernova 1987A pages 488–489

Right The interior of the Crab Nebula supernova remnant. The core survived as a pulsar (lower of the two bright stars just above and left of centre). Its rotation heats the surrounding gas, causing the blue glow. The supernova was witnessed by Chinese astronomers in AD 1054.

Random molecular evolution

Motoo **Kimura** 1924–94

For a century after Darwin published his theory in 1859, biologists could study evolutionary changes in the observable features of living creatures: in bodies, bones, even behaviour. By the 1960s they could also study molecular evolution: changes in proteins and (by the 1980s) in genes themselves. Molecular evolution turned out to have two unexpected properties: it is fast, and it proceeds at a relatively constant rate.

About 500 million years ago, the ancestors of modern humans looked like fish. Some modern fish are descended from those same ancestors and are almost unchanged in body form, while the evolutionary line leading to human beings has shown massive changes in body form – evolution in body form can be fast or slow. But the amount of molecular change up the lines to fish and to humans is almost identical: whether body form had been unchanged from fossil fish to modern fish or revolutionized from fossil fish to modern human, the molecules would have altered by the same amount. Both are governed by the same 'molecular clock'.

In 1968 the Japanese geneticist Motoo Kimura realized that the emerging facts about molecular evolution did not fit easily with the standard Darwinian theory of evolution by natural selection. Instead, he suggested that most (but not all) molecular evolution is 'neutral': the changes make no difference to the organism. Molecular evolution, then, proceeds by random drift. This was controversial as he was claiming that most evolution, at the molecular level, is not driven by natural selection, but his basic idea is now widely accepted. Kimura's neutral theory is the biggest development in evolutionary thinking since the 'modern synthesis' of the 1920s. It also underlies a major research programme in our own time: the attempt to reconstruct the history (or 'tree') of life from molecular evidence.

See also Darwin's *Origin of Species* pages 176–177, Neo-Darwinism pages 282–283, The double helix pages 374–375, Five kingdoms of life pages 426–427, Genetic cousins pages 442–443, Genetics of animal design pages 460–461, Ancient DNA pages 476–477, Out of Africa pages 492–493

Right The Scottish zoologist D'Arcy Wentworth Thompson, in a classic book of 1917, tried to show mathematically that different forms of fish were due solely to different growth rates.

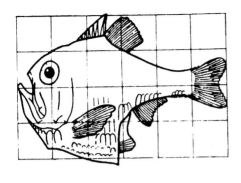
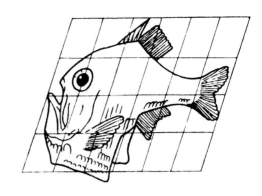
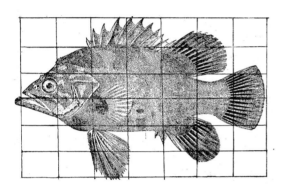
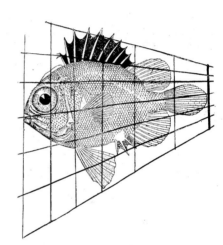
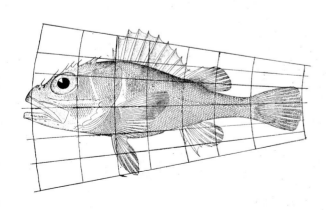
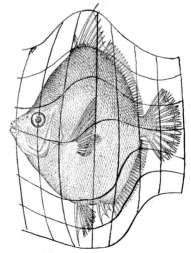

Apollo mission

Neil Alden Armstrong *b.*1930, Edwin Eugene Aldrin *b.*1930, Michael Collins *b.*1930

There has been no greater voyage of exploration in the history of mankind than the US Apollo mission to the Moon. Between July 1969 and December 1972, 12 men travelled 380,000 kilometres across space in six different craft. Lunar material weighing 381 kilograms was returned to the Earth's laboratories. A total of 79.4 hours was spent outside the Lunar Excursion Module and a lunar rover car was driven a distance of 30 kilometres around one of the sites. On 20 July 1969, Neil Armstrong and Edwin 'Buzz' Aldrin were the first humans to step out on the surface of the Moon, while Michael Collins remained in orbit.

An Apollo Lunar Surface Experimental Package was set up at each base. This monitored the lunar atmosphere, the flux of particles from the Sun, the effect on the Moon of the magnetosphere of Earth, the heat flow from the lunar interior, the shaking of the Moon as it is pulled by the ebb and flow of Earth's tides and impacted by small asteroids, and the Earth–Moon distance. The returned samples were used to date individual mare and highland regions of the Moon. Their mineralogy, magnetism and composition also gave invaluable clues to the history, origin and evolution of our satellite companion. But for most people the memorable parts of the missions were the astronauts' emotional descriptions of the magnificent desolation of the sterile, dry, lifeless lunar surroundings.

The programme was initiated by President John F. Kennedy in a speech to the US Congress on 25 May 1961. His commitment to 'landing a man on the Moon, and returning him safely to Earth... before this decade is out' was part of a Cold War space race against the Soviet Union. As soon as the Soviets stopped 'running', the Americans cancelled their remaining Apollo missions. Any hopes of extending manned space exploration by building a permanent base on the Moon or travelling onwards to Mars have been put on hold.

See also Heavens through a telescope pages 54–55, Origin of the Solar System pages 104–105, Sunspot cycle pages 158–159, Geomagnetic reversals pages 304–305, Solar wind pages 388–389, Water on the Moon pages 522–523

Right An hour into history's first moonwalk, Buzz Aldrin photographed his own boot and its imprint on the ancient lunar dust.

Five kingdoms of life

Robert Harding Whittaker 1920–80

'Animal, vegetable or mineral?' The question implies that living things are either plants or animals, and biologists have historically taken the same view. Biologists did encounter some creatures, such as mushrooms, that violated the distinction, but they forced them into the plant or animal group. Mushrooms are fungi, for instance, and biologists until recently classed fungi as plants – more accurately as 'plants that do not photosynthesize'.

Then there were the microbes. Biologists found increasing numbers of microscopic lifeforms in the wake of their discovery in the seventeenth century, and these were duly forced into the plant–animal distinction. Some microbes could photosynthesize: these were defined as algae and grouped with plants. Others seemed to be more like animals: they were defined as Protozoa and grouped with animals. In the nineteenth century biologists discovered bacteria – even smaller microbes – but these no one managed to define as either animals or plants.

By the twentieth century biologists knew that all life could not be divided into animals and plants, but the old idea was not finally laid to rest until 1969 when an American ecologist, Robert Whittaker, proposed his five-kingdom classification. He divided life into animals, plants, fungi, protists and bacteria. Animals, plants, fungi and protists are 'eukaryotes'; they are built of cells (or one cell, in the case of protists) with a distinct nucleus. Bacteria are 'prokaryotes'; their single cell has no distinct nucleus. Whittaker's classification struck a chord. Fungi have nothing to do with plants; indeed they are more closely related to animals.

Subsequent research has modified Whittaker's scheme. Some biologists prefer to divide the protists into more than one kingdom, but the most important development came when Carl Woese discovered that there are two groups of prokaryotes (archaeans and bacteria), not one. That has led to a 'three-domain' classification of life: archaeans, bacteria and eukaryotes (containing the other four kingdoms of Whittaker's scheme).

See also Birth of botany pages 18–19, Naming life pages 88–89, Viruses pages 232–233, Photosynthesis pages 344–345, Slime-mould aggregation pages 348–349, Oldest fossils pages 410–411, Symbiotic cell pages 418–419, Diversity of life pages 468–469

Right The green alga *Volvox aureus*. The photograph shows the release of daughter cells by a senescent mother colony.

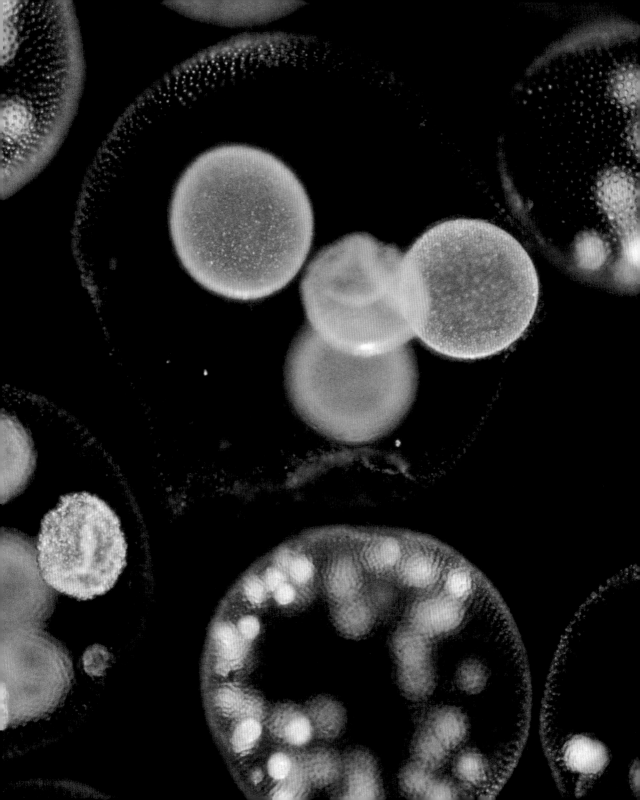

Green revolution

Norman Ernest **Borlaug** *b.*1914

The American plant breeder Norman Borlaug, who received a Nobel Prize for Peace in 1970 for his achievements as founding father of the green revolution, is often credited with saving more lives than anyone else in history.

Borlaug's work at Mexico's International Maize and Wheat Centre (1944–60) produced the high-yielding and disease-resistant semi-dwarf wheat that eventually increased crop yields worldwide. Wheat grows tall, a trait accentuated by the use of fertilizer, which makes it vulnerable to damage in bad weather. Borlaug added dwarfing genes, generating wheat cultivars that responded to nitrogen fertilizers with dramatic increases in grain yield rather than straw production. Then he persuaded the governments of Pakistan and India to grow them in the mid-1960s, when famine was rife on the subcontinent. As a result Pakistan became self-sufficient in wheat production by 1968 and India achieved similar status in 1974.

This quantum leap in yields – the green revolution – was repeated with rice in Southeast Asia in the 1970s and has kept famine at bay, but not without cost. Borlaug's detractors point to environmental damage caused by heavy dependence on fertilizers, pesticides and irrigation and deplore the socioeconomic impact on poverty-stricken farmers who cannot afford the costly new technology.

Current population growth rates point to the need for another green revolution. One strategy – favoured by Borlaug and some environmentalists such as Edward O. Wilson – would be to use genetic engineering to raise yields per hectare still further in existing fields, minimizing the need for more forest clearance. In 1999 Jinrong Peng and colleagues at the John Innes Research Institute in England cloned the dwarfing genes responsible for the semi-dwarf character of high-yielding wheat, raising hopes that they might be genetically engineered into other crops. Borlaug's green revolution may not yet have run its course.

See also Population pressure pages 110–111, Crop diversity pages 292–293, DDT pages 326–327, Genetic engineering pages 436–437

Right The new strains of wheat created by Borlaug greatly increased crop production, notably by tripling of Mexico's wheat yields.

Biological self-recognition

Niels Kai Jerne 1911–94

Our understanding of how the body protects itself from disease owes a lot to Niels Jerne, who was born in London in 1911 to Danish parents. After training in medicine in Copenhagen, he moved to the California Institute of Technology where, in 1955, he put forward one of the central concepts in immunology.

It had long been assumed that antibodies, the molecules that defend the body from infection, are produced only when an invader such as a virus or bacterium enters the body. Antibodies lock on to antigen molecules on the surface of an invader, marking it for destruction by the cells of the immune system. Jerne argued that the body already possesses all the antibodies it needs. One specific antibody out of millions attaches to its corresponding antigen and expands into a new population of antibodies that provide enough molecules to deal with the infection. The Australian immunologist Frank Macfarlane Burnet went on to suggest that antibodies were fixed to the surface of specialized immune cells and that the attachment of an antigen to the cell bearing the right antibody was the signal for the cell to start multiplying. This theory came to be known as 'clonal selection'.

But how does the immune system distinguish between 'self' (the body's own tissues) and 'non-self' (bacteria, viruses and transplanted tissue)? In 1971 Jerne proposed that the immune system 'learns' how to do this in the thymus gland, which is located in the chest. Here cells that make antibodies that attack 'non-self' matter multiply and those that make antibodies that would attack 'self' are suppressed. The most important of Jerne's ideas was his 'network theory', developed three years later. It describes how all the different cells of the immune system are in a state of balance. This allows the system to stay quiescent when not under attack, but to respond rapidly when there is a threat to the body.

See also Vaccination pages 102–103, Cellular immunity pages 204–205, Antitoxins pages 214–215, Blood groups pages 236–237, Transplant rejection pages 356–357, Monoclonal antibodies pages 448–449, AIDS virus pages 472–473

Right Computer model of immunoglobulin G bound to an antigen (red). This Y-shaped molecule is one of a group of related proteins that act as antibodies in the blood.

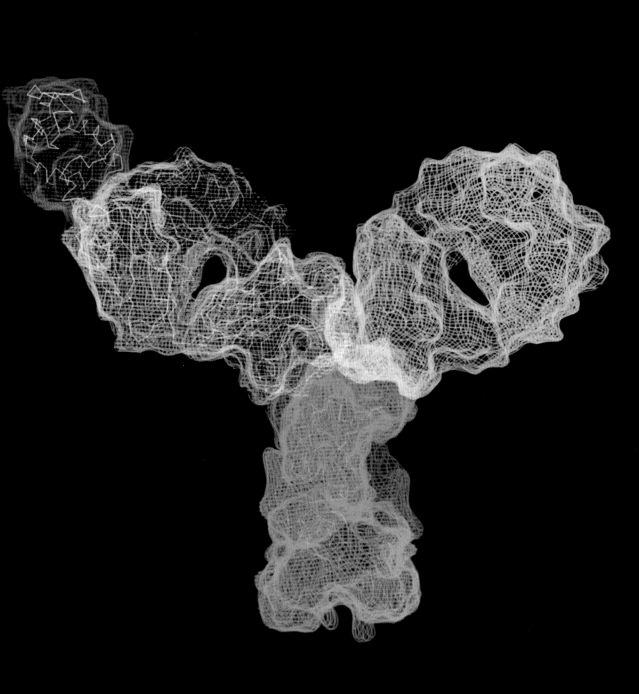

Gaia hypothesis

James Lovelock *b.*1919

The concept of Earth as a giant living organism can be traced back to Plato in around 400 BC, but not until the twentieth century did it gain scientific credibility. James Lovelock, an independent British scientist who had been employed by NASA (National Aeronautics and Space Administration) in the 1960s to explore the possibility of life on Mars, studied life on Earth just as he searched for it on distant planets – by analysing the atmosphere. He pointed out that our planet's atmosphere is a highly improbable mixture of gases maintained in balanced proportions by geochemical processes (such as rock erosion) and the activities of the organisms it supports (such as the removal of carbon dioxide and generation of oxygen by photosynthetic plants). His controversial 'Gaia hypothesis', named after the ancient Greek Earth-mother goddess, proposes that terrestrial biological and physical processes work together to produce and regulate conditions conducive to life's continued existence.

First put forward in 1972, the idea was rejected by mainstream scientists principally for its lack of rigour. But support came in 1981 when Lovelock produced 'Daisyworld', a computer simulation of a world populated by white or black daisies that either reflect or absorb solar radiation. As their relative numbers change according to the prevailing surface temperature, the daisy populations maintain a global temperature equilibrium. Later, more complex models with greater biodiversity improved the stability of the system.

Lovelock's Gaia hypothesis is now particularly relevant to human-induced changes in the global atmosphere that threaten the stability of our climate, ecosystems, food production and health. Without greenhouse gases, Earth's surface temperature would be –19 degrees Celsius, but if they increased uncontrollably above current levels, Earth's climate would resemble that of Venus. Ensuring that Gaia's greenhouse-gas composition remains stable has become one of the greatest scientific and political challenges of the twenty-first century.

See also Greenhouse effect pages 184–185, Nitrogen fixation pages 208–209, Climate cycles pages 276–277, Ozone hole 438–439, Life at the extreme pages 452–453, Eruption of Mount St Helens pages 462–463, Diversity of life pages 468–469

Right Earth as superorganism. The idea that life collaborated to keep our planet habitable was once dismissed as superfluous and misleading. But now even sceptics are taking a second look.

Gamma-ray bursts

US Department of Defense

In the late 1960s, US Vela military satellites were orbiting the Earth and monitoring the 1963 Nuclear Test Ban Treaty by looking for gamma-ray bursts from the explosion of atomic bombs. About one burst was detected every day. But these turned out to be of celestial rather than terrestrial origin. Confirmed by Thomas Cline and Upendra Desai in 1973, these intense flashes of photons appear and fade away again in a matter of seconds, but during its lifespan a 'burster' is likely to release more energy than the Sun will in its entire lifetime.

Only a handful of sources have produced repeat bursts. These have been found to coincide with supernova remnants, and the neutron star left behind after the supernova explosion. It has been suggested that all the photons start with an energy of 511,000 electron volts, and are produced by electrons and positrons annihilating one another near the surface of the neutron star. As predicted by general relativity theory, the radiation would then be 'redshifted' as the gamma rays 'climb out' of the neutron star's gravitational well, thus accounting for the broad range of burster energies.

The Compton Gamma Ray Observatory satellite was launched from the space shuttle Atlantis in April 1991. Its survey of gamma-ray sources showed that the distribution of bursters is the same in all directions. There are two possibilities. Either the sources are close, like the bright night-sky stars, or very distant, like neighbouring superclusters of galaxies. The first would have to be 10^{22} times less energetic than the second to achieve the same level of brightness.

Astronomers now favour the hypothesis that most bursters come from the far reaches of the universe. A burster seen in 1997 (GRB 970228) was also observed in X-rays by the Italian-Dutch Beppo-SAX satellite, and was seen optically near a very faint galaxy. Another burster (GRB 970508) appeared as a star-like point in the constellation Camelopardalis, with intergalactic spectral lines indicating that it was at least 4 billion light-years away. It may be that we are witnessing the merger of two neutron stars.

Above NASA's Compton Gamma Ray Observatory was launched in 1991 to study the universe in the gamma-ray part of the spectrum.

See also Stellar evolution pages 286–287, White dwarfs pages 310–311, The neutron pages 312–313, Antimatter pages 314–315, Pulsars pages 420–421, Supernova 1987A pages 488–489

Right Nobody knows what causes gamma-ray bursters or even whether sources lie close by or far away. Repeat bursts have been found to coincide with supernova remnants like the Crab Nebula shown here.

Genetic engineering

Paul **Berg** *b*.1926, Herbert Wayne **Boyer** *b*.1936, Stanley **Cohen** *b*.1935

Paul Berg and Herbert Boyer of Stanford University and Stanley Cohen of the University of California at Berkeley discovered how to transfer genes from one species to another in the early 1970s. Basically, genetic engineering is a 'cut, paste and copy' job done with stretches of DNA. First the gene of interest – let's say the gene for the blood-clotting protein factor VIII – is isolated by chopping up longer DNA molecules from human tissue with special 'restriction enzymes'. The factor VIII gene is then inserted into a carrier molecule – known as a 'vector' – which might be a virus, or a circular bit of bacterial DNA called a 'plasmid'. Finally the vector 'infects' the host cell, which could be a bacterium, a yeast or even a mammalian cell. The host cell now carries foreign DNA – in this case the human factor VIII gene – but treats it as its own. This kind of DNA – containing genes from more than one species – is called 'recombinant'. As the cells multiply, their DNA is active, turning out copies of host proteins, and copies of factor VIII as well. At the end of the experiment, the factor VIII molecules can be harvested.

Genetic engineering has many applications. It has been used to make several human proteins, such as insulin for treatment of diabetes and factor VIII for treatment of haemophilia, thus eliminating the use of human tissue. Many people with haemophilia have contracted HIV because the blood that was used to prepare their factor VIII was contaminated. Ideally they should now all be on the recombinant version.

More controversially, genetic engineering has also been applied to plants, transferring genes for resistance to herbicides and insects. The idea is to obtain higher-yielding disease-free crops – but the prospect of consuming food from genetically engineered plants has met with opposition from consumers.

Right The tools of genetic engineering can be used to splice together DNA molecules from two related bacterial viruses.

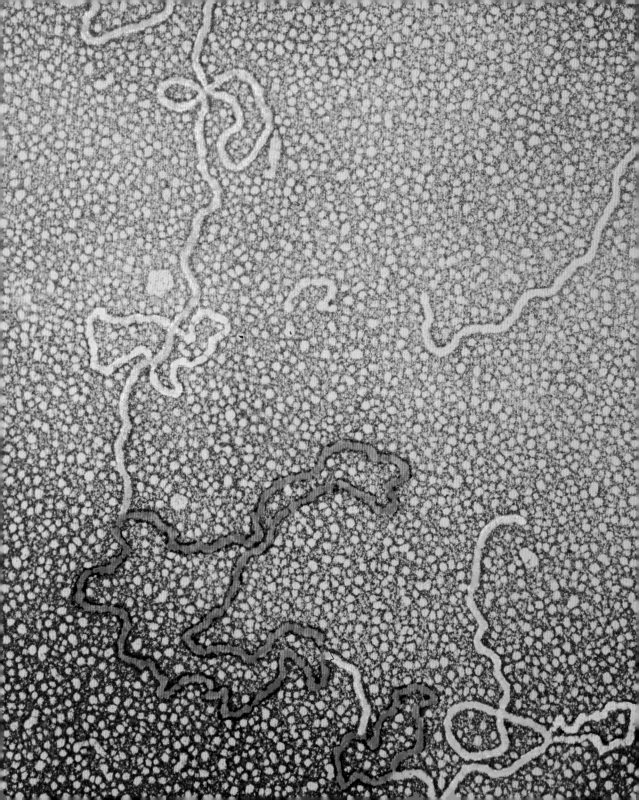

Ozone hole

Mario **Molina** *b.*1943, Sherwood F. **Rowland** *b.*1927

There is so little ozone in the atmosphere that if it were concentrated on the Earth's surface it would form a layer only three millimetres thick. And yet ozone – a molecule made up of three oxygen atoms – plays a hugely important role in our environment. It absorbs ultraviolet rays from the Sun before they reach the surface, where they would otherwise destroy the delicate molecules of life. Without this layer, land-based life could never have evolved.

In 1974, however, Mario Molina and Sherwood Rowland, then both at the University of California at Irvine, predicted that chlorofluorocarbons (CFCs), chemicals widely used in air-conditioning units, refrigeration plants and spray cans, destroy ozone far more quickly than it is created in the atmosphere. And because these chemicals were being widely released into the atmosphere, Molina and Rowland warned that the ozone layer could not survive the onslaught. Their announcement led to widespread debate but little action.

Then in 1985 Joseph Farman, a scientist with the British Antarctic Survey, discovered a huge hole in the ozone layer above the South Pole and linked the cause of the depletion to CFCs released by humans. The debate continued most forcefully in the southern hemisphere, where the increased exposure to ultraviolet rays as a result of the hole greatly increased the risk of skin cancer. Many researchers, including Molina and Rowland, lobbied governments to ban CFCs, arguing that they could easily be replaced with other, less damaging, chemicals.

After more than two decades their work paid off. With unprecedented international unity, the United Nations negotiated a ban on CFCs and other damaging chemicals. Known as the Montreal Protocol, the ban came into force in 1996. In effect the problem was solved. But CFCs take time to spread through the atmosphere, so the ozone hole is expected to persist for many years, perhaps even a century. In 1995 Rowland and Molina received the Nobel Prize for Chemistry for their work.

See also Ice ages pages 152–153, Greenhouse effect pages 184–185, Climate cycles pages 276–277, DDT pages 326–327, Gaia hypothesis pages 432–433

Right The ozone hole over Antarctica, 8 September 2000. First seen in 1980, the hole has grown each year since. It is shown as the blue area, spreading over some 28 million square kilometres.

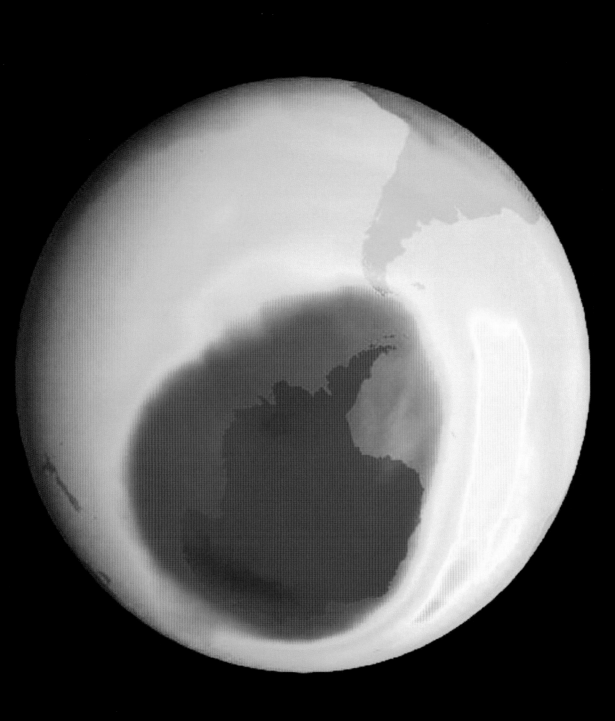

Black-hole evaporation

Stephen William **Hawking** *b.*1942

Olaus Roemer measured the speed of light in 1679 by timing the eclipses of the satellites of Jupiter. About a hundred years later, in 1784, the Reverend John Michell introduced the concept of the black hole. He suggested that the most massive objects in the universe might be invisible because 'all light emitted from such a body would be made to return to it by its own proper gravity'.

Evolved stars of more than three solar masses become black holes (the Sun's mass is 2 x 10^{30} kilograms). In the case of the binary star system Cygnus X-1, one of the pair of 'stars' is almost certainly a black hole, and matter from its companion star falls onto a flattened disc around it, producing prodigious amounts of X-rays as it is accelerated. Other candidates are active galactic nuclei. They are the most luminous objects known and they are only about as big as our Solar System. To be in equilibrium they must have a mass of about 10,000 million solar masses. Such a large mass in such a small space indicates that they are super-massive black holes.

As an object falls into a black hole it accelerates, reaching the velocity of light when it is at a distance from the centre of the hole known as the Schwarzschild radius. Here, at the 'event horizon', it disappears and is forever unobservable. Stephen Hawking discovered that the surface area of the event horizon could never decrease. But in 1974 he discovered a loophole. By combining black-hole theory, quantum mechanics and thermodynamics he found that black holes could evaporate. The production of an electron–positron pair just outside the event horizon could result in one of the particles falling into the hole and the other escaping. As the gravitational energy of the hole had been used to form the particles, the escaping particle essentially carries away some of the black hole's mass. These escaping particles are now referred to as Hawking radiation – black holes aren't black after all.

See also Origin of the Solar System pages 104–105, Laws of thermodynamics pages 164–165, The quantum pages 234–235, White dwarfs pages 310–311, Antimatter pages 314–315, Quasars pages 404–405

Right The gravitational pull of a suspected super-massive black hole forms a disk of hot gas, at the core of an energetic galaxy.

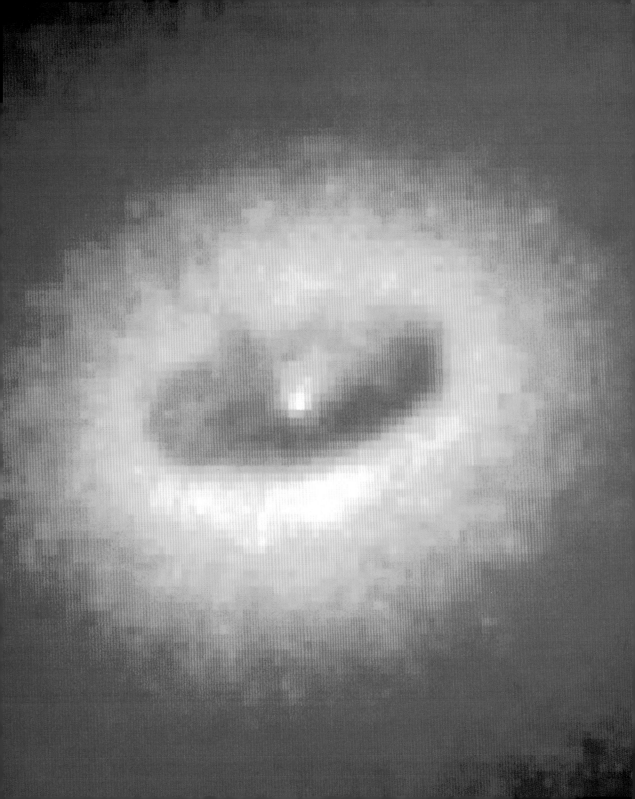

Genetic cousins

Mary-Claire King *b.*1946, Allan C. Wilson 1935–91

No one, when being serious, could confuse a human being with a chimpanzee. Humans walk upright, use language, have little body hair and have social lives based on pair-bonds. Chimpanzees are in all respects different. The biological attributes of all species are coded for in their DNA, and you might expect that the DNA of humans and chimps would be as clearly different as their bodies. But in 1975 two American biologists, Mary-Claire King and Allan Wilson, showed that the DNA of the two species is almost indistinguishable. Human and chimp DNA is 98.5 per cent identical: only about 1.5 per cent of the coding units in the DNA differ between the two species.

King and Wilson used the molecular difference between humans and chimps in two evolutionary deductions. Since 1960 it had been known that the molecular difference between two species increases at a roughly constant rate through time: a phenomenon referred to as the molecular clock. The molecular difference between two species can then (after calibration) be used to estimate the time back to their common ancestor. The 1.5 per cent difference between human and chimp DNA implies that they split about 5 million years ago. In 1975 this date for human origins seemed to contradict the (much earlier) date inferred from fossils, but it is now widely accepted.

The second deduction was that the big difference between the bodies of chimps and humans might have evolved because of changes in a small number of 'regulatory' genes. Some of our genes are 'structural' genes, coding for enzymes and bodily building blocks. Others are 'regulatory' genes, which control when the structural genes are switched on or off. Chimps and humans may differ little in their structural genes, but more in the regulatory genes that control the former's expression. This idea remains unproven, but it has been highly influential in current thinking about the genetic basis of evolutionary changes in bodily form.

See also The double helix pages 374–375, Chimpanzee culture pages 396–397, Random molecular evolution pages 422–423, Genetics of animal design pages 460–461, Out of Africa pages 492–493, Human genome sequence pages 524–525

442

Right A 'chimpanzee' from Angola, brought to Paris in 1738. The discovery of apes so similar to humans raised questions about humanity's relationship to the rest of the animal kingdom.

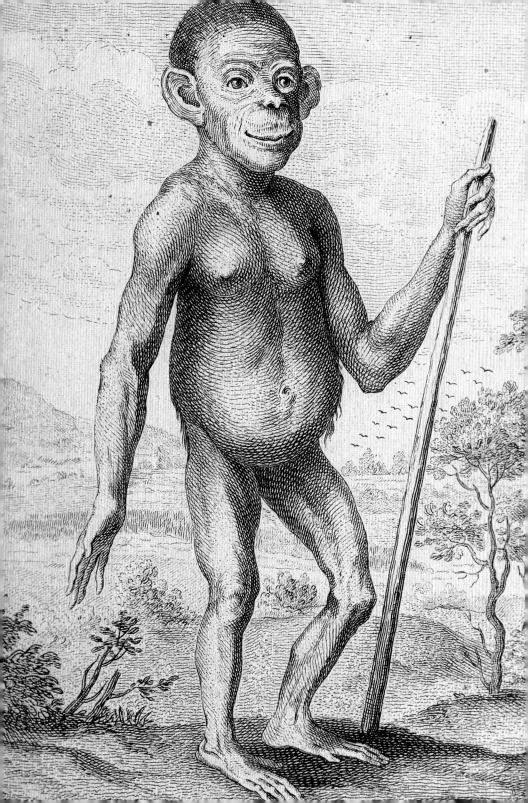

A unique species

Jared Diamond

Why did we evolve to be so different? This question becomes even more acute when we compare ourselves with our closest relatives among the world's mammal species, the great apes (as distinguished from the gibbons, or little apes). Closest of all are Africa's chimpanzee and bonobo, from which we differ in only about 1.6 per cent of our nuclear genetic material (DNA). Nearly as close are the gorilla (2.3 per cent genetic difference from us) and the orang-utan of Southeast Asia (3.6 per cent different). Our ancestors diverged 'only' about seven million years ago from the ancestors of chimpanzees and bonobos, nine million years ago from the ancestors of gorillas, and 14 million years ago from the ancestors of orang-utans.

That sounds like an enormous amount of time in comparison to an individual human lifetime, but it is a mere eye-blink on the evolutionary time scale. Life has existed on Earth for more than three billion years, and hard-shelled, complex large animals exploded in diversity more than half a billion years ago. Within that relatively short period during which our ancestors and the ancestors of our great ape relatives have been evolving separately, we have diverged in only a few significant respects and to a modest degree, even though some of those modest differences – especially our upright posture and larger brains – have had enormous consequences for our behavioural differences.

Along with posture and brain size, sexuality completes the trinity of the decisive respects in which the ancestors of humans and great apes diverged. Orang-utans are often solitary, males and females associate to copulate, and males provide no paternal care; a gorilla male gathers a harem of a few females, with each of which he has sex at intervals of several years (after the female weans her most recent offspring and resumes menstrual cycling and before she becomes pregnant again); and chimpanzees and bonobos live in troops with no lasting male–female pair bonds or specific father–offspring bonds. It is clear how our large brain and upright posture played a decisive role in what is termed our humanity – we now use language, read books, watch TV, buy or grow most of our food, occupy all continents, keep members of our own and other species in cages, and are exterminating most other animal and plant species, while the great apes still speechlessly gather wild fruit in the jungle, occupy small ranges in the Old World tropics, cage no animal and threaten the existence of no other species. What role did our weird sexuality play in our achieving these hallmarks of humanity?

Could our sexual distinctiveness be related to our other distinctions from the great apes? In addition to (and probably ultimately as a product of) our upright posture and large brains, those distinctions include our relative hairlessness, dependence on tools, command of fire and development of language, art and writing. If any of these distinctions predisposed us towards evolving our sexual distinctions, the links are certainly unclear. For example, it is not obvious why our loss of body hair should have made recreational sex more appealing, nor why our command of fire should have favoured menopause. Instead it may be that recreational sex and menopause were as important for our development of fire, language, art and writing as were our upright posture and large brains.

The key to understanding human sexuality is to recognize that it is a problem in evolutionary biology. When Darwin recognized the phenomenon of biological evolution in his great book *On the Origin of Species*, most of his evidence was drawn from anatomy. He inferred that most plant and animal structures evolve – that is, they tend to change from generation to generation. He also inferred that the major force behind evolutionary change is natural selection. By that term, Darwin meant that plants and animals vary in their anatomical adaptations, that certain adaptations enable individuals bearing them to survive and reproduce more successfully than other individuals, and that those particular adaptations therefore increase in frequency in a population from generation to generation. Later biologists showed that Darwin's reasoning about anatomy also applies to physiology and biochemistry: an animal's or plant's physiological and biochemical characteristics also adapt it to certain lifestyles and evolve in response to environmental conditions.

Evolutionary biologists have shown that animal social systems also evolve and adapt. Even among closely related animal species, some are solitary, others live in small groups and others live in large groups. But social behaviour has consequences for survival and reproduction. Depending on whether a species' food supply is clumped or spread out, and on whether a species faces risk of attack by predators, either solitary living or group living may be better for promoting survival and reproduction. Similar considerations apply to sexuality. Some sexual characteristics may be more advantageous for survival and reproduction than others, depending on each species' food supply, exposure to predators and other biological characteristics …

We can redefine the problem posed by our sexuality. Within the last seven million years, our sexual anatomy diverged somewhat, our sexual physiology further and our sexual behaviour even more, from our closest relatives, the chimpanzees. Those divergences must reflect a divergence between humans and chimpanzees in environment and lifestyle.

Fractals

Benoit Mandelbrot *b.*1924

In 1975 Benoit Mandelbrot published in French the landmark book *The Fractal Geometry of Nature*, the culmination of over 20 years of research that brought together a vast assortment of mathematical curiosities into a coherent framework. He coined the word 'fractal' from the Latin *fractus*, or 'broken', to highlight the fragmented and irregular nature of his computer-generated geometric landscapes.

A key feature of fractals is that they are self-similar at varying scales – a small part of the structure looks very much like the whole. If we look at the British coastline on a map we see that it looks fairly crooked and crinkly. If we could repeatedly zoom in we would see it in more detail but at each magnification the nature of the 'crinkliness' remains the same and is a fundamental part of the coastline's geometry. In fractals self-similarity is generated by a set of mathematical rules or 'algorithms'. Rather than using these to plot a graph, we use them to generate a sequence of numbers, each one fed back into the algorithm to generate the next. In the 1910s the French mathematicians Gaston Julia and Pierre Fatou had found what appeared to be random but bounded sequences of numbers. Not until Mandelbrot had developed a computer graphics program had anybody seen that far from being random the sequences yielded detailed intricate figures.

One other feature of fractals is their fractional dimension. The length of Britain's coastline depends on how accurately it is measured, and if we could theoretically zoom in further and further, the total length would get ever higher, approaching infinity, and yet the area of land would remain within a defined limit. The level of 'crinkliness' defines the 'fractal dimension' of the coastline, which is greater than one but less than two. This new geometry is all around us. Fractals and self-similarity are seen in the structure of plants, the formations of clouds, the fluctuations of stockmarkets, the distribution of galaxy clusters and, yes, coastlines.

See also Euclid's *Elements* pages 20–21, Chaos theory pages 238–239, Edge of chaos pages 490–491, Great Attractor pages 498–499

Right A three-dimensional 'spiral' fractal image derived from the 'Julia set'. Intimately linked to the Mandelbrot set, the Julia set was invented during the First World War by Gaston Julia and Pierre Fatou.

Monoclonal antibodies

César Milstein *b.*1927

When a bacterium or virus enters the body, antibody molecules bind to specific molecules called antigens on the invader's surface, tagging them for destruction. For years, researchers suspected that antibodies could have many applications, but it was not possible to get them in a pure form because different B lymphocytes in the immune system produce a complex mixture of antibodies to a particular antigen.

In 1975 César Milstein, working at the UK Medical Research Council in Cambridge with Georges Köhler, found a way of making pure, specific antibodies. In a cancer known as a multiple myeloma, one type of lymphocyte divides uncontrollably, producing a single, or monoclonal, antibody. Milstein found that if they immortalized the lymphocytes by fusing them with myeloma cells, they could culture the resulting hybrid cells and produce large amounts of pure monoclonal antibody of a predetermined specificity.

A monoclonal antibody can seek out a protein in the body by binding to a defined antigen on its surface. Since Milstein's ground-breaking discovery, many uses have been found for monoclonal antibodies. For instance, the test for HIV infection uses a monoclonal antibody to the virus, linked to a coloured molecule. If the virus is present in someone's blood, it binds the antibody, and a colour reaction indicates its presence. Monoclonal antibodies are also important in cancer therapy. Cancer cells are marked out from healthy cells by antigens on their surfaces. Monoclonal antibodies that bind to these cancer antigens are already used to treat breast cancer, and can act as 'guided missiles' to deliver anti-cancer drugs to a tumour, leaving healthy tissue unscathed. A similar technique can produce an image of a tumour – here the antibodies are linked to a dye or radioactive 'flag', allowing the tumour to be outlined in detail once they have reached their target.

See also Vaccination pages 102–103, Cellular immunity pages 204–205, Antitoxins pages 214–215, Blood groups pages 236–237, Transplant rejection pages 356–357, Biological self-recognition pages 430–431, Human cancer genes pages 456–457, AIDS virus pages 472–473

Right Monoclonal antibodies can now be produced in sufficient quantities to be used for some diagnostic purposes such as pregnancy tests and the microscopic study of diseased tissue.

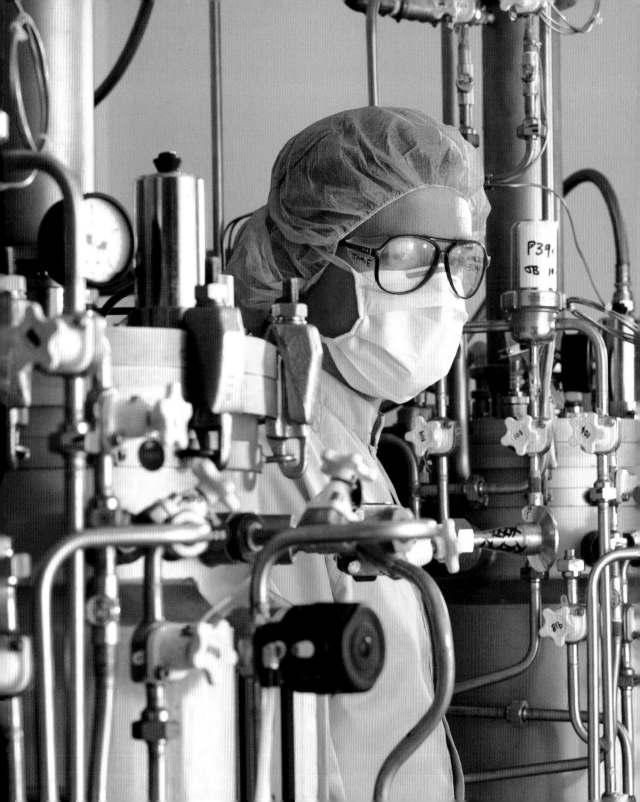

Four-colour map theorem

Kenneth Appel *b*.1932, Wolfgang Haken *b*.1928

In 1852 Augustus De Morgan, the first professor of mathematics at the newly founded University College London, was asked by a student to verify a seemingly innocuous conjecture: that four colours are the minimum number required to draw a map so that no two adjacent regions have the same colour. The problem piqued De Morgan's curiosity, and it soon became the subject of learned articles in the leading mathematics journals of the day.

With so many possible maps, mathematicians needed some form of classification to distinguish one layout from another, before checking that only four colours were indeed sufficient. In 1879 Alfred Bray Kempe, a London barrister and mathematician, published a proof in the journal *Nature*, and amid unanimous approval was elected to the Royal Society. Some ten years later, however, his method was found to be faulty, and by the twentieth century the problem had become a classic of topology – the branch of mathematics that deals with the configurations of regions of space and their relationships to each other, rather than their shapes or sizes. So attention shifted from the shape of the regions of the map to their configuration – how regions share a common border with other regions.

The four-colour conjecture finally became the four-colour theorem with the aid of a computer. In 1976 Kenneth Appel and Wolfgang Haken used 1,200 hours of computer time, supplemented with some 700 pages of hand calculations, to provide the first ever mathematical proof that nobody could read. The number of unique configurations analysed was so great that mathematicians accepted the proof, initially reluctantly, on the basis of an understanding of the algorithm that generated it rather than the results of the computation. The algorithm itself has since been further improved, but the question of what in mathematics constitutes a proof is still hotly debated.

See also Limits of mathematics pages 308–309, The computer pages 340–341, Fermat's last theorem pages 506–507

Right Many simple maps cannot be shaded using just three colours without two adjacent regions having the same colours. On the other hand, for all maps, such as the state map of New Jersey shown here, four colours suffice.

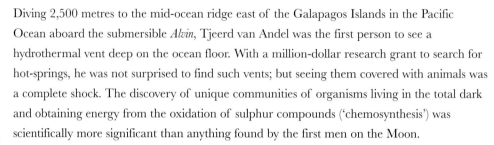

Life at the extreme

Tjeerd van Andel *b.*1923

Diving 2,500 metres to the mid-ocean ridge east of the Galapagos Islands in the Pacific Ocean aboard the submersible *Alvin*, Tjeerd van Andel was the first person to see a hydrothermal vent deep on the ocean floor. With a million-dollar research grant to search for hot-springs, he was not surprised to find such vents; but seeing them covered with animals was a complete shock. The discovery of unique communities of organisms living in the total dark and obtaining energy from the oxidation of sulphur compounds ('chemosynthesis') was scientifically more significant than anything found by the first men on the Moon.

Van Andel, a Dutch-born oceanographer, was working in the USA and professes to be 'a very lucky man to know exactly the most exciting moment of my life as a scientist. It was at 11.15 on the morning of 17 February 1977. The only other person I know who can say the same thing is the astronaut Neil Armstrong – he set foot on the Moon at exactly 9.28 on Sunday evening 20 July 1969.' With the *Alvin* dive, van Andel was following a hunch backed by a remotely sensed slight increase (0.01 degrees Celsius) in the temperature of the water at the ocean bottom.

At mid-ocean ridges, cold ocean water penetrates cracks in the hot young ocean-floor rocks. Heated, the water recirculates, scavenging chemicals from the surrounding rocks and streaming back into the ocean as hot-springs. These vent brines vary in chemistry and temperature with some reaching 350 degrees Celsius. Mixing with seawater precipitates manganese and iron minerals. Chemosynthesis by sulphur-eating bacteria (more than a million per millilitre) builds an ecosystem uniquely different from the photosynthetic one that dominates most life on Earth. Huge tube worms (*Riftia*, two metres high) and enormous mussels (such as *Calyptogenia* and *Bathymodiolus*, 150 millimetres long), which are hosts to endosymbiotic and chemosynthetic bacteria, thrive and are scavenged by crabs and fish.

Above View from the deep-submergence vehicle *Alvin* of black 'smoke' emitted by a hydrothermal vent on the Mid-Atlantic Ridge, 3,100 metres below sea level.

See also Photosynthesis pages 344–345, Origin of life pages 370–371, Oldest fossils pages 410–411, Plate tectonics pages 414–415, Galileo mission pages 514–515, Martian microfossils pages 518–519, Lake Vostok pages 520–521, Water on the Moon pages 522–523

Right Artist's impression of a hydrothermal vent on the ocean floor. Primitive forms of extremophile bacteria provide food for worms, molluscs, crustaceans and crabs.

Public-key cryptography

Ronald Rivest *b.*1947, Adi Shamir *b.*1952, Leonard Adleman *b.*1945

In cryptography, encryption and decryption keys must be kept secure. The problem is, the sender has to send not only the encrypted message but also the decryption key, either via another message or in person. In 1942 the British were able to decipher all of Germany's 'secret' naval communications after discovering the code-book in a captured U-boat. One option is that both the sender and the receiver are involved in the encryption process.

In 1976 Whitfield Diffie, Martin Hellman and Ralph Merkle at Stanford University, California, discovered a mathematical way of distributing encoded messages while keeping both keys private (the only disadvantage was the inconvenient sequence of events – sender encrypts, receiver encrypts, sender decrypts and finally receiver decrypts). Their second discovery was even more astonishing: that the encryption key could actually be made public without any loss of security. This counter-intuitive idea hinged on the fact that anybody can close a padlock but only the key-holder can open it. But it needed a mathematical padlock that was impossible to force open.

In 1977 another trio, Ronald Rivest, Adi Shamir and Leonard Adleman at the Massachusetts Institute of Technology proposed that very large prime numbers fit the bill. For a computer, multiplication of two primes is trivial, but the reverse procedure of finding the two original primes given their product is far harder. That year *Scientific American* ran a competition to find the two prime factors of a 129-digit public key – it took 17 years before anyone claimed the prize. Today RSA public-key encryption uses numbers so large that it would take all the computers on the planet longer than the age of the universe to crack.

And the covert history of public-key encryption? While working at GCHQ, James Ellis had discovered the concept of public-key encryption back in 1969, and by 1975 Ellis, Clifford Cocks and Malcolm Williamson had mastered its fundamental aspects. Unfortunately, the government then considered patenting to be a breach of security.

See also Deciphering hieroglyphics pages 136–137, The double helix pages 374–375

Right German signal troops during the Second World War communicate using a 'teletype' – a version of the Enigma coding machine.

Human cancer genes

Robert Allan Weinberg *b.*1942

Cancer is now thought of as a disease of the genes. This does not mean that it is always inherited (although sometimes it is). In the new view of cancer, certain genes, termed oncogenes, are 'hit' by carcinogens (chemicals, sunlight and some viruses) and the resulting damage upsets the delicate balance between cell division and cell quiescence (or cell death). When a cell – especially a damaged cell – starts to divide out of control, a tumour may result. Normally such cells should be prompted to commit suicide – a process known as 'apoptosis'. In many forms of cancer, apoptosis does not work properly.

In 1980 Robert Weinberg of the Massachusetts Institute of Technology discovered the first ever oncogene, known as *ras*. If this gene is mutated, cells divide continually, leading to the formation of tumours. Since then, mutated *ras* genes have been found in up to a third of all human cancers, and are especially common in colon, lung and pancreatic cancers.

Weinberg went on to discover in 1986 the gene that causes retinoblastoma, a rare eye cancer. Retinoblastoma arises in early childhood, and affects one child in 20,000. Some forms are inherited, and Weinberg showed that in these cases the child had lost the retinoblastoma gene (*Rb*) from chromosome 13. *Rb* was the first tumour-suppressing gene to be discovered; its role is to act as a brake on the cell-division cycle. Although retinoblastoma itself is rare, *Rb* is found in all cells. So studying the way it works has shed new light on how cancer progresses. Therapies are being developed that can restore the tumour-suppressing capacity to a cell that has lost it. Such targeted therapies, based on a knowledge of the genetics of cancer, are likely to be far more effective than the relatively crude anti-cancer drugs we rely on today.

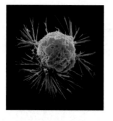

Above A spreading breast-cancer cell showing its uneven surface and cytoplasmic projections.

Right One renegade cell: a cancer cell in the last stage of division. Nuclear envelopes form around the duplicate chromosome sets and the two 'daughter' cells are connected by only a narrow bridge.

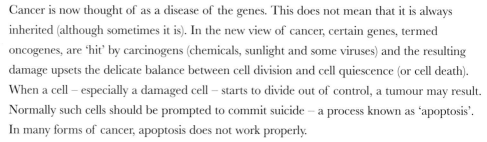

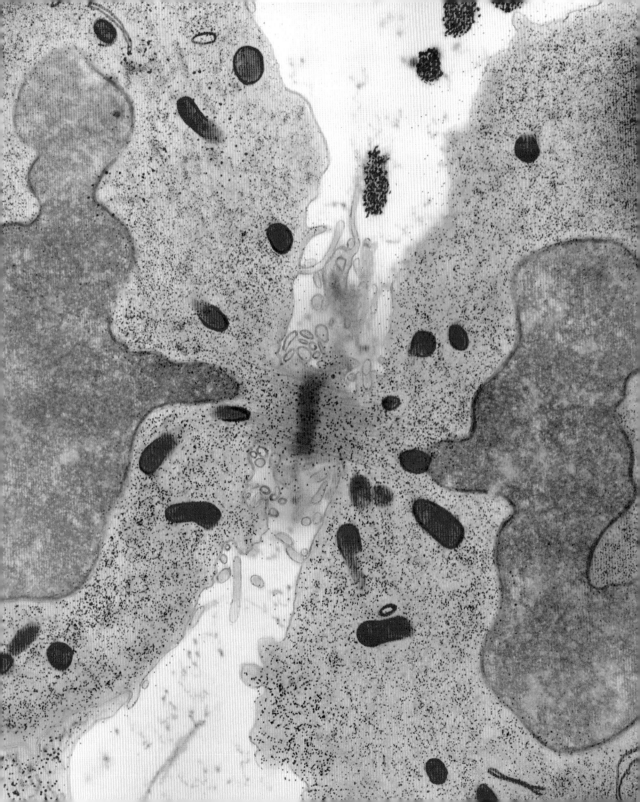

Extinction of the dinosaurs

Luis Walter Alvarez 1911–88, Walter Alvarez b.1940

In the late 1970s the American scientist Walter Alvarez, working with his father Louis Alvarez, Frank Asaro and Helen Michel, found a thin layer of clay at Gubbio in Italy that was rich in the rare element iridium. The layer was at the Cretaceous/Tertiary (K/T) boundary, a point marked by the mass extinction of the dinosaurs 65 million years ago. Louis suspected that the iridium came from outer space. This led the team to speculate in 1980 that a meteor or even a comet collided with Earth, causing the iridium-rich layer and the dinosaurs' extinction. The suggestion made international news.

There was no known crater of the right age or large enough to have caused the catastrophe, but dinosaurs were not alone in dying out at the K/T boundary. Many marine creatures such as the ammonites became extinct along with some 40 per cent of all other species living at the time. It took a further 10 years before the crater site was eventually pinned down to the Caribbean region and then revealed by sophisticated seismic measurements: it was buried beneath a kilometre of younger sediments at Chicxulub in the Yucatan peninsula of Mexico.

The impacting body was about 10 kilometres wide, travelled at about 30 kilometres per second and blasted an impact crater 100 kilometres wide and 12 kilometres deep. The impact pushed up a mountainous rim 8 kilometres high. With the energy of 100 million hydrogen bombs, 50,000 cubic kilometres of rock were blasted into the atmosphere in the form of dust, gas, droplets of molten rock and tiny diamonds, darkening skies and causing wild fires around the world. The mountainous rim collapsed, creating massive earthquakes, huge tidal waves and further crater rings up to 150 kilometres from the epicentre. Acid rain from the 800 million tonnes of sulphur thrown into the atmosphere devastated plant life and the whole basis of the global food chain. Earth history has been punctuated by such impacts in the past and will be again.

See also Comparative anatomy pages 106–107, Invention of the dinosaur pages 154–155, Burgess Shale pages 260–261, A cometary reservoir pages 360–361, Water on the Moon pages 522–523

Above Aerial view of Meteor Crater, Arizona. It is believed to have been formed by a meteor 50,000 years ago.

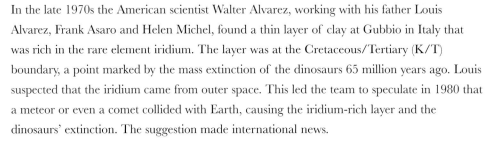

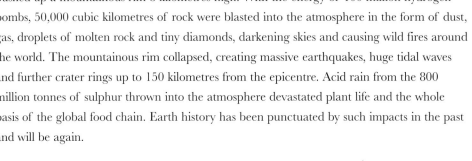

Right Gravity-anomaly map showing the extent of the Chicxulub impact crater, Yucatan peninsula, Mexico. The crater rim, seen as the larger green, yellow and red ring, is about 180 kilometres in diameter.

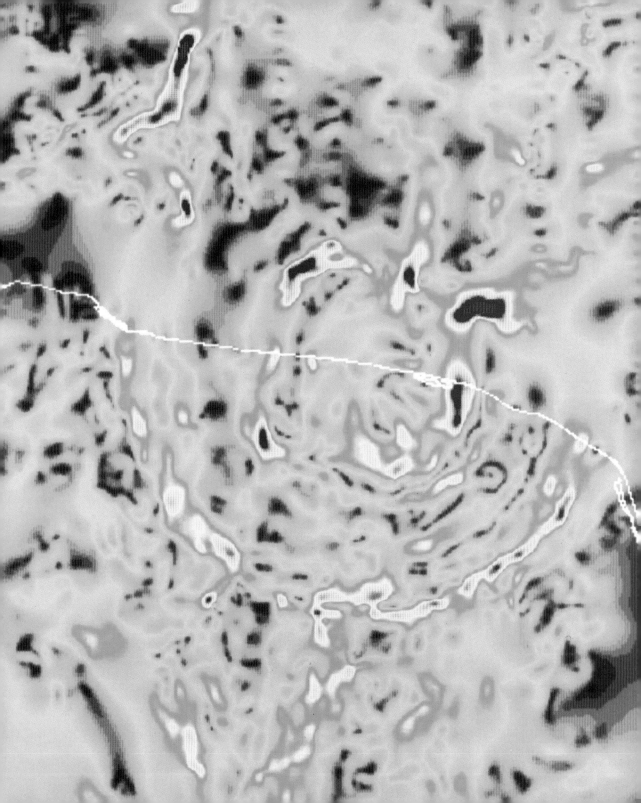

Genetics of animal design

Jani **Nüsslein-Volhard** *b*.1942, Eric **Wieschaus** *b*.1947, Ed **Lewis** *b*.1918

A single cell, the fertilized egg, gives rise to animals as diverse as humans and flies. Genes control how the cells behave during embryonic development and their future identity. The discovery of such genes and the key processes they control earned a Nobel Prize for Jani Nüsslein-Volhard, Eric Wieschaus and Ed Lewis in 1998.

In the late 1970s Nüsslein-Volhard and Wieschaus systematically searched for genes that control the early development of the fruit fly *Drosophila melanogaster*, whose genetics were already well known. By dosing flies with chemicals that cause mutations, they had, by 1980, identified key genes involved in setting up the fly's body plan, including the formation of the segments of the larva, each of which can be different. Surprisingly, these 'developmental genes' were found to act hierarchically, demarcating the embryo into ever smaller regions. At the top, for example, is a gene whose protein product forms a chemical gradient along the head–rear axis of the embryo, thus providing the cells with information about their position.

But cells also need to know what to become. This is controlled by 'homeotic' genes, so called because mutations in them can change one body structure into another – for example, in flies part of the balancing organ for flight can during early development be changed into part of the wing. Ed Lewis found that homeotic genes which pattern the body plan are arranged along the chromosome in the same order as the parts of the body they control. The character of the different regions is determined by a combinatorial code of these genes, later shown to contain a common 80-letter DNA region known as a homeobox. But the *coup de grâce* came when it was discovered that similar genes play a fundamental role in the development of most animals, be they sea-urchins, frogs, mice or humans. This breakthrough not only revolutionizes our understanding of the genetic control of development, but also highlights the common evolutionary ancestry of all living things.

See also Aristotle's legacy pages 16–17, Eggs and embryos pages 140–141, Genes in inheritance pages 264–265, Slime-mould aggregation pages 348–349, Chemical oscillations pages 364–365

Right As a fruit-fly embryo develops, genes are turned on and off in a pattern of stripes. This channels the cells of different regions into becoming different body segments.

Eruption of Mount St Helens

US Geological Survey

In March 1980 researchers with the US Geological Survey detected pre-eruption earthquakes beneath Mount St Helens in Washington State, USA. Small ash explosions punctured the snow cover of the summit. And in April a bulge nearly 2 kilometres across appeared on the north flank and grew rapidly (up to a metre a day). By early May this bulge was 150 metres above the rest of the ground and was clearly unstable.

At 08:32 on 18 May 1980, the northern side of the mountain began to slide down, turning into a huge avalanche. Although the immediate trigger was a magnitude 5.1 earthquake, the underlying cause was a mass of hot magma ascending beneath the volcano. Seconds later, a massive cloud of hot volcanic ash was released by the avalanche. The hot blast travelled at supersonic speeds, shot thousands of metres into the air and flattened millions of full-grown Douglas fir trees in an area of 600 square kilometres. The avalanche of rock, glacier-ice and soil debris, travelling at about 75 metres per second, then covered the region to a depth of 100 metres in grey mud with a peculiar hummocky appearance.

The event provided the first modern data on the main causes of death in explosive volcanic eruptions: trauma caused by flying rocks, burns, and lung damage from inhalation of hot gases. Even those rescued died later of their injuries – the likely fatality of such injuries explains why only two people out of 29,000 survived the 1902 eruption of Mount Pelee in Martinique.

Researchers carefully studied the hummocky terrain left by the Mount St Helen's avalanche, and began to recognize the dreadful origin of similar volcanic landscapes. And the eruption emphasized the importance of real-time geophysical measurements (especially of seismic activity and ground deformation): evacuation of the valley region that was predicted to be directly affected had saved countless lives (57 people died in total). But civil-defence authorities only learnt their lessons after the 1985 disaster at Nevado del Ruiz, Colombia, where 25,000 people died after an eruption was predicted, but no action was taken to evacuate.

See also Earth cycles pages 100–101, Catastrophist geology pages 108–109, Humboldt's voyage pages 114–115, Lyell's *Principles of Geology* pages 146–147, Mountain formation pages 206–207, Inside the Earth pages 250–251, Rocks of ages pages 252–253, Continental drift pages 270–271, Plate tectonics pages 414–415, Galileo mission pages 514–515

Right The eruption of Mount St Helens in 1980 was by far the best-documented event in the annals of volcanology. For the first time scientists could observe all the phases of a gigantic explosive cataclysm.

Quantum weirdness

Alain Aspect *b.*1947

What is reality? Common sense says that objects exist whether we look at them or not, but quantum mechanics takes a less comfortable view: that the world is made of uncertain possibilities which are realized only when a measurement is made.

Albert Einstein, Boris Podolsky and Nathan Rosen thought the quantum picture was absurd and devised a thought experiment in 1935 to prove it. If two particles fly out of a single reaction, their properties will be correlated, so by measuring one particle, it should be possible to make deductions about the other one. But quantum mechanics asserts that the particles have no real properties before measurement. So making a measurement on one particle must somehow instantly affect the other, even though you don't touch it – and even if it is, by now, a vast distance away. Einstein thought that this 'spooky action at a distance' was unacceptable. Surely each particle must have independent, real properties?

In 1965 the British physicist John Stewart Bell worked out that the correlation you see would be greater under quantum theory than it could be with any theory that gives the particles their own real properties. The delicate experiment needed to test this was finally done in 1982 by Alain Aspect and colleagues at the University of Paris, Orsay. They looked at the polarization of pairs of photons emitted by calcium atoms, and, sure enough, the level of correlation observed was too high to allow the realists' position, supporting instead the quantum theory. Reality isn't simple.

But that leaves the philosophers a lot of leg-room, and several interpretations of quantum mechanics exist. Does human consciousness turn quantum uncertainty into real measurement? Do all possible outcomes of a measurement exist in parallel universes? Is the world bound together by a web of 'non-local' connections? Or does quantum mechanics tell us only about measurements and experiments, not about reality itself?

See also The quantum pages 234–235, Wave–particle duality pages 300–301

Right Quantum incoherence: Salvador Dali believed that 'the key to today's art is just a quantified realism – a quantum of action in the iconography of the microphysics'.

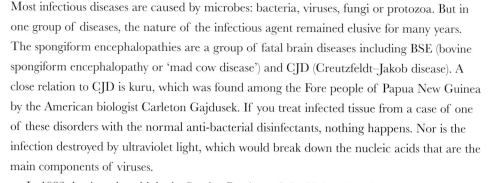

Prion proteins

Stanley **Prusiner** *b.*1942

Most infectious diseases are caused by microbes: bacteria, viruses, fungi or protozoa. But in one group of diseases, the nature of the infectious agent remained elusive for many years. The spongiform encephalopathies are a group of fatal brain diseases including BSE (bovine spongiform encephalopathy or 'mad cow disease') and CJD (Creutzfeldt–Jakob disease). A close relation to CJD is kuru, which was found among the Fore people of Papua New Guinea by the American biologist Carleton Gajdusek. If you treat infected tissue from a case of one of these disorders with the normal anti-bacterial disinfectants, nothing happens. Nor is the infection destroyed by ultraviolet light, which would break down the nucleic acids that are the main components of viruses.

In 1982 the American biologist Stanley Prusiner of the University of California published a controversial paper in which he suggested that the infectious agent of scrapie, a spongiform encephalopathy found in sheep, was a protein. He called the culprit a 'prion' – short for 'proteinaceous infectious particle'. This idea goes against the 'central dogma' of molecular biology which says that biological information can flow only from DNA to protein, not the other way around. Proteins, the argument holds, do not, unlike DNA, encode the necessary information for self-replication.

Above 'Mad cow disease' first appeared in Britain around 1985, with disastrous consequences.

Prions have now been isolated; they turned out to be an abnormal form of a protein found naturally in the body. Prusiner has proposed that the abnormal prion molecule attaches itself to a normal one and 'corrupts' it, turning it into the abnormal form. The two abnormal prion molecules can then corrupt further normal molecules – a kind of domino effect. The process eventually destroys brain tissue. The question now is whether ways of stopping the corruption can be found, so that people at risk of CJD might, for instance, be vaccinated against the ravages of prions. Prusiner won the 1997 Nobel Prize for Medicine or Physiology as a result of his work on prions.

See also Vaccination pages 102–103, Germ theory pages 202–203, Viruses pages 232–233, AIDS virus pages 472–473

Right Prions in the brain of a cow infected with bovine spongiform encephalopathy or 'mad cow disease'. The orange fibrils are thought to be aggregations of the protein that make up the infectious prion.

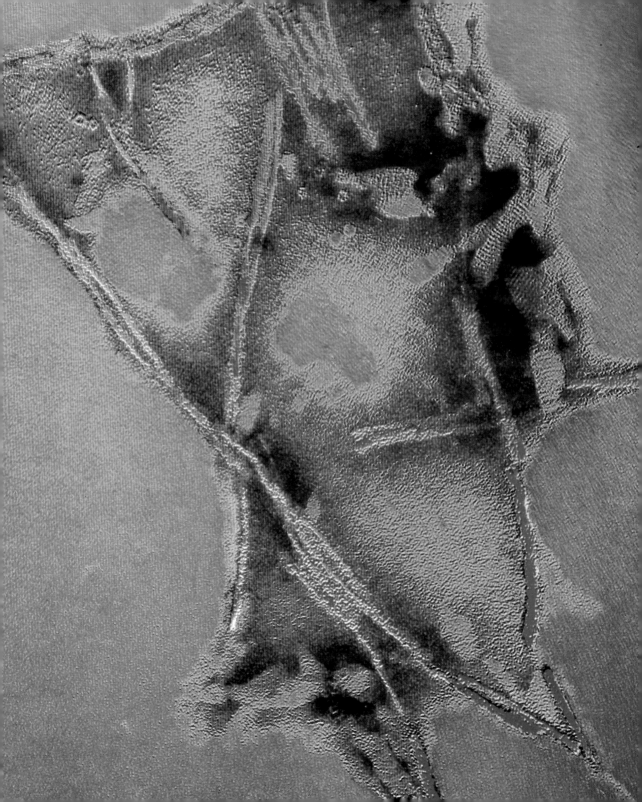

Diversity of life

Terry Erwin *b.*1940

Species are the standard units of biological diversity on Earth: human beings are an example of a biological species, as are gorillas, the common oak, the domestic cat and the garden robin. A biological species is a set of organisms that can interbreed with each other, but not with members of other species. One elementary fact about the ecology of our planet is the number of species that it contains. How many are there?

One estimate could come from the number of species that have been described: that number is about one-and-a-half million. But it is an underestimate of total biodiversity because there are many more undescribed species. In 1982 Terry Erwin from the Smithsonian Institution in the USA made an influential guess at the number of undescribed species. Beetles are the biggest group of animals on Earth. Erwin reasoned that most of the unknown beetle species probably live 30 metres up in the inaccessible canopies of tropical forests. He used a special method (known as the 'bug bomb') to make all the insects fall out of one such tree. He counted all the beetle species, known and unknown, and found 160 that were specific to that tree. Erwin reckoned that there are about 50,000 tropical tree species. Multiply up, and that gives an estimate of eight million tree-canopy beetle species. A further extrapolation implies that on Earth there are about 30 million arthropod species, and maybe 50 million species in all.

Erwin's estimate, based on one intensively studied tree, is obviously uncertain. Experts estimate that there are between 10 and 100 million species on Earth. But Erwin's research dramatized both the huge number of undescribed species that exist and our relative ignorance about global biodiversity, and made possible a reasoned estimate of what the total species count is.

See also Aristotle's legacy pages 16–17, Naming life pages 88–89, Burgess Shale pages 260–261, Neo-Darwinism pages 282–283, Gaia hypothesis pages 432–433, Five kingdoms of life pages 426–427, Life at the extreme pages 452–453, Lake Vostok pages 520–521.

Right Asked what could be inferred about the attributes of the Creator from a study of the natural world, the British scientist J. B. S. Haldane reportedly said, 'an inordinate fondness for beetles'.

Memory molecules

Eric Kandel *b.*1929

Eric Kandel was born in Vienna, Austria, in 1929, but emigrated as a child to the USA under the threat of Nazi oppression. After studying psychiatry at Harvard University, he became fascinated by the biology of the brain, in particular the molecular basis of learning and memory.

The human brain contains billions of nerve cells linked to each other in a complex network. Chemical neurotransmitters transfer information between nerve cells through specialized junctions called synapses. As the human brain is so complex, Kandel began in the 1960s to study the relatively simple nervous system of the sea slug *Aplysia* – a project that was to occupy him for the next quarter of a century.

Aplysia responds to noxious stimuli by withdrawing its gills. This gill reflex can be strengthened by learning. Kandel found that memory of a single noxious stimulus lasts only minutes. This short-term memory does not require gene activity or protein production, but it does involve chemical changes at the synapses between sensory nerves and the motor nerves mediating gill withdrawal. These changes involve enzymes called kinases that add phosphate groups to specific channels at the nerve terminal. This increases the influx of calcium ions into the nerve ending, which in turn boosts the release of neurotransmitter and so amplifies the gill reflex.

Stronger, repeated noxious stimuli lead to longer-term memory lasting days or weeks. Again the changes at the synapses involve the addition of phosphate groups to proteins, but now a signalling cascade is activated that leads to changes in gene activity and protein production. Resulting anatomical changes strengthen the connections between nerve cells. Kandel has since shown that a similar molecular mechanism also underlies short- and long-term memory in mice. So it may well form the basis of memory in humans too.

For this work Kandel won the 2000 Nobel Prize for Medicine or Physiology, along with fellow neuroscientists Arvid Carlsson and Paul Greengard.

See also Nervous system pages 210–211, Conditioned reflexes pages 242–243, Neurotransmitters pages 274–275, Behavioural reinforcement pages 322–333, Artificial neural networks pages 334–335, Nerve impulses pages 366–367

Right Kandel was drawn to the simple architecture of the sea slug *Aplysia's* nervous system, which consists of about 20,000 neurons, many large enough to be seen by the naked eye.

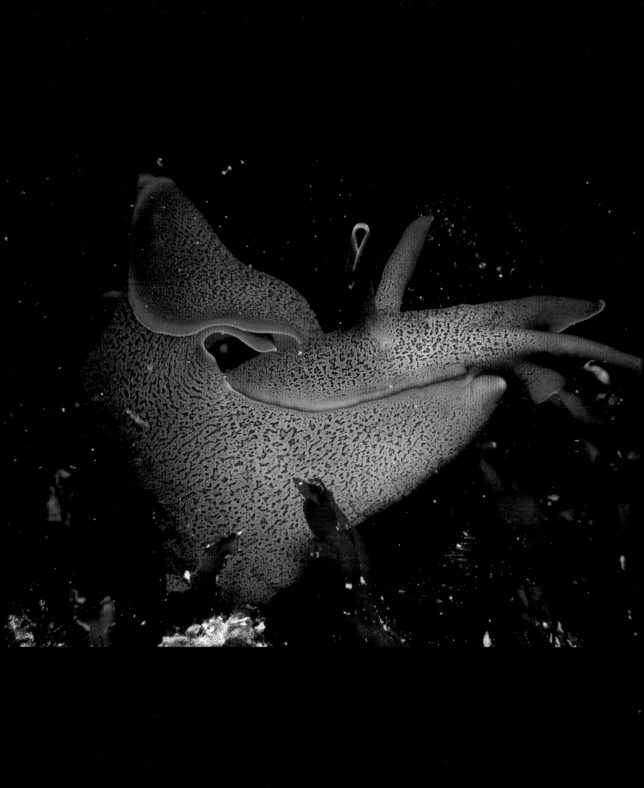

AIDS virus

Robert Gallo *b.*1937, Luc Montagnier *b.*1932

In 1982 a strange new disease caught the attention of biomedical researchers. Young homosexual men in California and New York were being afflicted by a rare form of pneumonia. Soon the handful of cases had become an epidemic – 750 people in the USA, 100 more in western Europe and an unknown number in Africa. All showed a profound decrease in T4 lymphocytes – cells that are key players in the immune system – and this allowed normally harmless infections, or a rare cancer called Kaposi's sarcoma, to take hold. The US Centers for Disease Control gave the disease a name: AIDS (acquired immunodeficiency syndrome).

But what caused AIDS? The answer was to come from two researchers – Robert Gallo of the US National Institutes of Health in Bethesda, Maryland, and Luc Montagnier of the Pasteur Institute in Paris – who in 1983 discovered the human immunodeficiency virus (HIV). There is still some controversy about Gallo's virus – whether it was actually a strain originally from the French laboratory or was an independent discovery – but officially both men are given joint credit for the work. It was an important breakthrough because it led to a test for HIV infection and, in time, to drugs that have turned AIDS from a sure killer into a chronic disease.

The origin of HIV remains unknown, although many researchers believe it may have arisen in Africa in the 1950s, when the original virus jumped species from monkey to human. HIV was the first known example of a retrovirus, so called because the flow of genetic information runs 'backward': its genetic material consists of RNA (ribonucleic acid) rather than DNA (deoxyribonucleic acid). It infects a range of white blood cells, the most important of which are the T helper cells. And that is the secret of its success: it cripples the immune system by knocking out vital components.

See also Viruses pages 232–233, The double helix pages 374–375, Monoclonal antibodies pages 448–449.

Right AIDS virus particles emerging as buds from the surface of a white blood cell. On breaking off, each bud reorganizes itself into a mature virus.

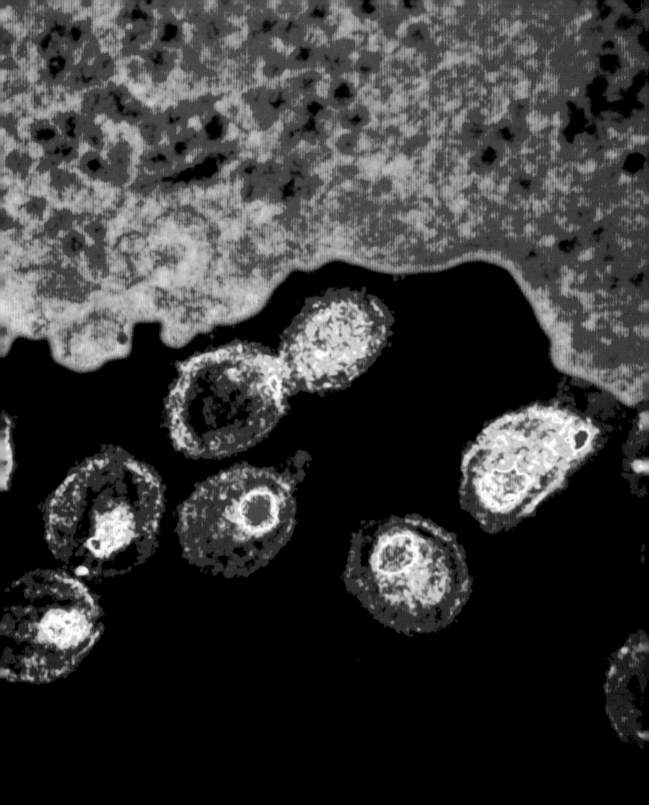

Superstrings

Michael Boris Green *b.*1946, John H. Schwarz *b.*1941

All of existence is simply notes plucked on a string. This is not mysticism or musicology; it is an assertion about the material world. String theory says that subatomic particles are not particles at all, but infinitesimally small one-dimensional loops. These 'superstrings' can vibrate, a little like a violin string – except that they are less than a millionth of a billionth the size of a proton, and they inhabit a space with six extra dimensions, all perpendicular to our familiar three dimensions, but curled up so tight we can't see them. The note of the vibration determines what properties the string has: one note gives you an electron, others give you a quark, a photon, a neutrino or anything else.

In 1984 the physicists Michael Green and John Schwarz showed that superstrings could unify nature. Electromagnetic and nuclear forces would be different sides of the same 'superforce' and there is even a kind of string vibration that could carry gravity. Not only would all forces and particles be manifestations of a single basic thing, the string, but the theory should dictate all the numbers that rule the universe. Why are electrons so light, why are there so many kinds of quark, why are neutrinos oblivious to electromagnetic forces? In theory, strings will explain it all.

But there are snags. Making calculations in string theory is so hard that few concrete predictions have yet come out, so it is untested. And some physicists object that the theory assumes the existence of space and time, whereas a true theory of everything should explain those too. There are several different versions of string theory, but Ed Witten and other theorists have now shown that they are all only facets of a barely glimpsed final theory called M-theory, which may be stranger still.

See also Electromagnetism pages 134–135, Maxwell's equations pages 186–187, The electron pages 228–229, The quantum pages 234–235, Model of the atom pages 272–273, The neutron pages 312–313, Quantum electrodynamics pages 352–353, A subatomic ghost pages 384–385, Quarks pages 408–409, Unified forces pages 416–417

Right Superstring theory holds that our universe has many more dimensions than meets the eye – dimensions tightly curled into the folded fabric of the cosmos.

Ancient DNA

Svaante Pääbo *b*.1955

Fragments of ancient DNA were first extracted in 1984 from the dried skin of a quagga, a zebra-like animal from southern Africa that was hunted to extinction over 100 years ago. Despite subsequent claims that fossil DNA had been recovered from dinosaur bone and insects trapped in amber, epitomized by Michael Crichton's novel *Jurassic Park*, it is not possible to extract DNA that is millions of years old. All attempts at the replication of such recoveries have failed: the supposed DNA was in fact contamination.

Complex cell molecules break down rapidly after death unless preserved by rapid freezing or dehydration, processes rare in nature. Nevertheless, careful use of new extraction and amplification techniques has allowed some research teams to recover DNA that is tens of thousands of years old. Foremost among these is Svaante Pääbo's group at the University of Munich.

Pääbo was a student of the late Allan C. Wilson of the University of California, Berkeley, who recovered the quagga DNA and showed its close, subspecies-level relationship to the zebra. In 1987 Wilson's team developed the 'African Eve' hypothesis to explain the origins of modern humans. Their global sample of DNA indicates that all modern humans originated from a single African population around 200,000 years ago.

Now Pääbo's team are using ancient DNA to improve our understanding of recent human evolution. Mitochondrial DNA from the 5,200-year-old 'Iceman' Ötzi found in the Austro-Italian Alps is surprisingly close to that of people living in the region today, suggesting a remarkable genetic stability. And three Neanderthal DNA samples, between 40,000 and 30,000 years old, are more similar to one another than to modern European DNA, which implies that the Neanderthals made no contribution to the modern human gene pool.

See also Neanderthal man pages 170–171, Java man pages 216–217, Taung child pages 298–299, The double helix pages 374–375, Olduvai Gorge pages 392–393, Random molecular evolution pages 422–423, Genetic cousins pages 442–443, Nariokotome boy pages 480–481, Genetic fingerprinting pages 484–485, Out of Africa pages 492–493, Iceman pages 504–505

Right A 35-million-year-old gall gnat trapped in Baltic amber. Despite claims that DNA of amber insects can be recovered, modern research shows this is not yet possible.

Images of the mind

Louis Sokoloff *b.*1921

Damage to particular areas of the brain often causes characteristic loss of function. For example, people with damage to Broca's area may comprehend language but fail to speak or write. Unfortunately, brain pathways are highly complex, so brain 'lesion' studies do not often reveal which parts of the brain do what. But with modern brain imaging we can see the workings of the intact brain in exquisite detail, be it engaged in processing language, thought, or memory; focusing attention or planning; or even registering emotion.

In 1984 Louis Sokoloff showed that positron emission tomography (PET scanning) could be used to monitor areas of the brain working hardest during particular tasks. PET scanning requires short-lived radioactive isotopes that emit positrons – positively charged electrons. If water containing such a radioactive isotope is injected into the bloodstream, the radioactivity will accumulate most quickly in parts of the brain with the greatest blood flow. Positrons emitted during radioactive decay collide with electrons, resulting in mutual annihilation. Each such event causes the release of energy in the form of two gamma rays travelling in exactly opposite directions. Gamma rays from many annihilation events are recorded by detectors surrounding the head, and a computer image is generated showing the concentration of radioactivity in slices through the brain. PET scanning is not only useful for measuring blood flow. By injecting a suitably labelled analogue of glucose, the technique can also be used to map areas of high fuel intake; and suitably labelled neurotransmitters can map out the distribution of the receptors that bind them.

Magnetic resonance imaging (MRI) can give even higher spatial and temporal resolution. When exposed to a strong magnetic field, many atoms behave like tiny spinning bar magnets, aligning themselves with the field. If they are now exposed to a pulse of radio waves, they emit detectable radio signals characteristic of the element and its environment. So MRI can be used to image the structure and composition of objects.

Right Window on the soul? PET scan of a human brain taking part in a stimulation test.

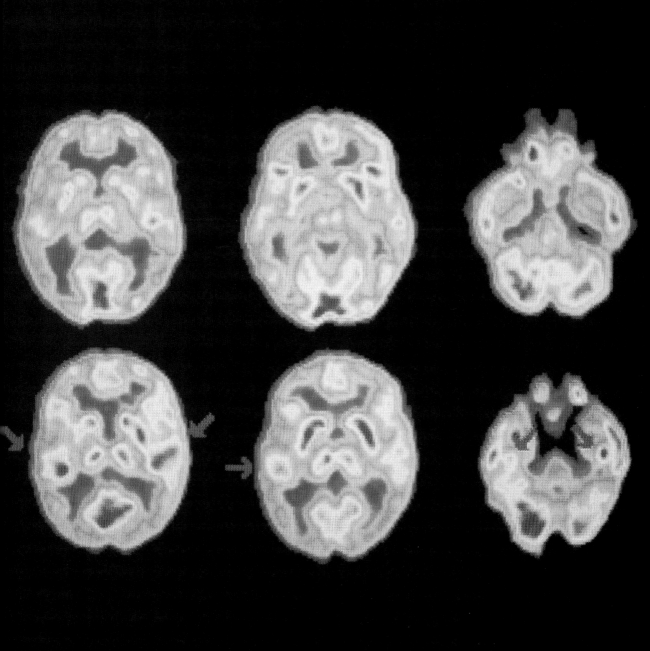

Nariokotome boy

Kamoya **Kimeu** *b.*1940

On 22 August 1984 the Kenyan palaeoanthropologist Kamoya Kimeu found a skull roof bone among black lava pebbles at Nariokotome, which lies west of Lake Turkana in northern Kenya – he had found the first remains of 'Nariokotome boy'. Around 1.5 million years old, the specimen became the most complete skeleton yet discovered that could be linked to our extinct ancestors *Homo erectus*.

Kimeu, foreman of the 'hominid gang' of Kenyans, has found more important fossils of our ancient relatives than anyone else in the world. The gang worked for Richard Leakey, son of Louis and Mary Leakey, and British-born Alan Walker, Professor of Palaeoanthropology at Pennsylvania State University, who have described and interpreted the finds.

More skull fragments were picked up and retrieved by sieving. Most hominid fossils are skull fragments and teeth. In the economy of nature, scavengers break up and disperse cadavers. It is exceedingly rare to find any of the post-cranial skeleton. 'Lucy', the 3-million-year-old australopithecine found by Donald Johanson in the Afar region of Ethiopia in 1974, was the only other partial skeleton (20 per cent preserved) then known to date from our ancient relatives' time.

But four years and some 1,500 cubic yards of sediment later, 67 bone fragments – around a third of the skeleton – had been recovered, including a skull that was spectacularly reconstructed. The Nariokotome skeleton includes a male pelvis and stands some 1.67 metres (5 feet 3 inches) tall, compared with Lucy's 1.07 metres (3 feet 6 inches). His slender build was well adapted to life out in the open tropical heat. Between 9 and 12 years old when he died, he was distinctly human but had a relatively small brain of 880 cubic centimetres, and almost certainly lacked speech.

His kin were the first really successful hominids. Long before Kimeu's find was born, *Homo erectus* were established as far away as southeast Asia and Russian Georgia. Modern humans are thought to have eventually evolved from African *Homo erectus* like the Nariokotome boy.

See also Prehistoric humans pages 148–149, Neanderthal man pages 170–171, Java man pages 216–217, Taung Child pages 298–299, Olduvai Gorge pages 392–393, Ancient DNA pages 476–477, Out of Africa pages 492–493, Iceman pages 504–505

Right Collection of hominid fossil skulls discovered during expeditions at East Turkana, Kenya, during the 1970s and 1980s. The skulls are, from left, *Homo habilis*, *Homo erectus* and *Australopithecus robustus*.

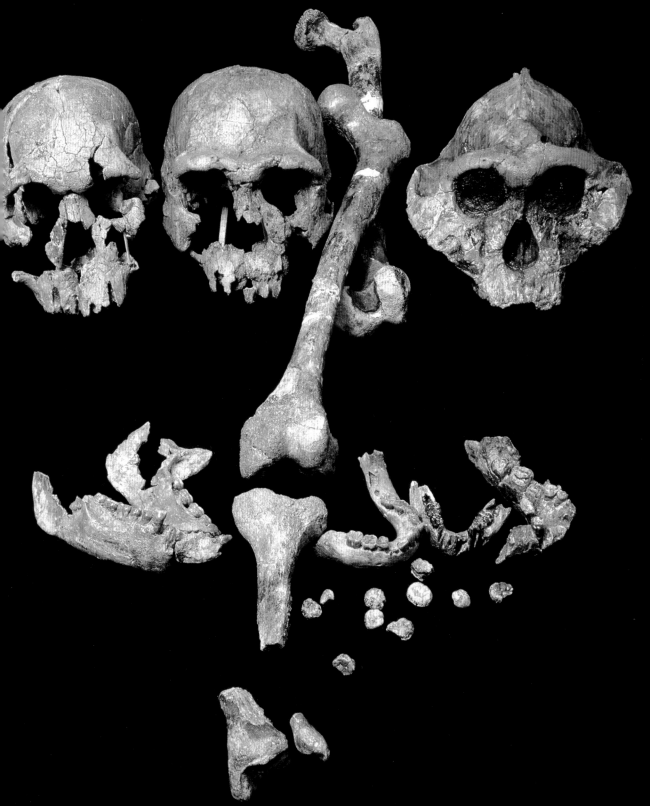

Quasicrystals

Dan Shechtman *b*.1941

Until the 1980s, one of the oldest tenets of physical chemistry was that all solids could be classified as either crystals or glasses. Rather like tiled floors, crystals are built up of highly ordered repeating patterns known as 'lattices'. Whereas the smallest repeated unit on a floor is an individual tile, the smallest periodic pattern in a crystal lattice is the so-called 'unit cell' of atoms or molecules. By contrast, the atoms and molecules that constitute a glass are amorphous, showing no long-range order at all.

But in 1984 a group at the US National Bureau of Standards in Gaithersburg, led by the Israeli crystallographer Dan Shechtman, made a shocking discovery: a material that didn't fit either classification. They found that by rapidly cooling an alloy of aluminium and manganese, they could produce a solid whose underlying structure had five-fold symmetry. Why was this a surprise? The most familiar shape with this kind of symmetry is a pentagon. Imagine a floor with tiles of this shape. No matter how they are locked together, they never completely cover the entire floor surface. Similarly, five-fold symmetry is totally incompatible with a periodic arrangement of identical unit cells. Amazingly, the paradoxical material that Shechtman and his co-workers discovered clearly had long-range order. So they named it a 'quasicrystal'. Many other such structures were soon reported.

Quasicrystalline alloys are harder than crystalline materials, and can have greater electrical resistance. Already they have been used in cookware, surgical tools and electric shavers. Yet only recently have scientists begun to explore the assembly of such complex arrangements. In 1997 the physicists Paul Steinhardt and Hyeong-Chai Jeong developed mathematical models to explain how quasicrystals could be made up from a single type of building block. This may in turn allow researchers to design and produce new quasicrystals with even more unusual properties.

See also Superconductivity pages 266–267, The transistor pages 350–351, Buckminsterfullerene pages 486-487

Right A jigsaw constructed from two types of tile fitted together without ever repeating the same pattern. Invented by the mathematician Roger Penrose, the tiling is a two-dimensional version of a quasicrystal.

Genetic fingerprinting

Alec Jeffreys *b*.1950

Our bodies are built from codes, or genes, carried in our DNA. But we have much more DNA than is needed for our genes. Indeed the 30,000 genes in a human being make up only one or two per cent of his or her DNA. The function of the other 98–99 per cent is uncertain, and is often called 'non-coding' or 'junk' DNA.

Some of the non-coding regions of our DNA are made up of repeats of various short sequences. DNA code is written with four kinds of molecule, symbolized by the letters A, T, C and G. So a stretch of repetitive DNA might consist of an eight-letter unit, such as GCAGGAGG, repeated a few dozen times. Some of these regions of repetitive DNA are highly variable between individuals: one person might have 10 repeats of the eight-letter unit, another 20 repeats, and another 100. The reason for the variation is that repetitive DNA has a high mutation rate, perhaps 10,000 times the standard mutation rate for DNA. The stretches of repeats expand and contract rapidly over evolutionary time. So every individual has a unique profile, or fingerprint, set by the size of his or her stretches of repetitive DNA.

Alec Jeffreys, working at the University of Leicester, England, discovered some of these highly variable regions of repetitive DNA in the 1980s. He also realized that the genetic fingerprints they provide could be applied to forensic problems, such as paternity disputes, identifying criminals and ruling out suspects in criminal prosecutions. In the UK, the national DNA database is currently linking criminals to evidence at the rate of over 500 per week. In the USA, genetic fingerprints have already shown 70 (one in eight) people facing the death penalty to be innocent.

See also Genes in inheritance pages 264–265, Jumping genes pages 362–363, The double helix pages 374–375, Random molecular evolution pages 422–423, Human genome sequence pages 524–525

Right Genetic fingerprinting reduces the information in each person's DNA to something like a bar code. A criminal's 'bar code' can be extracted from just one hair left at the crime scene.

Buckminsterfullerene

Harry Kroto *b*.1939, Richard Smalley *b*.1943, Robert Curl *b*.1933

If God were a geometer, carbon atoms would surely be His building blocks. In diamond's crystal lattice they link up in a four-cornered tetrahedral arrangement; in graphite they form hexagonal rings, joined edge-on in vast flat sheets; and in 1985 the British chemist Harry Kroto and his US collaborators at Rice University in Houston – Richard Smalley, Robert Curl and co-workers – found they could form a faceted, pseudo-spherical cage.

Kroto was keen to study linear chains of carbon molecules, which he believed might be formed in molecular clouds in space; Smalley was adept at making small clusters of atoms by blasting solid targets into vapour using a laser beam and letting the clusters condense from the cooled vapour. Together the researchers found that the large carbon clusters formed in this way all had an even number of atoms, and that by adjusting the experimental conditions they could create clusters almost exclusively containing precisely 60 atoms – 'C_{60}'.

Hours of haphazard experimentation revealed that the key to the unusual stability of C_{60} is its closed cage of carbon atoms, formed from five- and six-atom rings (pentagons and hexagons). The cage is a highly symmetrical polyhedron called a 'truncated icosahedron', like a soccer-ball made from pentagonal and hexagonal leather patches. It also resembles the 'geodesic domes' constructed in the 1950s and 1960s by the American architect Richard Buckminster Fuller; hence the name 'buckminsterfullerene'.

'Buckyballs' turned out to have some potentially useful properties, such as becoming superconducting when 'doped' with metal atoms. And in 1991 the Japanese scientist Sumio Iijima discovered a related hollow structure called the carbon 'nanotube' – a cylindrical tube, like a rolled-up graphite sheet, just a few nanometres in width and up to several micrometres long. Carbon nanotubes are extremely strong and stiff, and may have applications ranging from molecular-scale wires in ultra-miniaturized electronic circuits to electron-emitting antennae for making luminescent displays.

See also Benzene ring pages 190–191, Superconductivity pages 266–267, Quasicrystals pages 482–483

Right As well as their complex and beautiful shape, 'buckyballs' have novel physical and chemical properties that can be exploited to produce new catalysts, lubricants and superconductors.

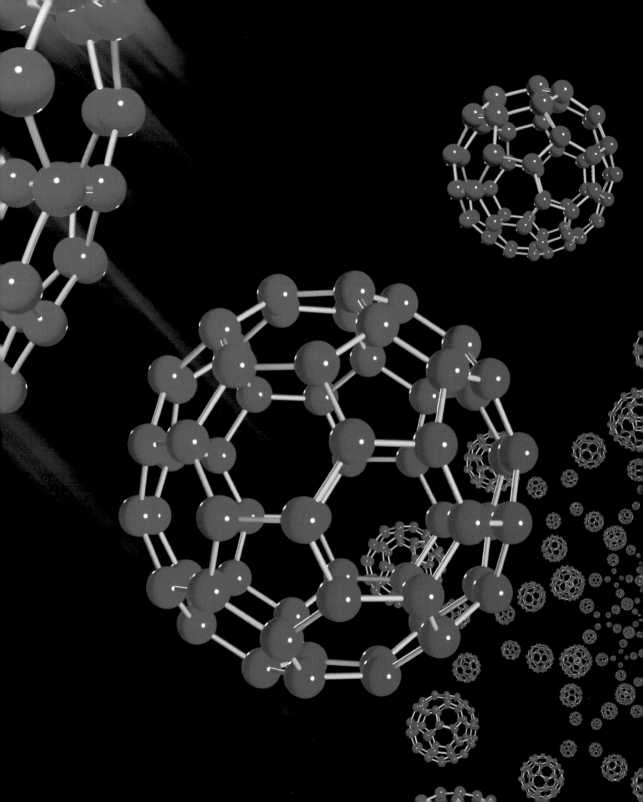

Supernova 1987A

Ian Shelton *b.*1958

A supernova in the Large Magellanic Cloud, 180,000 light years from Earth, was fortuitously discovered by Ian Shelton on 23 February 1987 using a small telescope at the Las Campanas Observatory in Chile. This exploding star caused threefold excitement. It was the first time since the development of modern astronomical instruments that a nearby supernova had been detected. The progenitor star, a 'small', 20-solar-mass, blue supergiant star known as Sanduleak −69 202, had been well studied before it blew up. And the supernova had been found before it reached maximum intensity.

Supernova 1987A is classified as a Type II supernova. The explosion was caused when the 1.5-solar-mass core of the star became unstable and collapsed, reducing in volume by a factor of a million in about one second, and became a neutron star a few tens of kilometres across. Material falling in towards the super-dense core bounced, sending a shock wave up through the silicon and carbon shells of the star and producing a frenzy of nuclear fusion. The stellar surface was heated to a temperature of nearly half a million degrees and was blasted into space at 30,000 kilometres per second (ten per cent of the speed of light). An enormous amount of light energy was produced, the supernova briefly becoming brighter than a whole galaxy. The neutrinos produced as the core shrank were volleyed through the star and into space, carrying away several hundred times the energy that the supernova radiated as visible light. The signals of just 19 of these neutrinos were picked up by large particle detectors in Japan and the USA.

See also A new star pages 48–49, Spectral lines pages 130–131, Stellar evolution pages 286–287, White dwarfs pages 310–311, A subatomic ghost pages 384–385, Pulsars pages 420–421, Gamma-ray bursts pages 434–435

Right Supernova 1987A is seen here as the bright star-like point to the lower right of the Tarantula Nebula (upper centre), an enormous area of glowing gas.

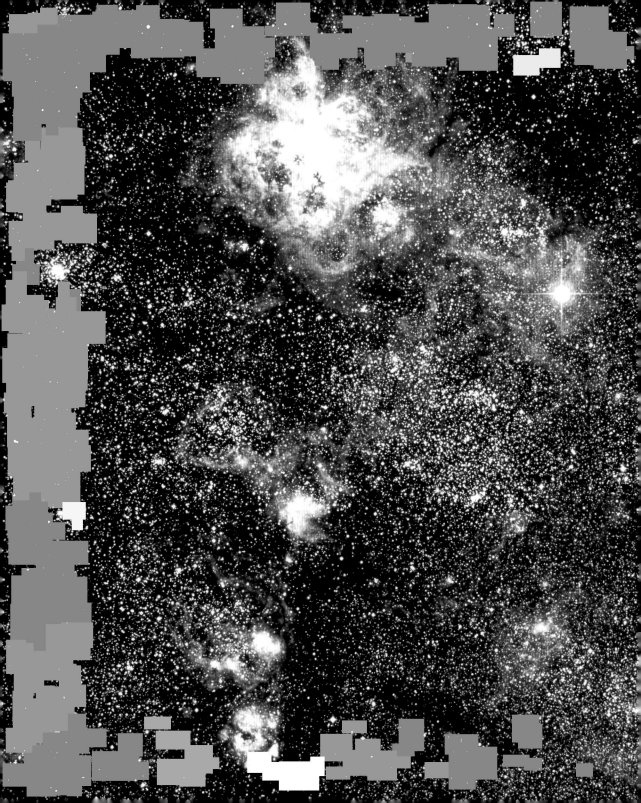

Edge of chaos

Per Bak *b*.1947, Chao Tang *b*.1958, Kurt Wiesenfeld *b*.1958

If the laws of physics are simple, why is the world complex? Here is one clue. In the 'game of life', a computer curiosity invented in 1977 by John Horton Conway, squares on a grid change colour from moment to moment following dumb, mechanical rules. Yet what emerges is an astonishing world of ever-changing novelty. Run this 'cellular automata' game on a computer and, as the screen begins to flicker, a thriving ecosystem comes to life. The colours never settle down, and 'creatures' even begin swimming about the screen, comporting themselves as if they were living, breathing beings. Conway's game is a mathematical marvel that teaches a lesson – dull and mindless rules on one level can give rise to spectacular, life-like complexity on another. It is a lesson with surprising implications.

Over the next two decades physicists discovered that systems as diverse as a pile of grains, the Earth's crust, its ecosystems and even the financial markets appear to work much like Conway's game. These systems and many others tend to 'self-organize' into what is known as a 'critical state' – a natural condition where ceaseless fluctuation and extreme instability is the norm. Anything in a critical state is always balanced on the edge of sudden, radical change, and it is next to impossible to predict what will happen next.

First proposed in 1987 by Per Bak, Chao Tang and Kurt Wiesenfeld, the notion of 'self-organized criticality' offers a one-size-fits-all explanation for the complex and tumultuous character of much of our world. In considering the outbreak of forest fires, mass extinctions that alter the course of biological evolution or even the character of human history itself, it may be something like a law of nature that the future will be punctuated necessarily by utterly unforeseeable upheavals.

See also Chaos theory pages 238–239, Fractals pages 446–447

Right Snapshot of a cellular automata universe. Starting from a randomly seeded universe, basic rules can create evolving systems with complex structures lying somewhere between order and chaos.

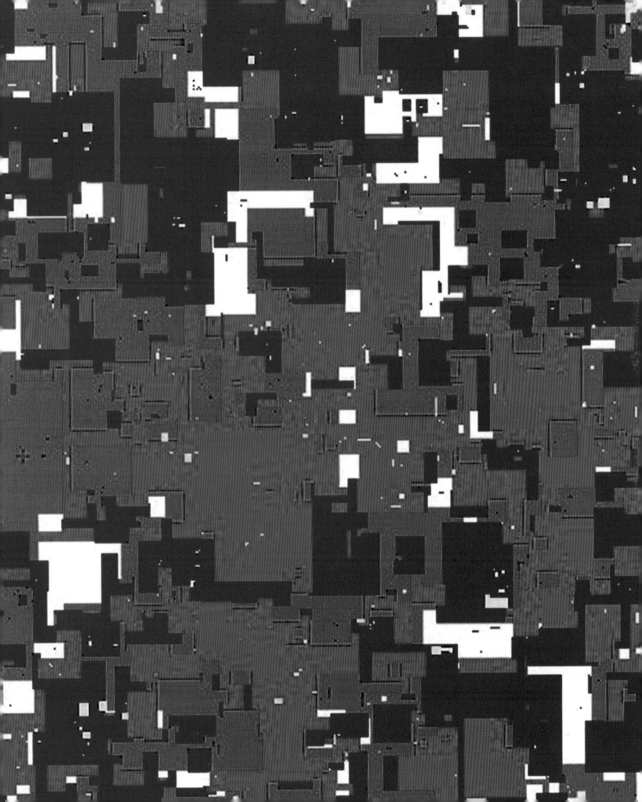

Out of Africa

Allan C. Wilson 1935–91, Rebecca Cann *b*.1951, Mark Stoneking *b*.1956

All human beings on Earth today have a distinctive, recognizable body form that sets us apart from our nearest fossil relatives. Anthropologists refer to us as 'anatomically modern humans'. Anatomically modern humans differ from the fossil humans that preceded them. For instance, anatomically modern humans first appeared in Europe about 40,000 years ago. They were preceded by Neanderthals, who had smaller chins, more prominent noses and longer, but flatter, brains than us. The origin of anatomically modern humans was the last stage in our evolutionary history, from the origin of life to the present time. When and where did anatomically modern humans evolve?

The fossil record gives ambiguous answers. Anatomically modern humans might have originated globally, a transformation of the existing human-like populations – including Neanderthals – that already occupied Europe, Africa and Asia. Alternatively, they might have originated in Africa, migrated out, and replaced the various local populations. In 1987 researchers in the laboratory of Allan Wilson at the University of California, Berkeley, used genetic techniques to answer questions about human evolution.

A group there, which included Rebecca Cann and Mark Stoneking, studied a special kind of DNA, called 'mitochondrial DNA', that is present in all our cells. They analysed mitochondrial DNA from about 150 people, selected because their ancestries spanned much of the globe. Wilson's group assumed a 'molecular clock': that mitochondrial DNA evolves at a roughly constant rate. They were able to infer that the common ancestor (sometimes called 'mitochondrial Eve') of all modern human beings lived in Africa, and as recently as 100,000 years ago. So modern Europeans are descended not from the Neanderthals, but from people who came to Europe from Africa. Since the Wilson group's work, genetic evidence has been used to trace in detail how the first anatomically modern humans migrated around, and populated, the globe.

See also Prehistoric humans pages 148–149, Neanderthal man pages 170–171, Java man pages 216–217, Blood groups pages 236–237, Taung child pages 298–299, Olduvai Gorge pages 392–393, Symbiotic cell 418–419, Ancient DNA pages 476–477, Nariokotome boy pages 480–481, Iceman pages 504–505

Right Making silent stones speak: East African stone tools found in deposits dating from 70,000 years ago at Olduvai Gorge, Tanzania.

Directed mutation

John Cairns *b.*1922

In 1988 the Harvard molecular biologist John Cairns reported on experiments which implied that bacteria can choose which mutations to produce in the face of environmental stress. Such 'directed mutation' flies in the face of the theory of evolution, which says that mutations are random events. Worse, it raises the spectre of Jean-Baptiste Lamarck's nineteenth-century argument that 'acquired' characteristics drive evolution – a theory that had been discredited.

In Cairns's experiments, bacteria were grown on a nutrient medium that lacked a vital component, such as the amino acid tryptophan. Bacteria mutate all the time, but in this case they seemed to produce advantageous mutations – those helping them to synthesize their own tryptophan – far more often than chance would allow. This suggests that the bacteria 'know' in advance what mutations are likely to benefit them. Since Cairns's controversial study, researchers have been looking for a mechanism that would explain directed mutation.

One suggestion – which Cairns himself goes along with – is that researchers are more likely to spot and count the beneficial mutations. However, this is not to imply any fraud on the part of the scientist. It may be that stress – and, to a bacterium, starvation is stress enough – causes bacteria to mutate a lot more, in an effort to get out of their state of crisis. The mutations are still random, in accordance with standard evolutionary theory. But of course the bacteria with harmful or neutral mutations do not survive to be counted – those with beneficial ones do. There is quite a lot of evidence for this 'hypermutation' – researchers have even discovered 'mutator' genes that drive the process along. But so far no one is sure whether hypermutation is a generalized phenomenon, or just confined to bacteria.

See also Acquired characteristics pages 128–129, Darwin's *Origin of Species* pages 176–177, Genes in inheritance pages 264–265, Genes in bacteria pages 332–333, Jumping genes pages 362–363, Random molecular evolution pages 422–423

Right The DNA of *Escherichia coli* is 1,000 times the length of the bacterium. Special treatment of the cell causes the DNA to be ejected, seen here as a gold-coloured fibre.

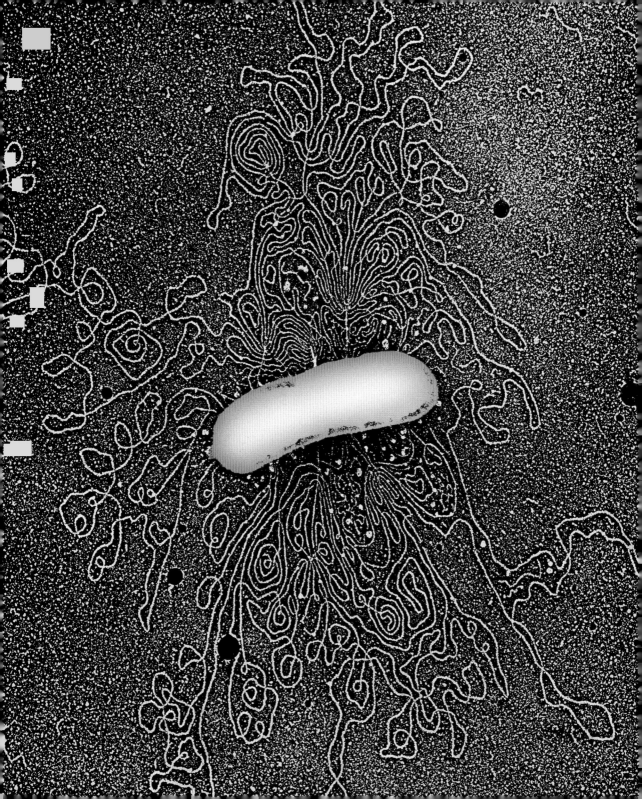

Nitric oxide

Robert **Furchgott** *b.*1916, Louis **Ignarro** *b.*1941

The explosive nitroglycerine was first used in the treatment of angina in 1870. It was even prescribed to its inventor, Alfred Nobel, who suffered from heart disease. But it was not until 1977 that the American pharmacologist Ferid Murad discovered that nitroglycerine releases the gas nitric oxide, which causes the coronary arteries to swell or 'dilate', thus improving the blood supply to the heart and relieving the pain of angina. Nitric oxide is a simple molecule, consisting of a nitrogen atom bonded to an oxygen atom. It is normally regarded as a pollutant and is found in cigarette smoke and car exhaust fumes.

In 1980 the American pharmacologist Robert Furchgott, working in New York, discovered an unknown signalling molecule in the body that, like nitroglycerine, also caused dilation of blood vessels. He called it 'endothelium-derived relaxing factor' (the endothelium is the layer of cells lining the blood vessels; a healthy endothelium is crucial for proper working of the heart and circulation). In 1986 Louis Ignarro discovered that this relaxing factor and nitric oxide were one and the same. This was the first time a gas had ever been shown to act as a signalling molecule in the body.

In heart disease the endothelium has a reduced capacity to produce nitric oxide. That is why nitroglycerine works as a treatment – it supplies the missing nitric oxide. By 1990 nitric oxide was shown to have many different roles in the body: it acts as a neurotransmitter in the brain (as a gas it can travel quickly and communicate with many brain cells at once); regulates blood pressure and blood clotting; and controls blood flow to different organs (including the penis – it is essential for normal erectile function). Researchers are now developing a range of new drugs based on nitric oxide for the treatment of heart disease and many other conditions.

See also Regulating the body pages 188–189, Dynamite pages 194–195, Aspirin pages 226–227, Neurotransmitters pages 274–275, Nerve impulses pages 366–367, Contraceptive pill pages 378–379, Chemical basis of vision pages 382–383

Right Cross-section of a small human artery containing red blood cells. When nitric oxide is released by cells in the walls of blood vessels, it relaxes nearby muscle cells and so lowers blood pressure. Nitric oxide was named 'molecule of the year' by *Science* magazine in 1992.

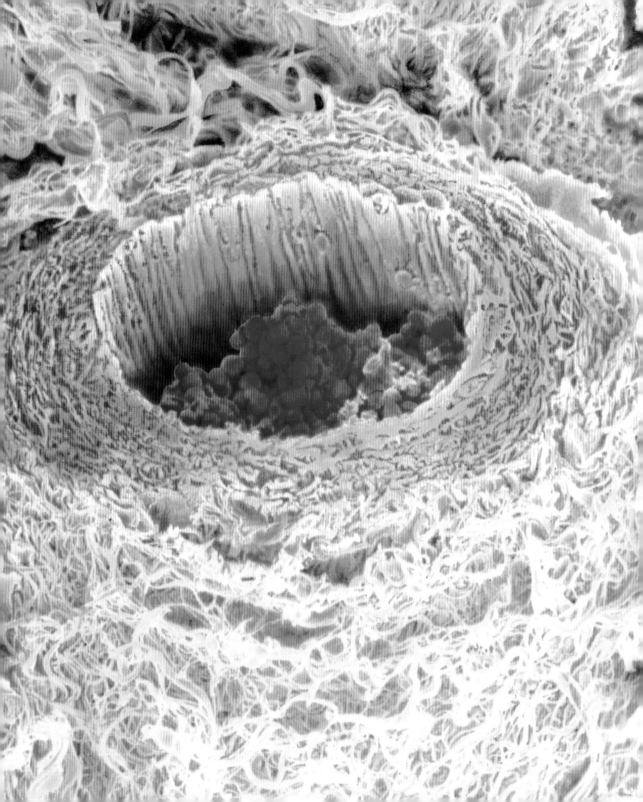

Great Attractor

Alan Dressler, *b.*1948 Sandra Faber *b.*1914

In an expanding universe, all observers see themselves as the centre of the expansion. All clusters of galaxies move farther apart from all other clusters of galaxies at a speed proportional to their distance from the observer – a large-scale pattern of motion known as the 'Hubble flow'. But it turns out that our local supercluster of galaxies – containing the Milky Way, the Virgo Cluster and many thousands of other galaxies – is deflected from the path it would follow in a pure Hubble flow. We are, in fact, moving at 600 kilometres per second in the direction of the constellation Centaurus.

In 1990 the American astronomers Alan Dressler and Sandra Faber suggested that we, and our neighbours, are flowing in that direction like a celestial river, pulled by the gravitational field of a great concentration of mass, which has been named the Great Attractor. This is estimated to be about 147 million light-years away, and unfortunately most of it is hidden behind dust in the Milky Way. It is also estimated to have a total mass of 5×10^{16} Suns.

Only about 7,500 galaxies have been found in the region occupied by the Great Attractor. So about 90 per cent of its mass must be in the form of unknown 'dark matter'. Dark matter controls certain aspects of galactic rotation as well as the gravitational interactions between galaxies. It cannot be 'seen' (as yet), but it certainly has a gravitational influence. Determining the composition of dark matter is one of the greatest challenges of modern cosmology.

Several suggestions have been made. Maybe the surrounding spherical regions or 'halos' of every galaxy contain MACHOs (massive compact halo objects). These might be ejected Jupiter-sized planets, faint brown-dwarf stars, cold white-dwarf stars or stellar-mass black holes. Maybe dark matter consists of cold WIMPS (weakly interacting massive particles) that flow through space, passing right through the Earth without interacting with normal atoms, electrons or radiation. Maybe it is hot and takes the form of massive neutrinos. Currently, there are more questions than answers.

See also Newton's *Principia* pages 78–79, General relativity pages 278–279, Expanding universe pages 306–307, White dwarfs pages 310–311, A subatomic ghost pages 384–385, Afterglow of creation pages 412–413, Black-hole evaporation pages 440–441, Fractals pages 446–447

Right Map of the universe showing galactic superclusters separated by voids. Centred on the Earth, the map covers a sphere 1.4 billion light-years in diameter, around five-thousandths of the observable universe.

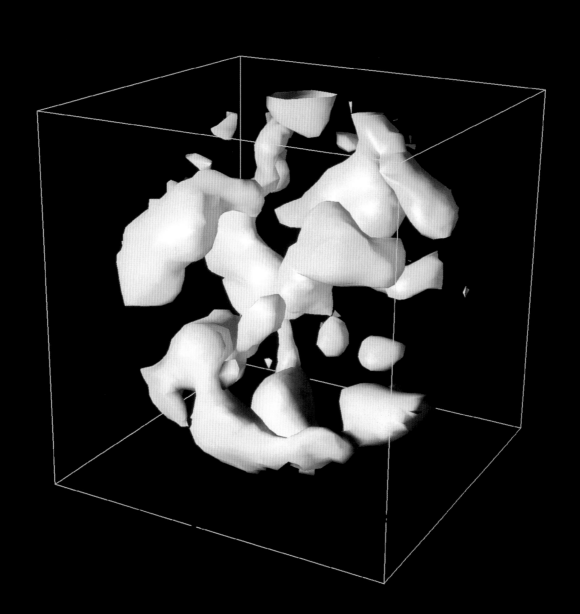

Bumblebee flight

Charles **Ellington** *b*.1952, Robert **Dudley** DATE UNKNOWN

Folk wisdom holds that bumblebees shouldn't be able to fly. With heavy bodies and short simple wings, they lack obvious aerodynamic refinement. But they do fly efficiently, thanks to complex adaptations, many of which were revealed by the work of Charles Ellington and Robert Dudley in 1990. Unlike bird wings, insect wings lack a permanent curved aerofoil section for generating lift. Instead they have a temporary one made of resilin, an elastic protein that, in combination with the stiffening of veins, enables the wing to fold partially on the upstroke and produce lift. Attachment musculature alters the wing pitch to allow hovering.

The muscle contractions that drive the wing beats need to be triggered faster than is possible through nerve impulses. Insects such as bumblebees get around this problem by using specialized muscles in their thoraxes. First, direct flight muscles attached to the wings and controlled by nervous impulses 'warm up' the flight mechanism. Then sets of indirect muscles attached to the thoracic walls come into play: the contraction of one set deforms the elastic thoracic exoskeleton, which, when it rebounds, tenses and triggers the automatic contraction of a second set. This oscillation of muscle contraction, operating beyond the limits of nervous control, creates the almost self-perpetuating system of high-frequency wing beats necessary for flight in heavy insects.

Temperature control is vital for the process. Insects shiver to get their muscles to the optimum temperature for lift-off. Once airborne, heat from energy expenditure is lost by convection from the surface of the insect, although bumblebees with full honey stomachs and pollen baskets sometimes need to land to prevent overheating. In wind-tunnel experiments, Ellington and Timothy Casey demonstrated that while a jogging human consumes the equivalent of a chocolate bar's worth of energy every hour, an equivalent-sized bumblebee would burn a similar amount of energy in one minute.

See also Animal instincts pages 318–319, Citric-acid cycle pages 320–321, Honeybee communication pages 338–339

Right The tawny bumblebee *Bombus pascuorum* in flight above bluebells.

Maleness gene

Robin **Lovell-Badge** *b*.1953, Peter **Goodfellow** *b*.1951

The chromosomal differences between men and women have been known for a long time. Usually, women have two X chromosomes (XX) and men have one X and one Y (XY). But both male and female characteristics are sometimes found in one individual, and there are also XY women and XX men. Study of such individuals has revealed some of the genes involved in sex determination. In cases of indeterminate or 'wrong' sex, one or more of these genes must be missing or defective. And the most important of these genes is the *SRY* (sex-determining region Y) gene, which controls the formation of the testes.

David Page of the Massachusetts Institute of Technology showed the way to *SRY* by developing the first gene map of the Y chromosome. Then, in the early 1990s, two British scientists – Robin Lovell-Badge and Peter Goodfellow – located *SRY* on this map in both men and mice. The protein encoded by *SRY* binds to DNA in the cell and alters its properties, with dramatic consequences for the embryo. The genital region develops into a penis and testes at around the twelfth week of gestation, while male hormones begin to act on the brain, and the body is shaped into a masculine rather than feminine form. Unlike other genes, *SRY* is remarkably similar in different men – and dramatically different in males of different species. It also appears to have changed little in 200,000 years of human evolution.

In other words we are all created as females at the moment of fertilization. It is whether or not we possess an *SRY* gene that sets us on the path to being male or female at birth. Lovell-Badge and his co-workers proved this in 1991 when they inserted an *SRY* gene into female mouse embryos. The animals changed sex, developing testes and other male characteristics.

See also Eggs and embryos pages 140–141, Genes in inheritance pages 264–265, The double helix pages 374–375, Genetic cousins pages 442–443, Human cancer genes pages 456–457, Genetics of animal design pages 460–461, Human genome sequence pages 524–525

Right We are all created as females at the moment of fertilization – it is whether or not we possess the maleness gene that sets us on the path to being male or female at birth.

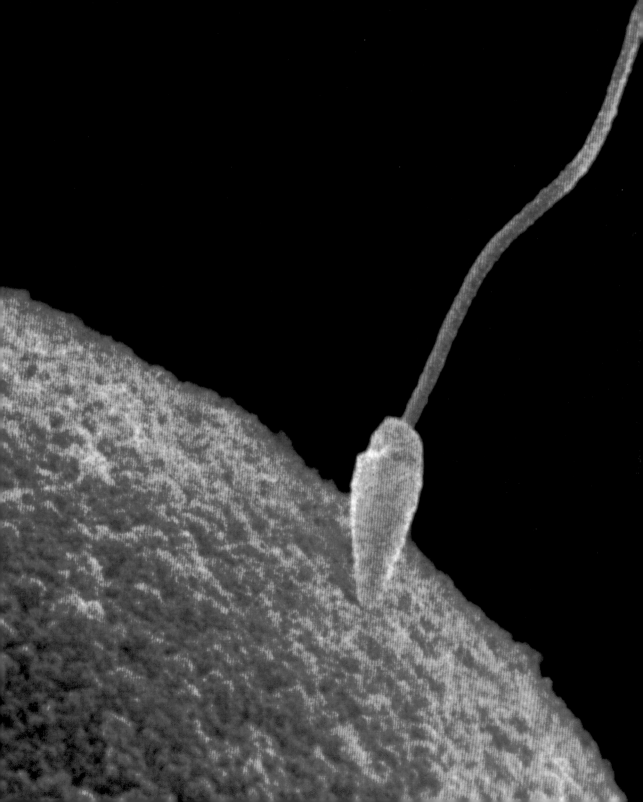

Iceman

Konrad Spindler *b*.1939

The 'Iceman', discovered by two German hikers, Erika and Helmut Simon, in September 1991 in the Ötztaler Alps of South Tyrol, Italy, is a unique and precious 'time capsule' of ordinary prehistoric life. He became the subject of intensive research by teams around the world, with much of the initial work headed by Konrad Spindler at the University of Innsbrück.

Dated to around 3300 BC, the Iceman is the oldest intact human corpse yet discovered. He was found with his clothes and equipment, including organic materials that normally disintegrate but survive here because of the frozen conditions. For the first time we can see the variety of materials used by late Stone Age people: clothes of leather, goat hide, bear- and deerskin, and woven grass; a fur backpack on a hazel and larchwood frame; a yew bow and arrows of viburnum and dogwood; sewn birchbark containers; and a copper axe with a yew haft.

The Iceman's body, intact albeit desiccated, tells much about his life and death. He was dark-skinned, in his mid- to late-40s and about 1.6 metres tall. His DNA confirms links with northern Europe, though plant material indicates he came from the Italian valleys to the south. The teeth are very worn, especially the front incisors, suggesting that he ate coarse ground grain or that he regularly used them as a tool. His internal organs are in excellent condition, though the lungs are blackened by smoke, presumably from open fires, and he has hardening of the arteries and blood vessels. His last meal seems to have consisted of meat (probably ibex), wheat, plants and plums.

One little toe has traces of chronic frostbite, and eight ribs had been fractured but were healed or healing when he died. Short parallel vertical blue tattoos on his spine and legs may have been therapeutic, aimed at relieving his arthritis – an early form of acupuncture perhaps? – and a fingernail hinted that he suffered periods of crippling disease, which may explain how he fell prey to adverse weather in the mountains and froze to death.

See also Prehistoric humans pages 148–149, Neanderthal man pages 170–171, Java man pages 216–217, Olduvai Gorge pages 392–393, Ancient DNA pages 476–477, Nariokotome boy pages 480–481, Out of Africa pages 492–493

Right The Iceman's body emerging from the ice, as he was first spotted on 19 September 1991 by a couple of hikers from Heidelberg at an altitude of 3,200 metres.

Fermat's last theorem

Pierre de Fermat 1601–65, Andrew Wiles *b*.1953

The French mathematician Pierre de Fermat gave his name to one of the most enduring mathematical riddles of the past 400 years. Yet in his lifetime he published virtually nothing, preferring to correspond in letters with a circle of mathematicians based in Paris. In fact the riddle might never have been discovered save for the efforts of his son, Samuel, who in 1670, five years after Fermat's death, took on the task of collecting together his father's scattered mathematical ideas. It was in a copy of Diophantus's *Arithmetica* that Fermat wrote the seemingly casual comment, 'I have discovered a truly remarkable proof which this margin is too small to contain'.

The theorem Fermat was alluding to is an extension of the Pythagorean theorem. For whole numbers, there is an infinity of so-called 'pythagorean triples' where the sum of two square numbers equals a third square number (such as $3^2 + 4^2 = 5^2$). Fermat said that this relationship did not hold true for numbers cubed, or to any higher power. From mere trial-and-error it seemed he was right, but proving this became a Herculean task. The mathematicians who tried their hand at solving 'Fermat's last theorem' reads like a roll of honour, but all failed in their attempts.

By 1993 computers had shown that it was true up to the four-millionth power. But a water-tight proof that it was always true was still elusive. Meanwhile mathematicians found that, far from being a mathematical curiosity, the truth or falsehood of the theorem was intimately linked to the nature of space. In 1993 Andrew Wiles, a British mathematician, gave a series of lectures at the Isaac Newton Institute in Cambridge, ending with his proof of the theorem. Unfortunately, after years of reclusive study, his adamantine proof concealed a slight but devastating flaw. Moving to Princeton, he renewed his efforts, and in 1995 published in the *Annals of Mathematics* his paper on 'Modular Elliptic Curves and Fermat's Last Theorem'. The riddle had been solved.

See also Four-colour map theorem pages 450–451

Right In addition to number theory, Fermat shared honours for probability theory and gave differential calculus its impetus. A magistrate, he was only a part-time mathematician.

Comet Shoemaker-Levy 9

Carolyn Shoemaker *b.*1929, Eugene Shoemaker 1928–97

The American astronomers Carolyn and Eugene Shoemaker started their photographic search of asteroids and comets in 1983 using the 46-centimetre Schmidt telescope at the Palomar Observatory in California. They were joined by David Levy in 1989. Among the 32 comets and 1,125 asteroids that they found, the most famous was Shoemaker–Levy 9. This was discovered close to Jupiter on 25 March 1993. Amazingly, it was not a single comet but a string of mini comets. By early April, Brian Marsden of the Smithsonian Astrophysical Observatory, Harvard, had collected enough observations to show that the comet was most unusual in that it was orbiting Jupiter and not the Sun. The astronomy community was greatly excited when, on 22 May, Marsden made the stunning prediction that the comet would actually hit Jupiter 14 months later.

Apparently, the one-kilometre-wide dirty snowball nucleus of the comet had been orbiting Jupiter since at least 1914. Unfortunately, it passed within 90,000 kilometres of Jupiter on 7 July 1992. The difference in Jupiter's gravitational force exerted on the near and far edges of the comet had been sufficient to break the fragile nucleus into pieces. The 22 fragments of the comet plunged into the tenuous upper atmosphere of Jupiter between 16 and 22 July 1994. Nearly every Earth telescope was trained on the planet at the time, even though the impacts took place on the face pointing away from us. In the visible part of the spectrum a black dusty region could be seen expanding away from each impact site. These dirty 'splodges' lasted for weeks.

Three flashes were seen in the infrared part of the spectrum: the first as each cometary fragment hit the planet, the energy being scattered off high dust clouds; the second as the impact fireball rose above the planetary limb; and the third, and by far the biggest, as the ejecta plume fell back and hit the cloudy atmosphere.

See also Halley's comet pages 84–85, A cometary reservoir pages 360–361, Extinction of the dinosaurs pages 458–459, Galileo mission pages 514–515, Water on the Moon pages 522–523

Above The comet was broken up following its close approach to Jupiter.

Right Impact of comet fragments left Jupiter with a temporarily 'bruised' appearance, caused by black debris tossed high above the giant planet's cloud-tops.

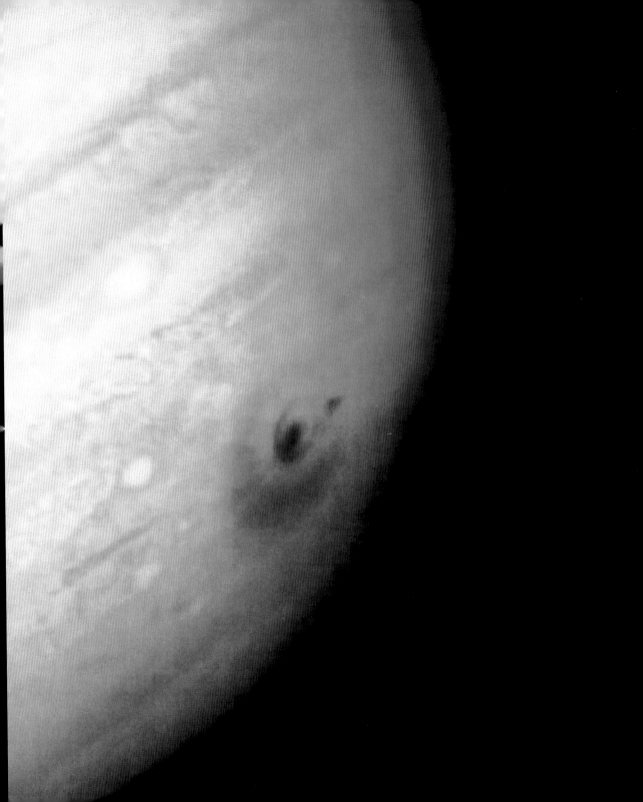

A new state of matter

Eric Cornell *b.*1961, Carl Wieman *b.*1951

Cold matter can go through a curious change in which its atoms lose their identity and merge into a sinister collective. This state can only be reached by about half of all materials. Particles of matter that have an integer number of quantum units of spin are called bosons; particles with $\frac{1}{2}$, $1\frac{1}{2}$ and so on are fermions. Two identical fermions cannot be made to overlap, and it is this antisocial tendency among electrons that stops ordinary matter collapsing. But bosons are less stand-offish – and they can be positively affectionate. Albert Einstein worked out in 1925 that a cold enough collection of bosons will all collapse into the same state. This Bose–Einstein condensate would be like a single superparticle.

This phenomenon is thought to cause the resistance-free transport of electricity in superconductors and the frictionless flow of superfluids, which can seep through the tiniest crack and will even 'magically' crawl over a barrier to find a lower level. But it was only in 1995 that a Bose–Einstein condensate was revealed in the laboratory.

The American physicists Eric Cornell and Carl Wieman first put a tenuous gas of rubidium-87 into a bottle made of magnetic fields. Rubidium-87 atoms are bosons, as are about half of all atoms. The physicists cooled their gas to less than a millionth of a degree above absolute zero, using evaporation and a battery of carefully tuned laser beams. The result was a condensate: a bizarre matter wave that subsumed the original rubidium atoms.

Now physicists are learning to use matter like light. Atom lasers, for example, turn a condensate into a beam of coherent matter waves. These waves can then be focused and scattered using lenses and gratings made of light – a reversal of ordinary optics. Condensates and atom lasers might be used to make ultrasensitive atomic clocks, or to beam tiny electronic components onto silicon wafers.

See also Changes of state pages 198–199, Superconductivity pages 266–267, Wave–particle duality pages 300–301

Right Computer-generated image of low-velocity rubidium atoms forming a Bose–Einstein condensate (blue and white peak). As the temperature drops, more and more of the trapped atoms occupy the same quantum state.

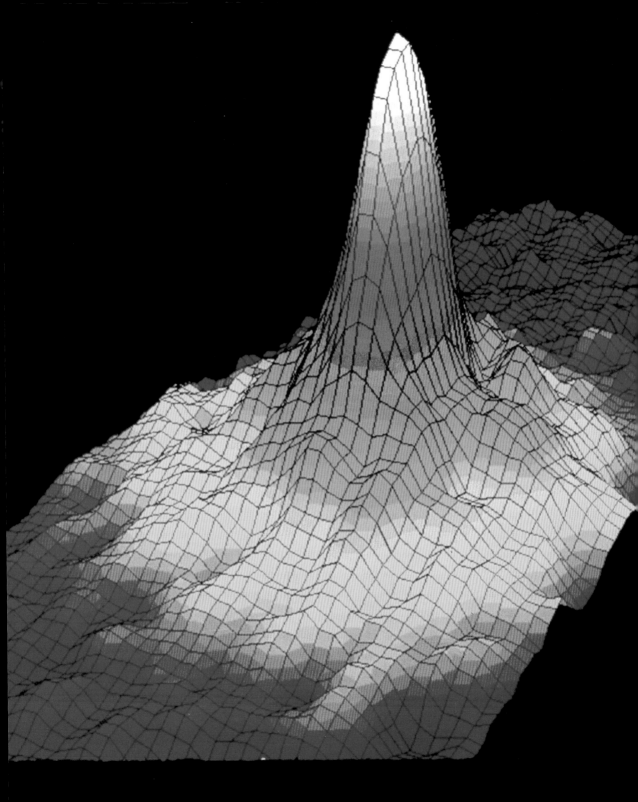

Planetary worlds

Michel **Mayor** *b.*1942, Didier **Queloz** *b.*1966

Imagine yourself on a planet orbiting the nearby star Proxima Centauri. Looking at the distant Sun in the night sky, how could you tell it had a planetary family? First, the Sun would wobble from side to side, as the centre of mass of the Solar System, and not the Sun's centre, moved against the background sky. This 1.5-million-kilometre wobble would have a period of about 12 years – the time it takes our most massive planet, Jupiter (0.001 solar masses), to orbit the Sun. Second, a 'Doppler shift' of the solar spectra would be produced as the Sun orbits the centre of mass at 12.5 metres per second. If you were lucky enough to be close to the orbital plane of the Solar System, this could be measured by spectroscopy. Third, Jupiter might cross the Sun's surface, and the partial eclipse could be detected as a brief change in brightness.

With today's astronomical instruments, method two is the most sensitive. Unlike method one, it is independent of stellar distance. Since the late 1980s, two groups of astronomers, Michel Mayor and Didier Queloz from the University of Geneva, Switzerland, and Paul Butler and Geoff Marcy from San Francisco State University, California, have been in friendly competition investigating spectra of nearby Sun-like stars for hints of new planets. On 6 October 1995 Mayor and Queloz announced that they had found a planet around 51 Pegasi. The planet was at least 0.47 times the mass of Jupiter and was (remarkably) orbiting every 4.229 days (the star–planet distance being 5 per cent of the Earth–Sun distance). And then, on 16 January 1996, Butler and Marcy announced that they had found planets around 47 Ursae Majoris and 70 Virginis.

Since 1996, planetary discovery has become a regular phenomenon. The selective way in which they are found dictates that the new planetary systems differ considerably from the system we live in.

See also Planetary distances pages 74–75, Discovery of Uranus pages 96–97, Spectral lines pages 130–131, Doppler effect pages 156–157, Discovery of Neptune pages 162–163

Right A suspected extrasolar planet thrown out of a double star system. The planet has a mass about three times that of Jupiter, and lies 450 million light-years away.

Galileo mission

National Aeronautics and Space Administration (Team Leader: Michel Belton)

The first stage in the space exploration of a planet is the flyby mission. Pioneer 10 passed Jupiter in December 1973, and Voyager 1 and Voyager 2 followed in 1979. The next step is the orbiter mission. Study time then increases from days to years.

The Galileo spacecraft started orbiting Jupiter on 7 December 1995 and was still returning data seven years later. The planet was observed in detail. And hundreds of passes over the four large inner moons of Jupiter – Io, Europa, Ganymede and Callisto – enabled these satellites to be accurately mapped without having been orbited individually. Investigation of the first two has been particularly fruitful.

Volcanic eruptions on Io feed gas into Jupiter's magnetosphere. Flying within 900 kilometres of Io's surface, Galileo measured complicated magnetic fields and plasma flows. The way in which the spacecraft's orbit was changed by Io's gravitational field indicated that the moon has a large metallic core, overlaid by a mantle of partially molten rock, topped with a thin volcanically active crust. Tidal heating has maintained much of Io in the molten state. Galileo took superb images of the bright, yellow, pockmarked surface. Hot spots were clearly visible, as was a volcanic plume 400 kilometres high.

Gravitational data from Europa indicate that it has an icy 'crust' 150 kilometres thick. The underlying rocky mantle is heated by both radioactive decay and tidal forces. Not only does Europa's surface show a variety of detailed structures suggesting ice-related tectonic activity, but the heating also ensures that the ice crust remains geologically active. The main fractures seen in the surface ice were probably produced at a time when Europa's rotation and revolution periods did not match. Europa may well have a deep extensive subsurface ocean. Unfortunately, the only way to prove this is to take radar measurements from the surface. What may ultimately demonstrate the real value of the Galileo mission is that it has led some scientists to propose that primitive life forms may exist in the Europan ocean.

Above Jupiter with the satellite Io in transit across the planet's disc.

See also Heavens through a telescope pages 54–55, Inside the Earth pages 250–251, Life at the extreme pages 452–453, Eruption of Mount St Helens pages 462–463, Lake Vostok pages 520–521

514

Right Europa, from Galileo. Linear and mottled terrains appear in brown and reddish hues, indicating the presence of contaminants in the ice. Icy plains are shown in bluish hues.

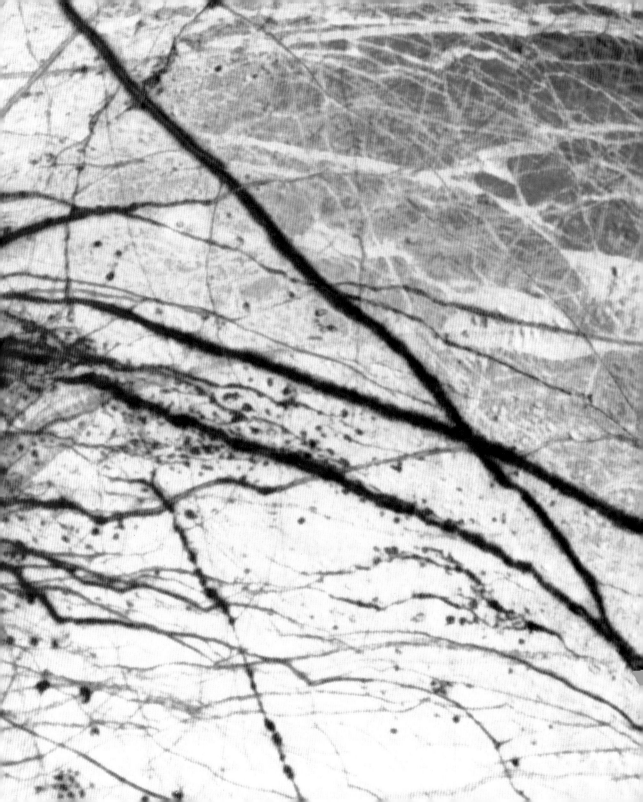

Dolly the cloned sheep

Ian **Wilmut** *b.*1944

On 5 July 1996 a very special lamb was born in the small village of Roslin, near Edinburgh. 'Dolly' had been cloned from a single cell from the udder of a six-year-old black-faced ewe. The year before, two other clones – Megan and Morag – had been created from sheep embryo cells. Dolly was unique because she was created from an adult cell. Instead of having a mother and a father and inheriting half her genes from each parent, she was the genetic double of the unknown ewe that donated the udder cell.

Cloning means making an exact copy, and it's really nothing new. Ian Wilmut and his team of researchers have been cloning DNA molecules, bacteria, plants and even frogs for many years. What's astonishing about the Dolly experiment is that it involved the reprogramming of the genes in the donor nucleus. Although we all have an exact copy of our genetic information in all our cells, different patterns of genes are active in different types of cell. Creating Dolly involved persuading the gene pattern in an udder cell to revert to that of the embryonic state – something that was previously believed to be impossible. Since Dolly, monkeys, sheep, cattle, goats, mice and pigs have all been cloned.

Reproductive cloning involves making a whole animal from a single cell. Understandably, there are ethical concerns about what could be done frivolously with humans. But cloning could help a man with no sperm to father a child by using one of his other body cells instead. It could also help to preserve endangered species and create flocks of genetically engineered animals for the production of new medicines.

Meanwhile, therapeutic cloning involves making just cells and tissues – not whole animals – from donor cells. It could allow the creation of bone marrow for people with leukaemia or neurons for brain repair in victims of strokes, Parkinson's disease and other neurological disorders. As yet scientists have only scratched the surface.

See also Eggs and embryos pages 140–141, Mendel's laws of inheritance pages 192–193, Hayflick limit pages 394–395, Genetic engineering pages 436–437, Monoclonal antibodies pages 448–449, Human genome sequence pages 524–525

Right The cloning of Dolly was published in *Nature* some eight months after her birth. The paper concluded with historic understatement: 'The implications of this work are far-reaching'.

27 February 1997

International weekly journal of science

£4.50 FFr44 DM175 Lire 13000 A$15

nature

A flock of clones

Extrasolar planets Fading from view

Climate cycles Eccentricity finds a role

Archaeology Hunting 400,000 years ago

New on the market
Genetics

Martian microfossils

David McKay *b.*1936

The Estonian astronomer Ernst Julius Öpik calculated in the 1930s that any large collection of meteorites would contain some blasted off the Moon and a handful from Mars. SNC meteorites, named after Shergotty (in India), Nakhla (in Egypt) and Chassigny (in France), are Martian. They crystallized as volcanic rocks between 1,300 million and 200 million years ago, and Mars was at the time the only extraterrestrial body nearby that was volcanically active. They also contain a variant form of nitrogen similar to that in the Martian atmosphere but different from that on Earth.

Another Martian meteorite was found in Antarctica on 27 December 1984. It had been resting on the blue ice in the Allan Hills region for the past 13,000 years. ALH84001 was special. It contained orange-brown carbonate globules, magnetite and polycyclic aromatic hydrocarbons. Even though these can be produced in many ways, they can be formed by the decay of life forms. The breakthrough came when the team, led by David McKay of NASA's Johnson Space Center, examined the meteorite at a magnification of 200,000 times under a scanning electron microscope and discovered an object that looked like a very short microfossil of an Earthly nanobacteria. This worm-like fossil had a segmented structure and its width was about a hundredth of that of a human hair.

The team wrote a paper on their evidence for relic biological activity in ALH84001. Details were discussed at a news conference on 7 August 1996. The media went wild, but scientists were more reserved. There was much talk of terrestrial contamination and water alteration, the likelihood of carbonate formation when the meteorite passed through Earth's atmosphere, and a sulphur isotopic composition that was more Earth-like than Mars-like.

More importantly, interest in the possibility of Martian life increased hugely. And Martian life-seeking space missions dedicated to digging and examining under the surface of the planet have been allowed to jump the queue.

See also 'Canals' on Mars pages 200–201, Origin of life pages 370–371, Alien intelligence pages 398–399, Oldest fossils pages 410–411, Five kingdoms of life pages 426–427, Life at the extreme pages 452–453, Extinction of the dinosaurs pages 458–459, Lake Vostok pages 520–521, Water on the Moon pages 522–523

Right Microscopic tubes within a meteorite from Mars were claimed as fossil evidence of past life on the 'red planet'. Researchers now admit that 'more data are needed'.

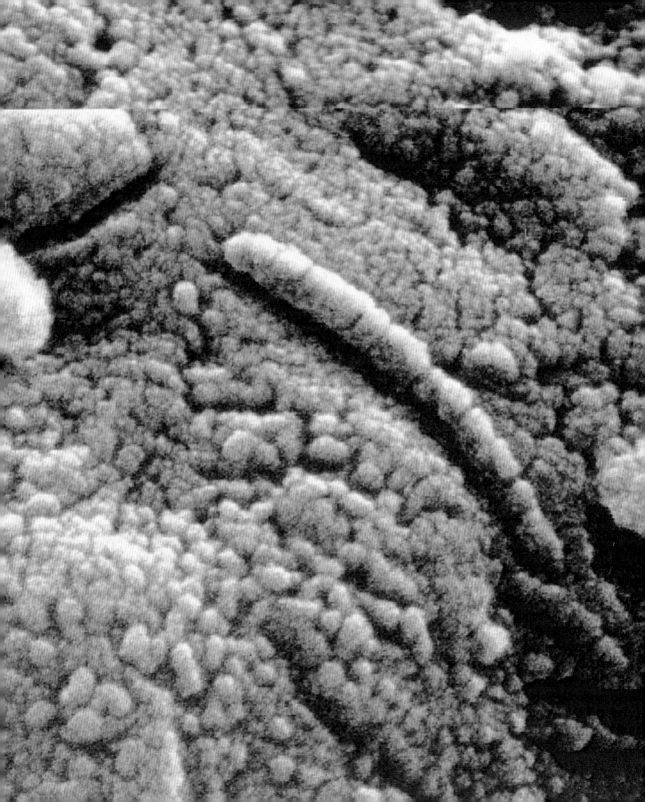

Lake Vostok

International research team

Lake Vostok, set in what must surely be the most inhospitable region of the world, is a daunting but irresistible challenge for scientific explorers. Yet nobody has ever set foot on its shores, and nobody ever will. The allure of Lake Vostok arises from the fact that, like the oceans of Jupiter's moon Europa, it is sealed from the atmosphere by a thick crust of ice. It lies 1,500 kilometres from the Antarctic coastline and 3,500 metres above sea level, near Russia's Vostok Station, a scientific outpost that has endured the lowest temperature ever recorded on Earth: −89.2 degrees Celsius.

The first hint of the lake's existence came in 1960. Flying over the Vostok area, the Russian geographer Andrei Kapitsa noticed an unusually large flat area on the ice sheet. But his suggestion that there was a subglacial lake underneath was not taken seriously. A British-led radar survey later showed that water was sandwiched between the ice and the bedrock. Vostok's ice is 4 kilometres thick, its water up to 500 metres deep. At 10,000 square kilometres (about the size of Lake Ontario in Canada), Vostok is the largest of about 70 similar Antarctic lakes and was sealed off at least a million years ago along with its plant and animal life. Today it is a dark, nutrient-starved place warmed by volcanic activity. Speculation about how this isolated biosphere might have evolved – if indeed it managed to survive – tests the imagination.

In 1998 drillers penetrated 3,623 metres into the ice, stopping 120 metres short of the lake to avoid contaminating it. They recovered the deepest ice core ever, along with a range of bacteria, fungi, algae and even pollen grains that had remained in place for as long as 200,000 years – the bottom of the core is in fact likely to be frozen lake water that formed beneath the glacier. Scientists now want to send a robotic explorer into the mysterious lake to search for signs of life. If this is successful, hopes will soar that life also exists millions of miles away, where ice-sealed oceans sheathe the moons of Jupiter.

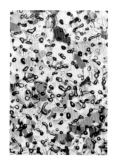

Above Bubbles of ancient air within Antarctic ice provide samples of the atmosphere from past times.

See also Oldest fossils pages 410–411, Life at the extreme pages 452–453, Galileo mission pages 514–515, Water on the Moon pages 522–523

Right A glaciologist prepares to use a hot-water drill to create holes for seismic tests. With satellite surveying and radar, these probes are delineating the extent of Antarctic subglacial lakes.

Water on the Moon

National Aeronautics and Space Administration (Principal Investigator: Alan Binder)

Water is essential for sustaining life. And the presence of warm liquid water is thought to be essential for generating life. Earth, fortunately, has plenty of it: if plate tectonics stopped forming mountains, and erosion then flattened our planet, the whole surface would be under 2.8 kilometres of water.

There are two obvious sources of terrestrial water. The rocks that formed Mercury, Venus, Earth, the Moon, Mars and the asteroid belt initially contained considerable amounts of water retained during their formation. If the temperature rose above 800 degrees Celsius, through, for example, radioactive heating, the rocks would have 'cracked' and water would have been released. More than half of the mass of comets is in the form of ice. So cometary impacts bring water to planetary surfaces too.

Venus and Mars were therefore wet on the surface in the past. In the case of Venus, the high temperatures and ultraviolet radiation flux have split the water into hydroxyl radicals (OH) and hydrogen (H), and these have slowly escaped from the clutches of the planetary gravitational field. Mars has also lost a lot of water but considerable amounts are probably buried just beneath the surface in the form of a permafrost.

In early 1998 the US spacecraft Lunar Prospector was placed in a 100-kilometre-high orbit around the Moon. A neutron spectrometer on board detected slow neutrons every time the spacecraft passed over the lunar north and south poles. These were produced by cosmic rays bouncing off hydrogen atoms. And the most likely 'home' of the hydrogen was thought to be water molecules. Some of the deep craters near the lunar poles are in perpetual darkness. Their temperatures are so low that any water released nearby is immediately frozen solid. Lunar Prospector found somewhere between 11 million and 330 million tonnes of ice at the poles. This might be extremely useful if ever the Moon is colonized.

See also Origin of the Solar System pages 104–105, A cometary reservoir pages 360–361, Apollo mission pages 424–425, Extinction of the dinosaurs pages 458–459, Galileo mission pages 514–515, Lake Vostok pages 520–521

Right The most conservative estimates of the amount of ice on the Moon would provide enough water for a colony of 2,000 people for at least 100 years, without recycling. This 1840 daguerreotype was one of the first astronomical photographs ever taken.

Human genome sequence

Human Genome Sequencing Consortium/Celera Genomics

In 1985 Robert Sinsheimer of the University of California dreamed of sequencing the human genome as biology's equivalent of putting a man on the Moon. On 26 June 2000 the completion of the 'working draft' of the human genome was announced – several years ahead of target. It was the biggest project ever undertaken in life science, involving thousands of scientists from both the private and the public sector in the US, UK, France, Germany, Japan and China.

Every cell of every living thing contains a copy of its instruction manual – the genome – written in the chemical code of DNA. There are four letters in this code: A, C, T and G. The draft now gives us the sequence of around 90 per cent of the three billion letters in the human genome.

The genome is divided between 23 pairs of chromosomes – structures in the cell nucleus that can be seen under the microscope. DNA from each chromosome was chopped up into more manageable fragments, which were then sequenced using chemical analysis. Powerful computers picked out overlapping fragments and pieced them together to generate the whole genome sequence. The technology advanced so fast that it took four years to sequence the first billion letters, but only four months to do the second billion.

The mapping of the genome makes it easier to find genes, the stretches of DNA that contain instructions for making proteins, which are the master molecules that control everything that happens in a cell. It looks as if there are 30,000 or so genes, accounting for just around two per cent of the entire genome. Mutations that lead to disease have already been discovered in around 1,100 genes, such as those responsible for Huntington's disease, cystic fibrosis and inherited breast cancer. Many more will now be discovered, accelerating the pace of research into common diseases with a genetic basis, such as cancer, heart disease, diabetes and asthma. The sequence itself contains vital clues to human evolutionary history.

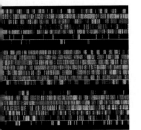

Above Human DNA sequence. Each colour represents a specific base – one of the four letters A, G, C, T – making up the genetic code.

See also Genes in inheritance pages 264–265, Sickle-cell anaemia pages 358–359, The double helix pages 374–375, Genetic engineering pages 436–437, Human cancer genes pages 456–457, Genetic fingerprinting pages 484–485, Maleness gene pages 502–503

Right Genetic cartography: the 46 human chromosomes – 23 of maternal origin, 23 of paternal origin. In 1986 the eminent molecular biologist Sydney Brenner said, 'The idea of trudging through the genome sequence by sequence does not command wide and enthusiastic support in the UK.'

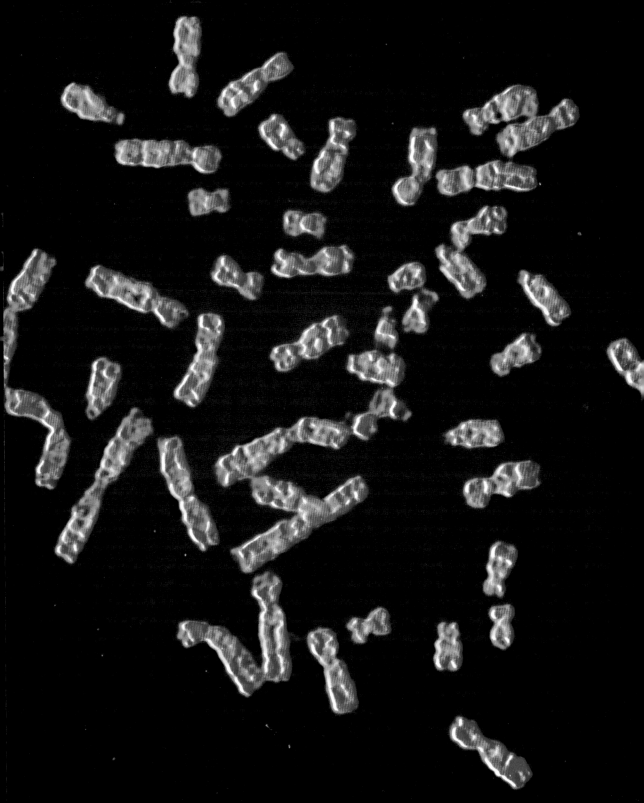

Index

Acknowledgements

Peter Tallack thanks the following people for their help and support: Peter Adams, Anthony Cheetham, Graham Farmelo, Tim Whiting and Eva Yemenakis. He is also grateful to the many contributors who advised him on the contents.

Picture credits Every effort has been made to trace the copyright holders. Cassell & Co apologizes for any unintentional omissions and, if informed of any such cases, would be pleased to update future editions. AKG London/ Erich Lessing p 11, 67, 95 Alfred Eisenstaedt/TimePix/Rex Features p308 © Anglo-Australian Observatory 1994 image by C Heisler/T Hill p508 Ann Ronan Picture Library p21, 25, 36, 39, 51, 74, 77, 105, 111, 117, 130, 131, 149, 160, 167, 169, 183, 195, 196, 197, 199, 202, 203, 204, 216, 217, 225, 229, 242, 249, 251, 263, 305, 355, 367, 387, 463, 507 Ann Ronan Picture Library/Photograph courtesy of The Nobel Foundation p262 Archives Photographiques Charmet p243 Associated Press p331, 337, 363 Berenice Abbott/Commerce Graphics Ltd, Inc p119 © Bettmann/Corbis p 2, 53, 61, 185, 201, 215, 221, 245, 271, 321, 327, 357, 381, 393, 397, 399, 429, 465 Bibliotheque Sainte-Genevieve, France/ Bridgeman Art Library p20 © Bob Goodale/Oxford Scientific Films p147 British Museum, London/ Bridgeman Art Library p137 British Society for the Turin Shroud p347 Cajal Institute - CSIC - Madrid, Spain p211 Carnival Collection, Special Collections, Tulane University Library p177 Chartres Cathedral/Giraudon/Bridgeman Art Library p15 © Corbis p43, 71, 83, 109, 120, 237, 401, 441. 455/Burstein Collection p373/Christel Gerstenberg p23/Farrell Grehan p294/Gary Braasch p469/Gianni Dagli Orti p85/Henry Diltz p379/Historical Picture Archive p87, 153/Hulton-Deutsch Collection p97, 295, 523/James A Sugar p79/Jerry Cooke p341/Lloyd Cluff p415/Michael Maslan Historic Photographs p451/Reuters NewMedia Inc p439/Roger Ressmeyer p371, 479/Rykoff Collection p293/Underwood & Underwood p213 D'Arcy Thompson Collection, courtesy of St Andrews University Library p423 Daily Herald Archive/NMPFT/Science & Society Picture Library p317 Daniel Heuclin/NHPA p291 Dave Watts/NHPA p411 David Tipling/Oxford Scientific Films p13 Department of Geology, National Museum of Wales p73 Edimedia p41 Equinox Archive p270 Frank Drake (UCSC) et al, Arecibo Observatory (Cornell, NAIC) p398 G I Bernard/NHPA p129, 443, 477 Hanny/ Frank Spooner Pictures p505 Harvest/Truth & Soul (Courtesey Kobal) p93 Henry E Huntington Library and Art Gallery p37, 307, 361 Institute for Cosmic Ray Research, The University of Tokyo p385 John Shaw/NHPA p304 L'Illustration/Sygma p163 Louvre, Paris/Giraudon/ Bridgeman Art Library p31 Mary Evans Picture Library p 55, 75, 171, 222, 241 Mitchell Library, State Library of New South Wales/ Bridgeman Art Library p63 Moebius Strip II by M C Escher c 2001 Cordon Art - Baarn - Holland. All rights reserved p 145 Moravian Museum, Brno p193 NASA p69, 434 NASA/Ann Ronan Picture Library p433 NASA/Galaxy Contact p405, 425 NASA/Oxford Scientific Films p207 Nationalgalerie, Berlin/Bildarchiv Steffens/ Bridgeman Art Library p115 Nature Magazine p517 Neil Bromhall/Oxford Scientific Films p259 Science, Industry & Business Library, The New York Public Library, Astor, Lenox and Tilden Foundations p158 Nick Birch/Ann Ronan Picture Library p303 Novosti (London) p292 Oxford Scientific Films p280, 281 Palazzo Farnese, Italy/Bridgeman Art Library p113 Peter Parks/NHPA p427 Photos by Mansell/Timepix/Rex Features p17 Physical Review, 1949 p352 Richard Mankiewicz p491 Roy Waller/NHPA p471 Royal Geographical Society, London/Bridgeman Art Library p49 Science Museum/Science & Society Picture Library p29, 57, 65, 91, 123, 135, 139, 164, 165, 172, 173, 219, 227, 234, 350, 351 Science Photo Library p125, 154, 198, 256, 289/A Barrington Brown p375/Alfred Pasieka p191, 374, 447/Anthony Howarth p483/B Murton/Southampton Oceanography Centre p452/Bernhard Edmaier p277/Biophoto Associates p334/Carlos Frenk, University of Durham p499/Celestial Image Co p287, 489/CERN p268, 353/Chris Madeley p389/CNRI p525/CSIRO p520/D Phillips p503/David A Hardy p453/David Parker/IMI/Univ. of Birmingham High TC Consortium p267/David Parker p150, 417, 458/David Scharf p349/David Vaughan p521/Dr Arnold Brody p205/Dr Gopal Murti p395, 495/Dr Jeremy Burgess p192, 209/Dr Kenneth R Miller p345/Dr L Caro p333/Dr Tony Brain p230/EM Unit, VLA p467/European Space Agency p84/Eye of Science p359/Geological Survey of Canada p459/J C Revy p431/James Holmes/Celltech Ltd p449, 525/James King-Holmes p464/Jean-Charles Cuillandre/Canada-France-Hawaii Telescope p161/Jean-Loup Charmet p19, 99, 103, 189, 309/John Reader p392, 481, 493/Juergen Berger, Max-Planck Institute p265/Ken Eward p391/Laguna Design p487/Lawrence Berkeley Laboratory p315/Manfred Kage p231, 366, 369/Mark Garlick p121/Martin Dohrn p47/Mehau Kulyk p239, 475/Michael Gilbert p273/Michael W Davidson p248/Nancy Kedersha/Immunogen p394/NASA p285, 286, 513, 514, 515, 519/National Cancer Institute p456/National Institute of Standards and Technology (NIST) p511/National Library of Medicine p223/National Optical Astronomy Observatories p159/Northwestern University p254/Pamela McTurk & David Parker p437/Pekka Parviainen p27/Peter Menzel p335, 485/Philippe Plailly p235, 301, 365/Philippe Plailly/Eurelios p255/Pr S Cinti/CNRI p275/Prof H Edgerton p157/Prof P Motta/Dept of Anatomy/University 'La Sapienza', Rome p383/Profs P Motta & T Naguro p419/Prof Peter Fowler p313/Quest p457, 497/Royal Observatory, Edinburgh p78/Sidney Moulds p143/Sinclair Stammers p175, 466/Space Telescope Science Institute/NASA p 279, 310, 311, 421, 435, 509/US Geological Survey p200/Volker Steger p176/William Ervin p407 Science Pictures Ltd/Oxford Scientific Films p232 Scott Camazine/Sharon Bilotta-Best/Oxford Scientific Films p233 © Scott Camazine/CDC/Oxford Scientific Films p473 Stephen Dalton/NHPA p329, 339, 501 Steve Hansen/TimePix/Rex Features p354 Ted Polumbaum/TimePix/Rex Features p323 The Art Archive/British Library p35 The British Library p48, 68, 89, 151, 187, 464 The British Library, London/Bridgeman Art Library p33 The Illustrated London News Picture Library p155 The Making of a Fly, published 1992 Blackwells Science by Peter A Lawrence p461 The Natural History Museum, London p101, 107, 133, 181, 253, 283 The Natural History Museum, London/ Bridgeman Art Library p299 The Natural History Museum/J Sibbick p261 The Washington Post p278 TimePix/E O Hoppe/Mansell/Rex Features p257 TimePix/John Florea/Rex Features p269 TimePix/Nina Leen/Rex Features p319 Wellcome Library, London p45, 59, 140, 141, 403